LANDFORMS AND LANDFORM EVOLUTION IN WEST GERMANY

Frank Ahnert (Editor)

LANDFORMS AND LANDFORM EVOLUTION IN WEST GERMANY

Published in Connection with the
Second International Conference on Geomorphology,
Frankfurt a.M., September 3–9, 1989

CATENA SUPPLEMENT 15

CATENA – A cooperating Journal of the International Society of Soil Science

ISSS - AISS - IBG

Cover photo (with friendly permission of the Verbandsgemeinde Loreley): The antecedent gorge of the middle Rhine near St. Goarshausen and the Loreley. Cf. paper by W. ANDRES in this volume.

CIP-Titelaufnahme der Deutschen Bibliothek
Landforms and landform evolution in West Germany / Frank Ahnert (Ed.). -
Cremlingen-Destedt: Catena, 1989
(Catena: Supplement; 15)
ISBN 3-923381-18-2
NE: Ahnert, Frank [Hrsg.]; Catena / Supplement

ISSN 0722-0723 / ISBN 3-923381-18-2

CONTENTS

Preface

Few publications exist in English on the landforms in the Federal Republic of Germany. It is hoped that this book, which is published in connection with the Second International Conference on Geomorphology (Frankfurt am Main, 1989) fills, to some extent, that gap.

Following a general introduction, four papers describe and discuss the major landform regions. They provide the background for the other papers and, in addition to covering basic factual information, also deal with major questions of landform development that have long been of particular interest to German geomorphologists. In the North German Lowlands, H. Liedtke discusses the chronology of the Pleistocene glaciations and the relationship between the age of glacial depositional landforms and their subsequent modification by non-glacial processes. In the Central Uplands north of the rivers Main and Nahe, W. Andres directs attention to the conflicting interpretations of the effects of tectonics and of past climates in the development of Tertiary planation surfaces and Quaternary river terraces. In the South German Scarplands, H. Bremer focuses on the role of structural and climatic controls in the long-term evolution of cuestas and valleys. K. Fischer discusses the effects of the complex nappe structure, of the lithology and of exogenic processes upon the landforms of the German Alps and the development of glacial, glaciofluvial and periglacial landforms in the Alpine Foreland.

Most of the remaining papers are studies of smaller regions or local areas. Their sequence in this volume is arranged more or less regionally; their geographical locations are shown in fig.1 (p.2). Several of these papers expand on themes in the general regional papers and examine, sometimes from differing points of view, particular aspects of landform development in detail. The wide variety of topics reflects in some measure the trends in geomorphological research in the Federal Republic.

A number of the papers deal with structural landforms and their relationship to planation surfaces. Two papers discuss Pleistocene glacial and glacio-fluvial landforms in type regions of particular interest. Karst and paleo-karst landforms and processes are investigated in three papers, Quaternary valley development and present-day sediment yields in several others. The relationships between geomorphology and hydrology, geoecology and soils are considered in papers on groundwater quality, interflow, soil formation, soil erosion and the diagnostic value of loess. A brief paper reviews the status of geomorphological mapping in the Federal Republic.

The English editing of the papers was undertaken by Bridget Ahnert, my wife.

Frank Ahnert

Department of Geography
RWTH Aachen
May 1989

CATENA SUPPLEMENT 15 p. 1–9 Cremlingen 1989

THE MAJOR LANDFORM REGIONS

Frank **Ahnert**, Aachen

1 Introduction

The Federal Republic of Germany is bordered in the north by the North Sea, the Baltic Sea and the German-Danish border across the Jutland peninsula. It extends about 800 km to the south, to the mountains of the Alps. Here on the border with Austria is the highest point in Germany, the summit of the Zugspitze (2963 m).

From north to south there is a rough symmetry in the distribution of the major types of structures, rocks and relief. The central part of the country, roughly between Hannover in the north and the river Danube in the south, is an area of uplands and basins. Its rocks are predominantly Paleozoic and Mesozoic and there is a local relief of, in general, several hundred meters. This central area of uplands is flanked by two areas of little-consolidated Cenozoic deposits: the lowlands in the north and the Alpine Foreland in the south. They lie at different elevations because of their different distances to baselevel. The local relief in both areas is low, for the most part in the order of a few tens of meters or less.

The north-south symmetry is accentuated further by the deposits of the Pleistocene glaciations in the northern lowlands and on the Alpine Foreland. Glacial and glaciofluvial landforms such as ground moraines, terminal moraines, lake basins and outwash plains occur in both areas.

These major landform units owe their spatial arrangement and their lithological and structural character to the events and phases of the tectonic and paleogeographic history of central Europe from the Paleozoic Era to the present. They owe their morphological physiognomy mainly to the work of the endogenic and exogenic processes on these rocks and structures during the Cenozoic Era. This brief review of the relevant major phases of the structural development and of morphoclimatic changes will give a general background for the papers in this volume (fig.1).

2 Development of the macrostructural pattern

From the Paleozoic to the present, three major orogenies have affected this part of central Europe: the Caledonian, the Hercynian and the Alpine orogeny.

Structures dating back to the Caledonian orogeny (Silurian) occur in Germany only in the Hohes Venn, the easternmost part of the Stavelot-Venn Massif that extends from SW Belgium across the border near Aachen. Remnants of Caledonian structures probably exist deep under the surface farther to the east

ISSN 0722-0723
ISBN 3-923381-18-2

D–3302 Cremlingen-Destedt, W. Germany
3-923381-18-4/89/5011851/US$ 2.00 + 0.25

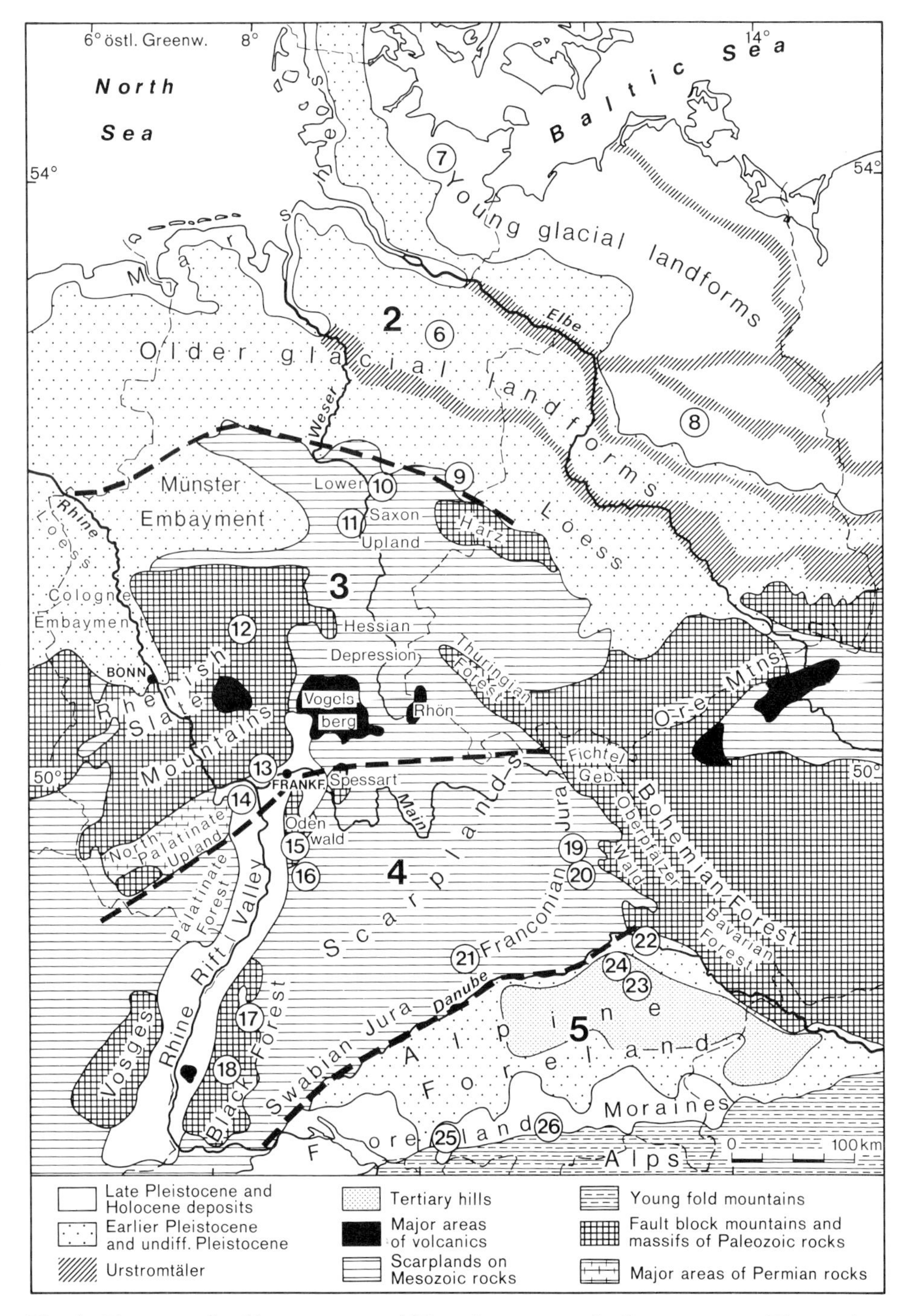

Fig. 1: *The major landform regions of West Germany and adjacent areas. The numbers 2–26 refer to the papers in this book, as indicated in the table of contents. Simplified from a map by LIEDTKE (1984).*

but are without geomorphological consequence in the present landscape.

The late Paleozoic Hercynian (Variscan) orogeny created a broad chain of fold mountains that stretched from the Central Massif of France across all of central Europe into Poland. Its folds strike, in general, southwest-northeast. From north to south the Hercynian fold belt is subdivided into three major zones (fig.1):

1. The Rhenohercynian zone (or Rhenohercynicum) which is represented by the Rhenish Slate Mountains (Rheinisches Schiefergebirge) and by the Harz. Lithologically this zone consists mainly of Devonian and Carboniferous sedimentary rocks, including slates, quartzites and limestones. The upper Carboniferous Ruhr coal beds lie along the northern flank of the Rhenish Slate Mountains. In the Harz, a granite intrusion forms the highest summit (Brocken 1142 m) just across the border in the GDR.

2. The Saxothuringian zone (or Saxothuringicum) which includes the North Palatinate Upland (or Saar-Nahe Bergland), the Odenwald, the Spessart, the Fichtelgebirge and, in the GDR, the Thuringian Forest and the Ore Mountains (Erzgebirge). In addition to some areas of crystalline rock, this zone consists of early Paleozoic sedimentary rocks and, in large interior synclinal troughs, Carboniferous and Permian sedimentary rocks (including the Saar coal beds). The Permian sediments are the products of the downwearing of the mountain range. Permian volcanics, such as the porphyries and melaphyres of the northern Palatinate, are also present.

3. The Moldanubian Zone (or Moldanubicum) of which the Black Forest in the southwest, the Oberpfälzer Wald and the Bavarian Forest (Bayrischer Wald) in the east are the morphologically significant remnants. The Oberpfälzer Wald and the eastern part of the Bavarian Forest along the border with Czechoslovakia are also called the Bohemian Forest (Böhmerwald). This area and the Bohemian Massif to the east also show evidence of at least one earlier, Precambrian, orogeny. Crystalline rocks predominate in the Moldanubian zone.

The mountain ranges of the Hercynian orogeny were worn down to landscapes of low relief by the early Permian. Troughs in the mountain system were filled by Lower Permian sediments (Rotliegendes formation). A long period of only slight crustal movements followed that lasted 150 million years until the Upper Cretaceous and during which deposition prevailed in most of central Europe. In the Upper Permian, large salt deposits were also accumulated in northern Germany (Zechstein formation).

The sedimentary rocks of the Triassic period consist of three formations of the Germanic Triassic: the terrestial sandstones and shales of the Bunter Sandstone (Buntsandstein), the marine limestones of the Muschelkalk and the partly terrestial, partly very shallow marine sandstones, shales, marls and gypsum deposits of the Keuper. Very distinct from the Germanic Triassic is the Alpine Triassic, a much more homogeneous series of marine limestones and dolomites that were deposited in the geosynclinal Tethys Sea that preceded the Alpine orogeny.

The Jurassic period is represented north of the Alps by the black shales of the Lias, the ferruginous brown limestones of the Dogger and the light grey to white limestones of the Malm formation. The Cretaceous sediments are also marine and were deposited mainly in the north. In southern Germany, the beginning of a broad upwarp in the southwest limited Cretaceous deposition to the east and southeast of the area.

The pre-Permian land surface, that is, the unconformity between the denuded rump of the Hercynian mountain system and the overlying Permian and Mesozoic strata, is not only a fossil denudation surface but also the most important structural boundary plane in central Europe north of the Alps. The two types of structure that it separates are termed the 'grundgebirge', the basement consisting of Hercynian and older structures, and the 'deckgebirge', the post-Hercynian sedimentary cover. They are of considerable geomorphological significance because the landforms on the two types of structure differ fundamentally.

During the Tertiary the northward-moving African plate collided with the Eurasian plate and caused the Alpine orogeny, which is not yet complete. This orogeny brought into being the complex young fold mountain system of the Alps of which Germany has only a small share, a strip about 30 km wide along the outer margin of the Northern Limestone Alps. These are composed of Mesozoic limestones and dolomites which are the forward parts of nappes that have been thrust northward from a root zone farther south. Some early Tertiary sediments have been included in the folding.

The Alpine Foreland between the foot of the mountains and the river Danube is a large marginal synclinc, a trough which subsided from the Upper Cretaceous onward throughout the Tertiary. While the Alps were being uplifted, the trough was filled with sediments derived from their denudation. Quaternary sediments form a superficial cover over wide areas.

The crustal stresses of the Alpine orogeny were transmitted northward where the 'grundgebirge', which had been folded in the Hercynian orogeny, was too rigid to be folded again and reacted mainly by warping and faulting. As a result, some areas were uplifted and others subsided. The Mesozoic 'deckgebirge' was largely stripped from the uplifted blocks and the old rocks and structures of the 'grundgebirge' were again exposed at the surface.

A major linear zone of subsidence extends from the vicinity of Basel in Switzerland and Mulhouse in France via Frankfurt, Kassel and Göttingen northward. It is part of an alignment of rifts and structural depressions that also continues southward through eastern France to the Mediterranean coast near Marseille. The Rhine Rift Valley, or Rhine Graben, between Basel and Bingen is its most pronounced component. Along the marginal faults of the Rift Valley are the uptilted blocks of the Vosges and the Palatinate Forest (Pfälzer Wald) in the west and the Black Forest (Schwarzwald) and the Odenwald in the east. The subsidence of the Rift Valley began in the early Tertiary and continued into the Quaternary.

North of Frankfurt the zone of subsidence continues as the Hessian Depression (Hessische Senke) between the Rhenish Slate Mountains in the west and the Thuringian Forest in the east. North of Kassel, near Göttingen, it becomes the Leine Rift Valley (Leinegraben). South of Hannover, it disappears under the Cenozoic sediments of the North German Lowland. North of Frankfurt, the surface rocks in the subsidence zone are predominantly Mesozoic sedimentary rocks.

In southern Germany, Mesozoic rocks occupy the large area between the uptilted eastern margin of the Rhine Rift Valley in the west and the uplifted blocks of the Bohemian and Bavarian Forests in the east. Along the Danube, the Mesozoic beds dip southward under the Cenozoic deposits of the Alpine Foreland.

An additional endogenic consequence of the tectonic stresses were the large number of volcanoes that erupted during the Tertiary and in the Pleistocene in the Central Uplands (Eifel, Westerwald, Vogelsberg, Rhön) and in southern Germany (Swabian Jura and Kaiserstuhl). In the Eifel the volcanism continued throughout the Pleistocene.

Subsidence and sedimentation predominated in northern Germany throughout the Cenozoic. Near the base of the sediments, however, are the Zechstein salt deposits (Upper Permian) which reacted to the crustal stresses of the Alpine orogeny with halokinetic deformations. The development of salt domes and elongated salt diapirs was accompanied by warping of the overlying Mesozoic strata. On the northern margin of the Central Uplands, the warped beds form cuestas, hogbacks and intervening vales. Farther north they are covered for the most part by Cenozoic deposits. An exception is the island of Helgoland in the North Sea which is formed of Bunter Sandstone.

3 Major climatic changes and their effects

Apart from altitudinal differences in temperature and the effects of relief on precipitation, the climate was probably fairly uniform over all of central Europe at any given time. The Mesozoic was a period of warm climate which continued, with fluctuations, into the early Tertiary. During the Oligocene and Miocene, the mean annual temperatures also varied but decreased overall from about 20°C to about 18°C. The temperature decreased more rapidly in the latter half of the Miocene and during the Pliocene. Towards the Quaternary, the temperatures may not have differed much from the present conditions, with a mean annual temperature of about 10°C, a January mean around 0°C and a July mean of about 18°C.

An estimate of the moisture supply during the Tertiary is more difficult. Paleopedological evidence indicates intensive chemical weathering and therefore long periods of humid or subhumid climate; these alternated repeatedly, however, in the Eocene, Oligocene and early Miocene with drier periods, shown by the occurrence of evaporites and by the plant and animal remains in the sediments.

The Tertiary climates in central Europe have often been compared to the subhumid and semiarid climates of the marginal tropics and subtropics in which planation processes are common. Even though such a comparison may not be justified in terms of annual temperature ranges, seasonal variations of precipitation and the frequencies of morphoclimatic events of particular magnitudes, it is very probable that the climatic conditions of the early and middle Tertiary generated geomorphic process systems that were conducive to planation. On many of the German Uplands there arc denudation surfaces that truncate the underlying structures and that represent landform generations older than the Pleistocene valleys by which they are dissected. Particularly good examples occur in the Rhenish Slate Mountains and in the Harz.

Compared to the peneplains in the subhumid tropics, these denudation surfaces are quite small in extent and are, perhaps, locally developed pediments and pediplains rather than remnants of extensive peneplains. Areas of resistant rock, such as the quartzitic zones of the Hunsrück and the Taunus in the southern Rhenish Slate Mountains and of the Harz, stand several hundred meters above the denudation surfaces that surround them. They must have been present as inselbergs while the denudation surfaces wcrc bcing formcd. Such a spatial association indicates pedimentation. The development of pediments would have been possible during both the semiarid phases and the subhumid phases of Tertiary climate.

In the tropics, denudation surfaces occur next to well-dissected uplands. Similarly, there is evidence of Tertiary fluvial dissection in some German upland areas not far from denudation surfaces that were formed at the same time. This applies particularly to the marginal uplands of the Rhine Rift which were already partially uplifted during the Tertiary. Although a climatic-morphological tendency to planation was undoubtedly present in all of central Europe, it has led to the development of denudation surfaces only where the tectonic movements were weak enough for planation to take place.

The general cooling trend during the Pliocene fluctuated. Warmer periods occurred during which intensive chemical weathering and planation processes seem to have taken place. Further cooling, with fluctuations, led to the alternating glacials and interglacials of the Pleistocene.

The Pleistocene epoch lasted over two million years. The first million years contained several 'cold' and 'warm' periods but no identifiable glacials. These occurred only in the second half of the epoch. During the glacials, continental glaciers advanced from the Scandinavian Shield southward into the lowlands of north central Europe. Simultaneously, glaciers flowed from the summit regions of the Alps into the valleys and out into the Alpine Foreland where they spread out as large piedmont glaciers. The moraines left by the Alpine glaciers indicate that there have been at least four major glacials, from oldest to youngest, the Günz, Mindel, Riss and Würm. In north Germany, the moraines of the continental ice sheets show evidence of only three glaciations (Elster, Saale and Weichsel). The equivalent of the oldest Alpine glaciation appears to be represented only by a cold period (Menapian) in north Germany, although if there was a glaciation, its traces must have been obliterated by the ice of the glacials that followed.

In the area between the northern glaciated area and the Alps, only some of the higher summits in the uplands had small glaciers. Evidence of glaciation has been found in the Black Forest, the Harz and the eastern Bavarian Forest.

The unglaciated regions of central Europe were tundra in which periglacial process systems predominated. Frost weathering produced debris which was transported downslope by nivation, frost creep, solifluction and, from bedrock cliffs, by rockfall. Where the stream gradients and discharges were high enough, erosion occurred. Elsewhere the rock waste was accumulated in valley fills, alluvial fans and depositional plains. Much of the fine-grained fraction of this material was blown from the alluvial flats by the wind and deposited as loess.

During the interglacials, many valley fills were eroded to form river terraces. It is not always certain, however, whether these phases of erosion were caused by climatic factors, by uplift, or both.

The Pleistocene was, however, undoubtedly a period in which valley formation and valley deepening predominated in the German Uplands. Tertiary planation surfaces became deeply dissected. In many areas they were destroyed altogether by denudational lowering of the interfluves. Only the major divides have not yet been reached by the effects of Quaternary headward erosion.

There have also been several short climatic fluctuations during the Holocene, but only of relatively small amplitude. The present morphoclimate of central Europe is humid temperate, with a moderate number of freeze-thaw cycles in winter, year-round precipitation and only rare occurrences of high-intensity rainfall. The natural or quasi-natural vegetation is dense. Consequently, present-day geomorphic processes are not very active and confined mainly to chemical weathering and to the breaking down and removal of waste materials left behind by the last glacial. Greater process intensities are found only where high local energy concentrations occur, for example, on seashores and high mountains and in stream channels with steep gradients, or where the process system reacts to man- induced changes of the landscape.

4 The major landform regions

On the basis of the morphostructural pattern, of the existing landform associations and of their morphogenetic development, the area of the Federal Republic of Germany can be subdivided into five major landform regions:

The North German Lowlands
The Central Uplands
The Rhine Rift Valley and Southern Scarplands
The Alpine Foreland
The German Alps

Of these, the North German Lowlands and the Alpine Foreland delimit themselves as areas of young sediments and of low relief in contrast to the uplands where older rocks and structures with erosional and denudational landforms of moderate to high relief predominate. The North German Lowlands are, of course, a segment of the depositional plain that stretches across the northern part of central Europe from NW Belgium and the Netherlands to Poland. The upland area can be divided by a line that runs from the lower Saar valley near Merzig east-northeast to Wiesbaden and Frankfurt and from there eastwards to the Frankenwald in the northeast of Bavaria.

The area north of this line, the Central Uplands region, includes two mountain blocks of the Rhenish Slate Mountains and the Harz, made up of Paleozoic rocks and structures, which were uplifted and exhumed during the Cenozoic. Their highest summits, particularly those on resistant rocks, lie from 800 m to more than 1100 m above sea level. The remnants of their Tertiary denudation surfaces are at about 500–600 m.

Between these two mountain blocks are the Hessian Depression, the Leine Rift Valley and the uplands of Lower Saxony (Niedersächsisches Bergland) where there are cuestas and hogbacks on Mesozoic rocks.

The Hessian Depression is a tectonically complex area in which there are remnants of Tertiary volcanism, including the large lava shield of the Vogelsberg and the summits of the Rhön. The northern boundary of the Central Uplands region lies near Hannover where the Mesozoic rocks and their warped structures have subsided below the Cenozoic cover of the North German Lowlands.

South of the line that divides the uplands is the region of the Rhine Rift Valley and the South German Scarplands. The landforms developed at the same time as those of the Central Uplands but in a different spatial pattern and with different geomorphological characteristics. The Rhine Rift Valley is much larger and deeper than the Hessian Depression and has also been of much greater consequence for the morphological development of the adjacent uplands. These uplands are tilted away from the Rift Valley and are, therefore, structurally and lithologically asymmetrical. Near the edge of the Rift Valley the Paleozoic 'grundgebirge' is exposed but farther back it is covered by the Mesozoic 'deckgebirge'. The Mesozoic strata, of varying resistance, dip more or less continuously away from the Rift valley so that eastwards, progressively younger rocks lie at the surface. The more resistant beds form cuestas scarps and the weaker beds underlie the intervening dip slopes.

In the east, the region is bordered by the crystalline upland blocks that form the margin of the Bohemian Massif: the Fichtelgebirge, the Oberpfälzer Wald and the Bavarian Forest. The southern border is marked by the river Danube which flows approximately along the line where the limestone dipslope of the Swabian/Franconian Jura cuesta disappears under the younger sediments of the Alpine Foreland. West of the Rhine Rift Valley, the scarplands form the structural eastern rim of the Paris Basin. The summits of the uplands in this southern region are higher than those in the north. Both the Black Forest and the eastern Bavarian forest have summit elevations of more than 1400 m, and the highest cuesta scarp in Germany, the Swabian Jura, exceeds 1000 m above sea level in its western part.

The two regions of low relief, the North German Lowlands and the Alpine Foreland, are areas of Cenozoic deposition on a generally subsiding base and both have glacial and glaciofluvial landforms. There are several differences due mainly to their locations. The North German lowland borders on the seas. Coastal processes have destroyed some of the original glacial and glacialfluvial landforms by erosion, redistributed the eroded material and formed Holocene marshes and islands, especially on the North Sea coast. The river systems of the North German Lowland drain mostly to the North Sea. Their northwesterly alignment is, in part, determined by the Pleistocene urstromtäler or glacial spillways.

By contrast the morphological development of the Alpine Foreland is controlled mainly from the Alps, formerly by the Alpine glaciers and now by the rivers that, with the exception of the Rhine, flow from the mountain valleys across the Foreland to the Danube.

The German Alps extend from Lake Constance (Bodensee) in the west to the

Salzach valley near Salzburg in the east. Along their northern margin there is a narrow pre-Alpine zone of flysch mountains behind which rise the Northern Limestone Alps with summits that are generally over 2000 m high.

The courses and drainage basins of the four large trunk rivers of Germany, the Rhine, the Weser, the Elbe and the Danube, have little relation to the five major landform regions. Only the Danube, which has its baselevel in the Black Sea, flows for any distance along a major regional boundary. The Rhine flows from the Swiss Alps through all five regions to the North Sea. The Weser rises in the Central Uplands and the Elbe in the Bohemian Massif in Czechoslovakia and both also flow to the North Sea. All these rivers are essentially older than the uplift of the blocks they cross so that they flow for long stretches in antecedent courses, of which the gorge of the middle Rhine between Bingen and Bonn is the most spectacular (see cover photo).

References

KNETSCH, G. (1963): Geologie von Deutschland und einigen Randgebieten. Enke, Stuttgart, 388 pp.

LIEDTKE, H. (1984): Geomorphological mapping in the Federal Republic of Germany at scales 1:25000 and 1:100000 - a priority program supported by the German Research Foundation (Deutsche Forschungsgemeinschaft). Bochumer geographische Arbeiten 44, 67-73.

LOTZE, F. (1971): Dorn-Lotze, Geologie Mitteleuropas. Schweizerbart, Stuttgart, 4th. Edition, 388 pp.

SCHWARZBACH, M. (1974): Das Klima der Vorzeit. Enke, Stuttgart, 3rd. Edition, 380 pp.

SEMMEL, A. (1980): Geomorphologie der Bundesrepublik Deutschland. Steiner, Wiesbaden, 4th Edition, 192pp.

Address of author:
Prof. Dr. Frank Ahnert
Geographisches Institut der RWTH Aachen
Templergraben 55
D-5100 Aachen
Federal Republic of Germany

CATENA SUPPLEMENT 15 p. 11–24 Cremlingen 1989

THE LANDFORMS IN THE NORTH OF THE FEDERAL REPUBLIC OF GERMANY AND THEIR DEVELOPMENT

Herbert **Liedtke**, Bochum

1 Introduction

The northern part of the Federal Republic of Germany is characterized by landforms that have been developed almost exclusively in, or on, glacial deposits. The only exceptions are the coastal strips on the North Sea and the Baltic Sea, a few salt domes where Upper Permian or Mesozoic rocks have been pressed up to the surface and a few other areas of Mesozoic rocks, most of which lie near the southern margin of glaciation. The North German Lowland is defined as the area that stretches from the coast and the German-Danish border to the margin of the uplands. The uplands include not only the Paleozoic basement complexes of the Rhenish Slate Mountains and the Harz but also the Lower Saxon Upland which forms a spur into the North German Lowland between the river Weser near Minden in the east and Osnabrück in the west and which has also been overrun by the Pleistocene inland ice. The Münster Lowland Embayment is bounded on the northeast by the Lower Saxon Upland and is connected in the west with the lower Rhine plain.

ISSN 0722-0723
ISBN 3-923381-18-2

D–3302 Cremlingen-Destedt, W. Germany
3-923381-18-4/89/5011851/US$ 2.00 + 0.25

The forward edge of the Pleistocene inland ice moved across the present course of the lower Rhine river for a distance of about 10 to 20 kilometers. Immediately to the south, the Cologne Lowland Embayment extends into the Central German Uplands. This embayment was not glaciated but was covered by extensive gravel deposits of the Rhine and Meuse (Maas) rivers and also by loess over wide areas (see ANDRES in this volume).

2 The coastal areas of the Federal Republic of Germany

The coasts of the North Sea and the Baltic Sea differ greatly although both were developed by the drowning of moraine landscapes after the eustatic rise of sea level in the last few thousand years of the Holocene. The previous landscape along the North Sea coast had been characterized by middle Pleistocene moraines of low relief, in contrast to the Baltic coast where there had been a landscape of late Pleistocene moraines with a higher local relief. The frequent storm floods on the North Sea erode the shore year after year whereas the occasional storm floods that occur in the Baltic Sea are usually much less damaging and generally affect only parts of the southern Baltic Sea.

Tides on the North Sea coast vary. The tidal range near Borkum is 2.2 m, at the mouth of the Elbe near Cuxhaven 2.8 m and near Sylt 1.7 m. The highest tides occur in the Jade Gulf near Wilhelmshaven where there is a maximum range of 4.1 m. When high tidal ranges coincide with strong northwesterly storms of long duration, considerable damage can result. For example, during the storm flood of February 1962, which caused 315 deaths in Hamburg, a northwestly wind blew for two days at 40-70 km per hour. Fortunately it did not coincide with spring tides. The dates of storm floods that have eroded large areas of coast in the past are known from historical records. In 1164, the Jade Gulf was formed and in 1287, the Dollart. In January 1362, 7600 persons drowned in a storm flood in North Friesland and East Friesland, and in October of 1634, also in North Friesland, 9000 people lost their lives in a storm flood.

The German North Sea coast has three different morphological segments:

1. the chain of islands off the coast of East Friesland;
2. the inner part of the German Bight off which there are hardly any islands and into which the rivers Elbe and Weser flow and
3. the indented coast of North Friesland with numerous irregularly distributed islands.

The origin of the East Friesian Islands is not entirely clear. From the first century A.D. onward, the level of the sea began to rise again very slowly and the construction of the Wurten, artificial settlement mounds, became necessary. From the ninth century onward, the building of dykes also became essential. Towards the end of the Middle Ages the sea level had risen eustatically by about 1.3 m. Since that time it has remained more or less constant (HANISCH 1980). Because the western coasts of the East Friesian Islands, which lie at the outer edge of the 3–15 km wide Wadden Sea, have been more strongly exposed to erosion, it was assumed previously that they were migrating eastward. However, historical maps do not show any clear evidence of such a movement and it is now thought that the islands are more or less stationary and lie on morainic mounds. The tops of these mounds are 3–5 m below sea level. The islands are separated by "Seegats" through which tidal currents flow. Sand is also transported parallel to the coast eastward into the inner part of the German Bight.

Some of the North Friesian Islands have a morainic core but others are formed of marsh remnants (Halligen) which continue to exist only because of protection by dykes or because of high artificial mounds. The amount of land lost by marine erosion has increased greatly in the past one thousand years. Not only have the storm floods combined with the slow but progressive rise of sea level to cause more rapid erosion, the coastal area has also subsided by up to 1.6 m following the compaction of older marine deposits and peat layers. The severity of storm floods is also related to fluctuations in the atmospheric circulation, the causes of which are not yet known (LINKE 1984).

In contrast to the North Sea, the tides in the Baltic range from 10–20 cm. The Baltic Sea coast of Schleswig-Holstein is made up of drowned subglacial channels or tunnel valleys and broader embayments formed in glacial deposits. Since

Federal Republic of Germany	GDR		Poland	European USSR
GRUBE et al.	CEPEK		RZECHOWSKI	VELICHKO & FAUSTOVA
Holocene	Holocene		Holocene	Holocene
		-10[2]		
Weichsel Glacial	Weichsel Glacial		Vistula	Waldai
		- 115 -		
Eem Interglacial	Eem Interglacial		Eem	Mikulino
		- 128 -		
Warthe Stadial	S III (Lausitz = Warthe) S II		Warta	Moscow
		- 195 -		
Treene Interstadial[1] Drenthe Stadial (Saale II)	Treenethermomer Saale I		Polichna Oder-, Central Polish Glacial (S I)	Odincovo Dnepr
		- 297 -		Romny
Saale Sands (Saale I) Holstein Interglacial	Dömnitz Interglacial Fuhne Cold Period		(S) Mazovian	Pronya Lichwin Wilga
		- 502 -		
Elster Glacial	Elster Glacial		South Polish Glacial	Oka
		- 688 -		
Cromer Interglacial Menapian Cold Period	Mahlis Unstrut Cold Period		Podlasia Narew Cold Period	

[1] STREMME (1982) Treene Warm Period 230–250,000 B.P.
[2] RZECHOWSKI (1986); Ages in 1000 years B.P.

Tab. 1: *Subdivisions of the Nordic Glaciations in northern Germany.*

the prevailing westerly winds blow offshore, the coastal erosion, transport and deposition are of much less significance than on the coast of Pomerania farther to the east. Locally shoreline retreat can be considerable. An annual rate of cliff retreat of about 0.5 m has been measured, for example, near Travemünde. Along many coastal stretches retreating cliffs and prograded beaches alternate (Ausgleichsküste). It is probable that 2000 years ago the sea level was similar to that of the present time. In contrast to the North Sea, there appears to have been a regression to about 1.0–1.3 m below the present sea level around 1000 A.D. but since then, there is a slow rise which is still continuing (KLUG 1980, 1985).

3 The subdivisions of the glaciations in north Germany

There are major geological and geomorphological differences between young moraines and the old moraines in north Germany. Young moraine landscapes were covered by an ice sheet during the last glaciation but the old moraine landscapes were ice-covered only during previous glaciations. In north Germany three ice ages and four glaciations can be distinguished (tab.1). The Saale ice age consists of at least two major ice advances, the (older) Drenthe and the (younger) Warthe advance. They are separated by the Treene warm period. However, evidence based on warm period deposits in Poland and the Soviet Union, indicate that this warm period should be termed an interglacial.

4 The young moraine landscape in eastern Schleswig-Holstein

In the north of the Federal Republic of Germany only the eastern area of Schleswig-Holstein has been covered by the last inland ice sheet; the other formerly glaciated areas were covered by the ice from the Elster, Drenthe or Warthe glaciations.

In Schleswig-Holstein, the furthest extension of the Weichsel (Vistula) glaciation is, in general, apparent either from the large terminal moraines or from the boundary between moraine and outwash plain (fig.1). Where the last ice sheet bordered higher terrain formed of old moraines, or where a large amount of outwash deposition occurred as, for example, south of Rendsburg (STREHL 1986), its boundary is not always visible in the landscape. The terminal moraine deposited by the furthest advance of the last glaciation, terminal moraine A of the Brandenburg stage, runs from the Danish border near Flensburg southward to Rendsburg (fig.1), turns southeastward near Bad Segeberg and continues in a large semi-circle to the northeastern edge of Hamburg and to the Stecknitz Canal near Büchen, from where it continues eastward as the Brandenburg end moraine in Mecklenburg (GDR). In some areas the Weichsel ice sheet advanced across the surface (STEPHAN 1981) without any disturbance of the substrate. Elsewhere strong push effects have occurred. In eastern areas considerable differences in the relief have been produced by the advancing inland ice, (photos 1 and 2), such as the Bungsberg (164 m). The ice advance also disrupted the marine connection between the North Sea and the Baltic which had existed during the Eem interglacial. Marine Eem deposits lie today at several different elevations.

The Weichsel ice sheet reached its maximum extent in Schleswig-Holstein about 20,000 years ago. Beyond this limit lies the strongly denuded older moraine relief of the Warthe stadial, where differences in elevation have been further reduced by the sedimentation of the Weichsel outwash plains. In the glaciated area to the east of the ice margin, long, deep subglacial tunnel valleys and a landscape of young moraines with numerous knobs and kettles characterize the surface. Ancient and modern traffic routes to Denmark ("Ochsenweg", railway line, auto-

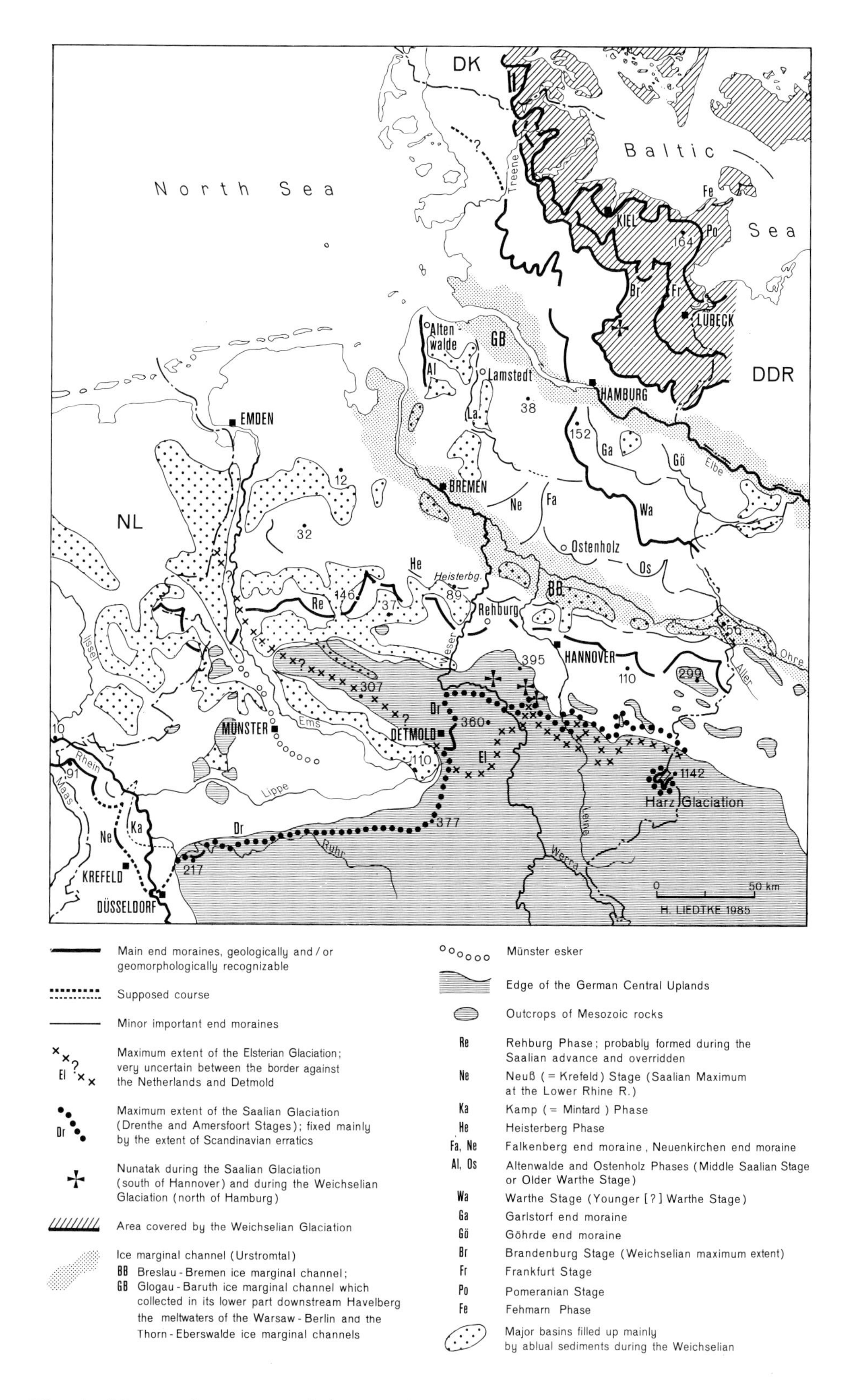

Fig. 1: *Main end moraines of the Scandinavian glaciations and related landforms in the Federal Republic of Germany.*

bahn) avoid the rough young moraine topography and run northwards along its outer margin.

The area to the west that lies beyond the limit of the last glaciation is subdivided into the "high Geest" made up of deposits of the Warthe glacial and the "low Geest" which is intercalated in the high Geest and consists of outwash valley floors and outwash plains of Weichsel age. The meltwaters that drained westward flowed directly into the then ice-dammed North Sea. Those that flowed southward were tributary to the lower Elbe river which served as the ice-marginal channel (urstromtal) for all glacial spillways during the Weichsel ice age.

The retreat of the Weichsel ice is difficult to date. A large number of ice margin positions have been identified which are often traceable only over short distances so that precise linkages to the widely-spaced major ice margins in Brandenburg and Poland are not possible. Groups of ice margin positions have been distinguished, "A" moraines (outer), "M" moraines (middle) and "I" moraines (inner), and been correlated with the three Weichselian ice margin positions farther east, the Brandenburg, the Frankfurt (on the Oder), and the Pomeranian moraines. These correlations have been disputed. The stratigraphy of the young moraines has not yet been solved completely (GRUBE et al. 1986).

In Schleswig-Holstein, the three ice margin positions were very close together. In two areas, near Flensburg and near Schleswig, the Brandenburg margin ("A" moraine position) was reached or overtaken by the Pomeranian stage ("I" moraine position; HOUMARK - NIELSON 1981) because during this period the ice no longer came from the northeast but from the Baltic basin in the east-southeast and could, therefore, advance easily and reach the previous extreme position of the ice margin.

The terminal moraines of the young moraine landscape in Schleswig-Holstein are relatively linear and often also inconspicuous features. Some end moraines are only a few meters high in one location but rise to 70 meters in another where the ice movement had been impeded by higher relief (GRIPP 1964). Nevertheless, the terminal moraines are important for the determination of the temporal and stratigraphic subdivisions of glacial retreat. Drowned glacial channel valleys (förden) subdivide eastern Schleswig-Holstein into several distinct landscapes that are characterized by ground moraines, occasional drumlin fields and a few eskers. The landscape of Angeln lies between the Flensburg Förde and the Schlei, that of the Danish Wohld between the Schlei and the Bay of Eckernförde. East of the Kiel Förde is the Wagrien landscape in which there are many large lakes.

The complex pattern of landforms in Schleswig-Holstein developed in two stages. First, with the melting of the last ice sheet extensive areas of outwash were deposited. Secondly, after the melting of the contiguous ice sheet isolated masses of dead ice of varying sizes remained covered by moraine and outwash material. This dead ice did not melt until several thousand years later and, as a result, the previous flow patterns of glacial meltwaters were obliterated and kettle holes and lake basins of various sizes were left. In some places the outwash channels of the younger ice margins can be reconstructed from the presence of lakes (fig.2). The lower terraces on the lakes were formed

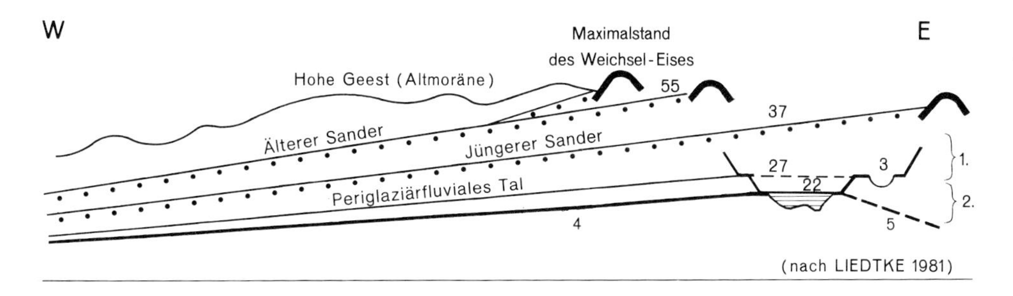

Fig. 2: *Outwash plains and lake terraces in Holstein (after LIEDTKE 1981).*

1. Phase of deep thawing (until end of Alleröd period)
2. Phase of post-glacial thawing-out
3. Lake level position of the younger Tundra period with a broad lake terrace at the eastern shore and subsequent development of hollows due to thawing out.
4. Present valley floor which follows the former centrifugal drainage direction.
5. Present valley floor of centripetal drainage. Because of the deep thawing and the thawing out of dead ice the original centrifugal drainage at the maximum of the Würm glaciation was changed at the beginning of the Holocene to, for the most part, an eastward centripetal drainage.

in the younger Tundra period between 10,900 B.P. and 10,000 B.P., when a period of cold climate interrupted the thaw of the dead ice masses and, as a result, shoreline terraces were formed, especially on the eastern shores of the lakes where the surf could erode a cliff.

Periglacial modification of landforms in the young moraine area is of minor importance. There are occasional cryoturbations and ice wedges and also periglacial cover sands, but the increase in temperature during the melting period and the extensive effects of late-melting dead ice have destroyed many periglacial features.

By the time the ice had retreated to the vicinity of the present coastline of the Baltic Sea, the meltwaters could no longer drain off across the higher elevations of the marginal moraine belt and were impounded as ice marginal lakes in which varved clays were deposited. Some of these lake basins persisted for a long time. At Bad Oldesloe, for example, 1016 varves have been counted (CIMIOTTI 1983). Further retreat, interrupted by a short readvance (Fehmarn advance), exposed the low relief ground moraine surface near Oldenburg south of Fehmarn. The area was traversed by a narrow meltwater channel which ran from Dahme west-northwest past Oldenburg to Weissenhaus. During the postglacial rise in sea level, this channel filled with peat. From excavations near Rosenhof it is known that at the time of the oldest settlements, about 3900-3700 B.C., the sea level was about 3.3 to 3.5 m lower than at present. Since then the coastline of the Baltic has remained close to its present position. There have been some local changes, such as the retreat of cliffs or the formation of spits and barrier beaches. One example is the closing off of the embayment which now forms the Hemmelsdorf Lake north of Lübeck. At extreme high storm levels the low barrier beach is crossed by the sea and the water becomes brackish. After about 60 years the lake water is again fresh. Large numbers of shallow lakes and kettles have been filled with sediments during the Holocene, some disappeared entirely.

5 The landforms of the old moraine landscape

Several glaciations and their accompanying climatic conditions have affected the old moraine landscapes. The climatic effects varied depending on how far south the ice sheets advanced. In the Federal Republic of Germany, the Drenthe glaciation covered the largest area, reaching the lower Rhine, the northern margin of the Rhenish Slate Mountains and the Paderborn plateau. Only east of the Paderborn plateau did the ice of the Elster glaciation reach a little farther south. The margins of the inland ice sheets are shown in fig.1. The ice margins shown on the map are the extreme positions of each glaciation. In general, each successive advance of the Saale and Weichsel glacials did not reach as far as the previous one. The Elster age of the moraines northwest of the Harz has recently been questioned. They could possibly be the southernmost moraines of the Saale glaciation.

The number of landforms and of deposits remaining decreases progressively towards the southern limit of the inland ice advance. Three zones of relief development can be distinguished:

1. The area of young moraines with strongly developed glacial landforms, lakes and hollows. There are considerable elevation differences within short distances.

2. The area of old moraine, now an area of low relief, in which there was strong glacial landform development and deposits of 50–100 m thickness but where considerable downwearing of the higher areas by periglacial processes and depositional filling of the lower areas has taken place.

3. The formerly ice-covered area on the northern edge of the uplands and some other marginal areas of resistant rock which have been affected by periglacial processes but only minimally by glacial forming processes.

The effectiveness of periglacial processes depends on the bedrock resistance which affects the intensity of denudation, on the frequency and duration of periglacial conditions and on the local relief.

The old moraine landscape previously resembled the present young moraine landscape of Schleswig-Holstein. However, periglacial processes during the last glaciation, and also at the end of the Warthe glaciation, altered the surface forms so that the alignment of terminal moraines cannot be reconstructed with certainty. Often no specific landforms are visible. The former moraine relief has been obliterated and changed into a monotonous landscape (Geest) dominated by sands (photo 3).

The few traces left by the inland ice in the areas underlain by consolidated bedrock are usually covered by loess. Terminal moraines cannot be identified at all and urstromtäler and diversion channels are absent. The inland ice did not remain in this extreme position long enough for large end moraines to accumulate or for diversion channels to be cut into the bedrock. In a few areas there are short stretches with accumulations of

Photo 1: *Young moraine landscape, view northwestward from the Bungsberg (164 m).*

Photo 2: *Young end moraine near the Westensee, Schleswig-Holstein.*

Photo 3: *Old moraine landscape, Lüneburg Heath.*

coarse erratics in near-level terrain that have been interpreted as stationary positions of the inland ice. The arca covered by the inland ice has been reconstructed from remnants of ground moraine and from erratics that have been preserved, often in stonelines.

In the old moraine landscape the local relief ranges from 10 m to 25 m per square kilometer, with the exception of very few higher hills (Dammer Berge north of Osnabrück 146 m, Harburger Berge south of Hamburg 169 m, Wilseder Berg 164 m).

The present-day low relief is due in part to the low pre-glacial relief, as this was an area of sedimentation, and in part due to the occurrence of permafrost during Pleistocene cold periods. It has been estimated that during the coldest phase, with mean annual temperatures of about -5°C, the permafrost may have had a thickness of at lcast 150 m (FRÄNZLE 1988). At this time permafrost was continuous over all of north Germany, with the exception of a few river courses. During the interstadials the mean annual temperature rose to >0°C. and the permafrost became discontinuous or disappeared entirely.

Fossil ice wedges and ice wedge polygon networks provide evidence of the former existence of permafrost. During the summer, the permafrost thawed to a depth of 0.5–1.5 m and solifluction and surface runoff occurred. The runoff transported silts and the fine sand fractions, even on slopes of as little as 3–5 degrees, into the adjacent hollows. The sand and coarse silt remained in the hollows and the fine silt and clays were transported as suspended load to the sea. This areally effective periglacial-fluvial removal from the slopes and subsequent deposition at lower elevations is termed abluation. On loose fine-grained material and in the absence of a dense vegetation cover, abluation has been more important than solifluction, filling closed hollows and aggrading broad basins and lowlands. Because of the low gradients in these areas during the Holocene, deep peat bogs developed which have, however, been cut and drained during the past few hundred years. At the present time numerous straight ditches and canals traverse these areas. A small proportion of the fill in the hollows was deposited near the end of the Warthe stadial. During the following interglacial (Eem) organic material was deposited and then covered in the early Weichsel glacial by thick deposits of sand and occasional clay and gravel layers (BEHRE & LADE 1986, LIEDTKE 1983). The hollows were filled to about two-thirds or three-quarters of their total volume, the remainder being filled during the middle and late Weichsel glacial.

Not all hollows were filled up entirely and in the old moraine area of northern Germany there is a large number of very shallow pans (kaven). In northern Lower Saxony, about 10,000 of these forms with a diameter of 50 to 200 meters and depth of 0.5–3.0 m have been recorded. Most are dry and two-thirds are the result of deflation. The origin of the others is often uncertain but they are easily distinguishable from the dead ice kettles of the young moraine landscape because they all occur in very flat areas. Many of them lie along the southern land ridge (Südlicher Landrücken), a line of low hills from south of Hamburg to Berlin, that formed the ice margin of the Warthe glaciation. South of Berlin on the Fläming between Magdeburg and Dahme in the German Democratic Republic, 592 of this type of hollow have

Young moraines	Old moraines
Weichsel Irregular, chaotic stream patterns with areas that are not connected to a stream system (interior drainage, lakes without surface outlets).	Saale: Drenthe, Warthe; Elster? Landscape of valleys; in very flat areas, extensive peat bogs
Many lakes, or former lakes filled in during the Holocene	No glacial lakes remaining; they have disappeared Weichsel glacial due to periglacial denudation
Numerous kettle holes (dry) or water-filled small depressions in usually hilly terrain	Numerous "Kaven": very shallow depressions most often filled with water in the very flat ground moraine area or on nearly horizontal lowlands covered with ablual and periglacial fluvial deposits and in some cases caused by deflation
Silty ground moraine of boulder marls with a small decalcification depth up to 1 m	Sandy ground moraine (boulder loam) that is deeply decalcified and has lost its silt fraction by ablual denudation
Knobby terrain with considerable local relief	Low expressionless terrain with very low relief except for a few gravelly ridges
Numerous glacial channels with ungraded longitudinal profiles	Channels for the most part filled with sediments, barely recognizable in most areas
Glacial series of landforms identifiable	Glacial series of landforms rarely recognizable
Terminal moraines frequently clearly visible	Terminal moraines most often identifiable in outcrops
Few periglacial valleys without climatic-caused asymmetry	Extensive networks of dry valleys with occasional asymmetry
Occasional occurrence of stone lines and ventifacts especially in the region of the Brandenburg and Frankfurt stadials	Very frequent occurrence of stone lines and ventifacts on the Geest surfaces
Dunes on sandy substrates especially east of the Elbe in the urstromtäler; additional aeolian deposits on adjacent ground moraine areas	Extensive areas of aeolian sand sheets (cover sands)
No loess	Loess belt along the northern foot of the Central German uplands. Very occasionally areas of coarse loess (Flottsand)

Tab. 2: *Geomorphological characteristics of the young and old moraine landscapes in the north of the Federal Republic of Germany.*

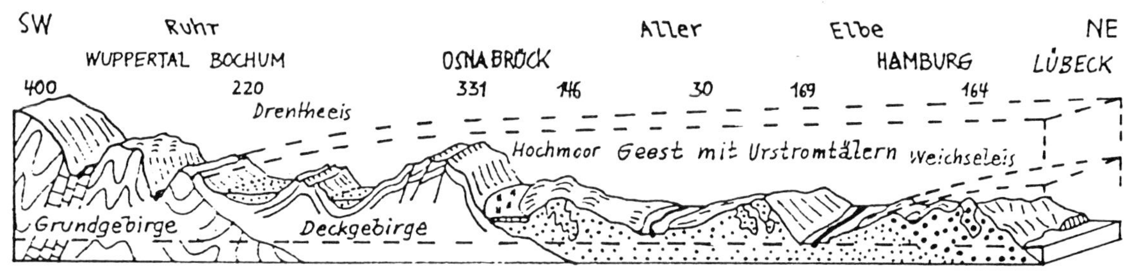

Length of profile approximately 400 km (not true to scale)

Paleozoic	Upper Cretaceous	Mesozoic	Quaternary	Holocen
Former periglacial area with solifluctional denudation and terrace formation	Old moraine landscape		Young moraine landscape	Baltic S
	Elster glaciation and/or	Saale glaciation	Weichsel glaciation	Holocen
	Drenthe stage; Rehburger phase (overrun)	Warthe stage	Fehmarn advance Pomeranian stage Frankfurter stage Brandenburger stage	Tempor. ice-dam lake
Basement complex with marginal ice cover	Thin Quaternary cover on Mesozoic beds which in places are at the surfaces	Usually 50-100 m Quaternary sediments with urstromtäler, kaven and occasional hills of Quaternary material	Knobby relief with numerous lakes, 50-150 m Quaternary sediments	Marine transgre sion ac glacial landfor
Some erratics	Sandy ground moraine in lowlying areas. At the southern lowland margin loess	Sands and sandy ground moraine (boulder loams)	Boulder marls	
Landforms hardly affected by glaciation	Landforms little affected by glaciation because substrates are at low angles	Strong ice push in some areas; strong periglacial modification; resistant gravel ridges as denudadational remnants	Landforms strongly affected by glaciation. Relief accentuated by late thawing of dead ice	Cliffs a barrier beaches tendenc develop "neutra shorelir

Fig. 3: *Generalized profile from the Baltic Sea to the Rhenish Slate Mountains.*

been identified in an area of 4000 sq., a mean density of about three hollows per twenty square kilometers. These natural hollows, which often lie on outwash deposits, have not been filled because of the grain size of the deposits and the low surface gradient. Others lie on divides, indicating that the area has not yet been entirely converted into a fluvial valley landscape.

The amount of periglacial denudation has been determined in only a few areas. For example, in the Uelzen Basin, a periglacial denudation of 20 m has been estimated and in several areas of the Dammer Berge, a denudation of 30 m (GALBAS, KLECKER & LIEDTKE 1980). After the meltwater flow ceased in the urstromtäler, several alluvial fans were deposited by periglacial tributary streams on the former channel bottoms.

Aeolian sand deposits, in some instances with a typical dune form but more often irregular in shape, occur on the floors of the urstromtäler and on the outwash plains. They were developed towards the end of the late Weichsel glaciation. Ventifacts, stone pavements and stone lines in the former ground moraine areas are another indication of the deflation that occurred during the last ice age.

In the lowland along the northern margin of the central German uplands a zone of loess up to 50 km in width was deposited, primarily during the last glacial. The loesses are from one to ten meters thick and cover an undulating relief, so that the elevation differences have been still further reduced.

Tab.2 summarizes the differences between the young moraine and old moraine landscapes.

6 The Cologne Lowland Embayment and the lower Rhine

There are extensive deposits of loess in the Cologne Lowland Embayment from the last glaciation and from earlier Pleistocene cold periods. These loesses cover the gravel terraces of the Main Terraces of the Rhine and the Meuse (Maas). The terraces are older than the Elster glaciation but have been tectonically modified in the late Quaternary.

A push moraine area occurs on the lower Rhine between Düsseldorf and Kleve. This is the only area along the southern boundary of the Pleistocene glaciation in which large end moraines occur. Since there was no permafrost under the channels of the large rivers the unfrozen gravels were pushed into moraines by the Drenthe ice sheet (THOME 1980). The Main Terrace gravels, located east of the Rhine at a higher elevation, were overrun by the ice but were little affected by glacial erosion.

The terraces of the Rhine become lower in elevation towards the Dutch border and eventually disappear under the Holocene deposits. The Rhine delta is an area of subsidence and the Pleistocene Rhine sediments have, therefore, been preserved underneath.

In the north of the Federal Republic, therefore, only 6,700 km^2 is young moraine landscape in contrast to 53,000 km^2 of lowland with older moraines, including an area of 7,300 km^2 along the coast. In addition, there is an area of about 5,900 km^2 in the Central German Uplands. An additional 5,490 km^2 in the Cologne Lowland Embayment and area of the lower Rhine has not been covered by ice during the Pleistocene. Fig.3 shows a schematic profile of the major landscape types between the Baltic Sea and the Rhenish Slate Mountains.

References (since 1980)

BEHRE, K.-E. & LADE, U. (1986): Eine Folge von Eem und vier Weichsel-Interstadialen in Oerel/Niedersachsen und ihr Vegetationsablauf. Eiszeitalter u. Gegenwart **36**, 11-36.

CEPEK, A.G. (1986): Quaternary Stratigraphy of the German Democratic Republic. In: SIBRAVA, V., BOWEN, D.G. & RICHMOND, G.M. (eds), Quaternary Glaciations in the Northern Hemisphere. Quaternary Science Reviews **5**, 359-372.

CIMIOTTI, U. (1983): Zur Landschaftsentwicklung des mittleren Trave-Tales zwischen Schwissel und Bad Oldesloe, Schleswig-Holstein. Berliner Geographische Studien **13**, 92 pp.

FRÄNZLE, O. (1988): Periglaziäre Formung der Altmoränengebiete Schleswig-Holsteins. Berliner Geographische Abhandlungen **47**, 23–35.

GALBAS, P.-U., KLECKER, P.M. & LIEDTKE, H. (1980): Erläuterungen zur Geomorphologischen Karte 1:25,000 der Bundesrepublik Deutschland, GMK 25 Blatt 3415 Damme. 48 pp.

GRUBE, F., CHRISTENSEN, S., VOLLMER, T., DUPHORN, K., KLOSTERMANN, J. & MENKE, B. (1986): Glaciations in North West Germany. In: SIBRAVA, V., BOWEN, D.Q. & RICHMOND, G.M. (eds), Quaternary Glaciations in the Northern Hemisphere. Quaternary Science Review **5**, 347-358.

HANISCH, J. (1980): Neue Meeresspiegeldaten aus dem Raum Wangerooge. Eiszeitalter u. Gegenwart **30**, 221-228.

HOUMARK-NIELSEN, M. (1980): Glacialstratigrafi i Danmark ost for Hovedopholdslinien. Dansk Geologisk Förenig, 61-76.

KLUG, H. (1980): Art und Ursachen des Meeresanstiegs im Küstenraum der südwestlichen Ostsee während des jüngeren Holozäns. Berliner. Geogr. Stud. **7**, 27-37.

KLUG, H. (1985): Küstenformen der Ostsee. In: NEWIG, J. & THEEDE, H. (eds), Die Ostsee, Natur.- und Kulturraum.

LIEDTKE, H. (1981): Die nordischen Vereisungen in Mitteleuropa. Forsch. z. dt. Ldkd. **204** (2. Aufl), 307 pp.

LIEDTKE, H. (1983): Periglacial slopewash and sedimentation in Northwestern Germany during the Würm (Weichsel) Glaciation. Permafrost, Fourth Intern. Conf. in Fairbanks, Proceedings, 715-718.

LINKE, G. (1981): Ergebnisse und Aspekte der Klimaentwicklung im Holozän. Geol. Rdsch. **70**, 773-783.

RZECHOWSKI, J. (1986): Pleistocene Till Stratigraphy in Poland. In: SIBRAVA, V., BOWEN, D.Q. & RICHMOND, G.M. (eds), Quaternary Glaciations in the Northern Hemisphere. Quaternary Science Reviews **5**, 365-3

STEPHAN, H.-J. (1981): Eemzeitliche Verwitterungshorizonte im Jungmoränengebiet Schleswig-Holsteins. Verh. naturwiss. Ver. Hamburg N.F. **24**, 161-175.

STREHL, E. (1986): Zum Verlauf der äußeren Grenze der Weichselvereisung zwischen Owschlag und Nortorf (Schleswig-Holstein). Eiszeitalter u. Gegenwart **36**, 37-41.

STREMME, H.E. (1982): Pedostratigraphie in Schleswig-Holstein. In: EASTERBROOK, D.J. et.al. (eds), Quaternary glaciations in the northern hemisphere. IUGS - Unesco International Correlation Program, Proj. 73-1-24, 7, 223-231.

THOME, K. (1980): Der Vorstoß des nordeuropäischen Inlandeises in das Münsterland in Elster- und Saale Eiszeit. Westfäl. Geogr. Studien **36**, 21-40.

VELICHKO, A.A. & FAUSTOVA, M.A. (1986): Glaciations in the East European Region of the USSR. In: SIBRAVA, V.. BOWEN, D.Q. & RICHMOND, G.M. (eds), Quaternary Glaciations in the Northern Hemisphere. Quaternary Science Reviews **5**, 447-461.

Address of author:
Prof. Dr. Herbert Liedtke
Geographisches Institut der Ruhr-Universität Bochum
Universitätsstraße 150
D-4630 Bochum 1

CATENA SUPPLEMENT 15 p. 25–44 Cremlingen 1989

THE CENTRAL GERMAN UPLANDS

Wolfgang **Andres**, Marburg

1 Introduction

The Central German Uplands are a region of great geological and geomorphological complexity (see AHNERT in this volume). Because it is not possible to discuss the area in detail within the space available, emphasis is placed on the Rhenish Slate Mountains (Rheinisches Schiefergebirge) and adjacent areas. SPÖNEMANN, BRUNOTTE and GARLEFF discuss other parts of the Central Uplands in this volume and SCHMIDT describes areas of limestone in the Rhenish Slate Mountains in detail. The geomorphology of the Rhenish Slate Mountains has been intensively researched, particularly the development of valleys, of Tertiary planation surfaces and of Quaternary river terraces. A number of papers on the geomorphology of the area have appeared in FUCHS et al. (1983).

2 The major morphostructural units of the Central Uplands

The most important structural feature of the Central Uplands is the Rhenohercynian zone of the Variscan Fold Mountains. Its rocks were folded in the Carboniferous. Where they lie at the surface in, for example, the Rhenish Slate Mountains and the Harz, they are part of fault block mountains, the boundaries of which follow the tectonic pattern of the Variscan orogeny, either SW-NE (Ore Mountain or Variscan orientation) or SE-NW (Hercynian orientation).

Between the Harz and the Rhenish Slate Mountains, and also at the western margin of the latter, the old folded basement is overlain by the Mesozoic sedimentary cover and cuestas and tablelands predominate (fig. 1). The structural influence of the underlying Variscan basement is, nevertheless, apparent at the surface in the orientation of faults, joints and deformations of the strata.

During the Alpine orogeny the Central German Uplands were located in the forward marginal zone of the collision belt (ILLIES & FUCHS 1983). Both the basement rocks and the Mesozoic cover were divided into a large number of fault blocks. Paleozoic faults were reactivated and new faults formed. In most cases they were normal faults aligned mostly either NNE-SSW (Rhenish orientation) or NNW-SSE (Egge orientation). The pattern of tectonic lines has a major influence on the development of the surface forms.

Along the northern margin, in southern Lower Saxony, the Mesozoic cover rocks have been folded and faulted (Saxonic tectonics) so that a combination of cuestas and hogbacks occur in this area instead of the tablelands and cuestas

ISSN 0722-0723
ISBN 3-923381-18-2

3-923381-18-4/89/5011851/US$ 2.00 + 0.25

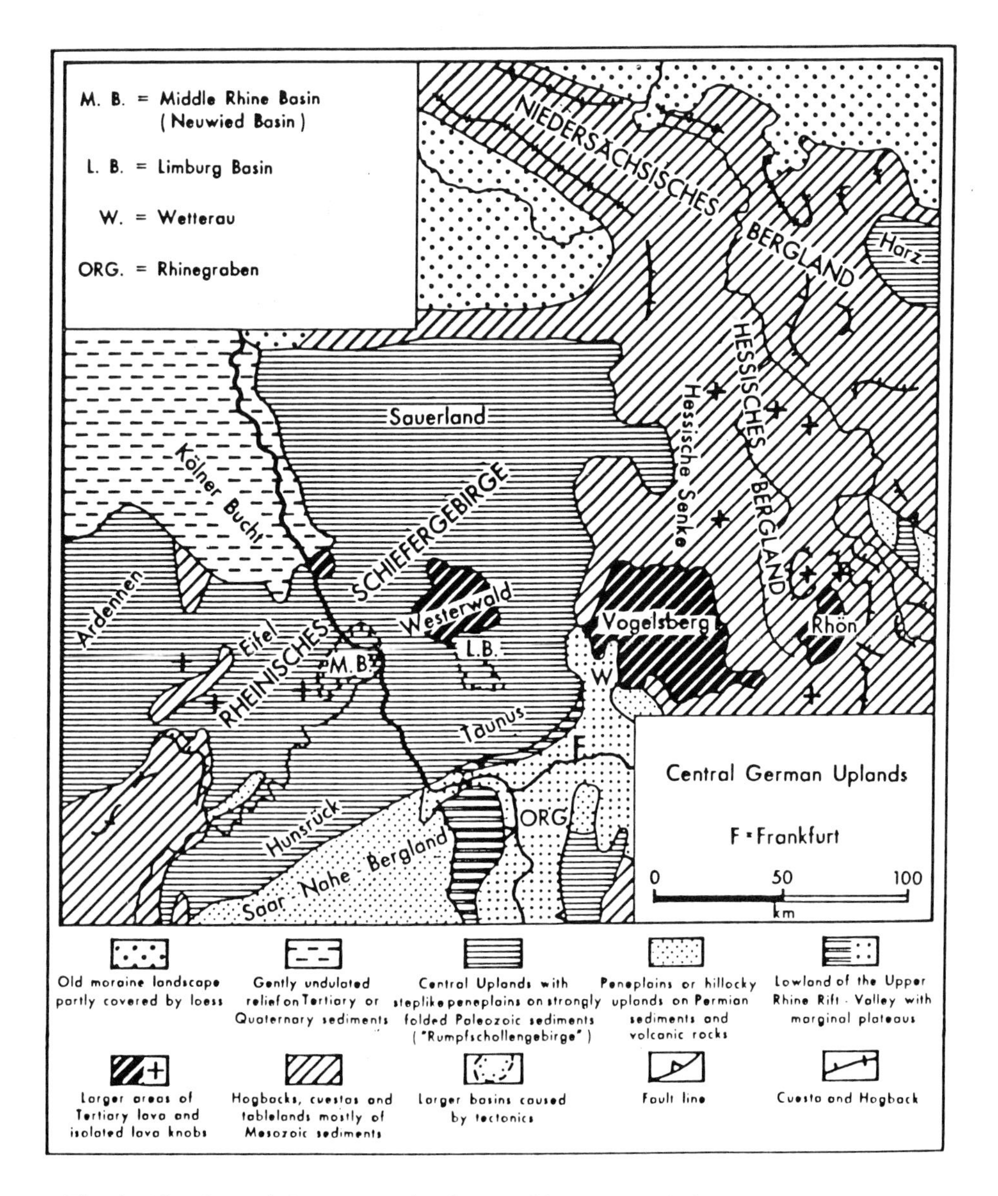

Fig. 1: *The Central German Uplands. Landform types (after LIEDTKE 1984).*

found elsewhere. The Cenozoic faulting was accompanied by volcanism which affected the Eifel, the Siebengebirge and the Westerwald in the Rhenish Slate Mountains and the Vogelsberg, the Rhön and, locally, several parts of northern Hessen in areas of the Mesozoic cover rock (fig.1). At the western margin of the middle Rhine basin (Neuwied basin) and in the southern Eifel the volcanism continued until the late Quaternary. The most recent eruptions took place less than 10,000 years ago.

The North Palatinate Upland (Saar-Nahe Bergland, fig.1) is composed of Permian sedimentary rocks that were derived from the denudation of the Variscan fold mountains and deposited in a synclinal intramontane basin of that mountain system. The sedimentation was complemented by post-orogenic volcanism which left large massifs of porphyry and extensive sheets of melaphyr.

All rivers in the Central Uplands drain to the North Sea. Most of the area of the Rhenish Slate Mountains is tributary to the Rhine; the northern part of the region of Mesozoic cover rock is tributary to the Weser and the greater part of the Harz is tributary to the Elbe.

3 The Rhenish Slate Mountains and adjacent areas

The Rhenish Slate Mountains are a remnant of the Variscan fold mountains the central part of which was an upland, and therefore an area of denudation, during much of the Mesozoic and the entire Cenozoic. Devonian and Carboniferous slates and greywackes are the predominant rocks. More resistant beds of quartzite support prominent ridges, particularly in the Hunsrück and the southern Taunus. Middle Devonian limestones occur in synclines (see SCHMIDT in this volume) and locally the stratigraphic sequence includes predominantly basic Devonian volcanic rocks.

In all parts of the Rhenish Slate Mountains there are extensive denudation surfaces at elevations of more than 400 m above sea level. Only a few summits and ridges are higher. The highest local relief occurs along the deeply incised valleys. A similar association of landforms is found in many other parts of the Central Uplands.

3.1 Morphological development of the Rhenish Slate Mountains in the early Tertiary

It has long been recognized that the Variscan folds are truncated by peneplains at different levels. The work of W.PENCK (1924) greatly stimulated discussion and the different levels were interpreted as having different ages. Although it was also known that on these peneplains remnants of a deep regolith occurred, relics of terrestial weathering under tropical to subtropical conditions, no connection was made between the peneplanation process and climate. The peneplains were thought to be the final stage of long-lasting cycles of erosion as postulated by W.M.DAVIS. JESSEN (1938) and, particularly, BÜDEL (1963) first suggested a predominantly climatogenetic interpretation. The summits and ridges that rise above the peneplains were assumed by some to be areas of greater uplift and by others to be due to greater rock resistance (SEMMEL 1984, RATHJENS 1977). The development of the quartzitic ridges of the Hunsrück and Taunus Mountains is probably due to both causes.

Photo 1: *The upper middle Rhine valley near Kaub (view southward).*
Above the narrow valley with the small town of Kaub and the castle (Pfalz) in the river, the Main Terrace is marked by the fields on the right side of the photo. The middle ground is part of the Tertiary "Trogfläche". The background is dominated by the forested quartzite ridges of the Taunus (left) and the Hunsrück (right).

Photo 2: *Meanders of the Moselle near Minheim (view northward).*
On the gentle slip-off slopes the entire sequence of Pleistocene terraces is present. The steep undercut slopes extend from the river directly up to the Main Terrace.

Photo 3: *The valley of the lower Lahn near Cramberg (view northwestward).*
The deeply incised valley has a broad Low Terrace here, above which the slopes rise to the flat early Pleistocene Main Terrrace, the gravels and sands of which are frequently covered with loess.

Photo 4: *The Pulvermaar near Gillenfeld in the Eifel (view northward), a classical example of a volcanic explosion crater.*
The nearly circular crater, about 700 m in diameter, lies in the Paleozoic bedrock and is rimmed by volcanic ash that is only a few meters thick. The lake is over 70 m deep. In the foreground is the forested cinder cone of the Römerberg, with a small maar on its flank. All these volcanic landforms are of late Pleistocene age.

The development of the highest denudation surfaces began in most areas of the Rhenish Slate Mountains in the Mesozoic. The early Tertiary peneplains in the Rhenish Slate Mountains had a gently undulating relief with inselbergs (BREMER 1978). Towards the margins, the gradients were probably considerable. Tectonic movements occurred both within and at the margins of the peneplains so that the denudation levels were uplifted and deformed. The reconstruction of continuous denudation levels over great distances on the basis of the elevations of peneplain remnants is, therefore, not possible.

In contrast to the peneplains on which there are relics of deep weathering, the younger planation surfaces (Trogflächen or trough surfaces) follow the main lines of the present-day drainage systems and are covered by fluvial quartz gravels in some areas. The Trogflächen indicate a phase of increasing uplift which resulted in the removal of the deep regolith cover on the old peneplains and the initiation of valley development. The quartz gravels were for the most part derived from the weathering of resistant quartz dikes in the Devonian rocks. The finer material of the old regolith was largely transported into the adjacent basins where it is the major constituent of the widespread clayey deposits found there. However, thick layers of clay also occur locally within the Rhenish Slate Mountains, often interbedded with sands and gravels. These fluvial deposits allow dating of the end of the peneplanation process and of the renewal of erosion due to uplift. The deposition of the quartz gravels, the Vallendar gravels, probably took place during the Oligocene and, locally, also in the upper Eocene, particularly the finer-grained facies (LÖHNERTZ 1978), suggesting that the peneplanation was largely completed by the early Tertiary (GLATTHAAR & LIEDTKE 1984). In some areas, such as the Hohes Venn, the Eifel and the Ardennes, marine abrasion by the Cretaceous sea probably also played a role in the planation process.

The Vallendar gravels were deposited in several independent drainage basins. Their deposition within the upland, their spatial distribution and the relative coarseness of their grain sizes indicate that a well-developed valley system must have existed before the middle of the Oligocene. In the middle and upper Oligocene, the marine transgression from the Mainz Basin and the Lower Rhine Embayment (Cologne Embayment) advanced far into the Rhenish Slate Mountains; marginal regions and zones of subsidence were buried under deposits of gravel that attained great thicknesses in some areas (SONNE 1970, LÖHNERTZ 1978, ZÖLLER 1985, GLATTHAAR & LIEDTKE 1984). Some of the limestone areas, which had been lowered by solution denudation (SCHMIDT 1975), were inundated and the karst landforms were filled with Oligocene sediments (ANDRES & PREUSS 1983).

The varying elevation and thickness of the Vallendar gravels were interpreted by LOUIS (1953) to indicate that a pre-Oligocene relief had been buried. Later BIRKENHAUER (1972) attempted to link the peneplanation with these gravel accumulations whereby the planation was thought to have taken place by lateral erosion that started from the surface of the gravel fill. An older, middle Oligocene fill was supposed to have reached to the present 360 m level and an Oligocene/Miocene fill to the 400 m level and to have caused planation at these levels. This has not, however, been proven conclusively (ANDRES, BIBUS & SEMMEL 1974, BIBUS 1980). For the following reasons it seems unlikely that such a burial occurred:

1. Tectonic uplift and subsidence of the Rhenish Slate Mountains as a whole, as postulated by BIRKENHAUER, has been disproved by other geological and geomorphological investigations (ILLIES & FUCHS 1983, MURAWSKI et al. (1983).

2. Even several hundred meters of regolith removal could not supply the quantity of quartz gravels necessary for the postulated burial of the relief (SEMMEL 1984).

3. An accumulation surface which has a given elevation for only a short time cannot provide the impulse for extensive truncation surfaces in bedrock.

It is reasonably certain that the lower position of the Oligocene quartz gravels is due to the local burial of a valley relief and to synsedimentary and post-sedimentary tectonic displacement (SEMMEL 1984, HÜSER 1973, BIBUS 1983).

The development of the present drainage system, and particularly the Rhine valley through the Rhenish Slate Mountains, was probably initiated during the regression of the sea from the Mainz Basin and the Cologne Embayment in the upper Miocene. The drainage followed the linear tectonic depressions that were formed in the Oligocene and in the lower Miocene and that are believed to have connected the Mainz Basin, the Neuwied Basin and the Cologne Embayment in the middle Miocene (BOENIGK 1982, QUITZOW 1974, SEMMEL 1984). Superposition of streams by erosion into local Oligocene gravel fills has also taken place in some valleys (ZÖLLER 1985).

The spatial and temporal differentiation of Tertiary crustal movement is also reflected in the age and distribution of volcanic events. The first eruptions took place in the Eifel in the Eocene and continued into the Oligocene. During this period the centre of activity shifted eastward to the Siebengebirge and the Westerwald. In the Miocene it moved to the Hessian Upland and the Vogelsberg, east of the Rhenish Slate Mountains. The easternmost volcanic area, the Rhön, was, however, active first in the Oligocene (LIPPOLT 1983).

3.2 Geomorphological development in the Rhenish Slate Mountains during the Pliocene

Pediment-like surfaces extend from the older denudation levels down to broad valley zones into which the present large river valleys have been incised. Usually two separate levels can be distinguished (ANDRES 1967, BIBUS 1980, KAISER 1961). In the central Rhenish Slate Mountains these levels lie at about 300 m above sea level and are covered by a thin layer of quartz gravels. The high proportion of quartz pebbles and the heavy mineral components show that these deposits are Pliocene rather than Pleistocene. Quartz oolite pebbles that have been transported from the upper Rhine indicate that the Rhine river was already flowing from south Germany to the Cologne Embayment (KAISER 1961). In the Mainz Basin to the south, the Pliocene sediments are much thicker than farther north, indicating that the Rhine river was aggrading before flowing into the Rhenish Slate Mountains. The lower Pliocene deposits of the Rhine in the mountains have been displaced upwards by about 50 m above those in the Mainz Basin (ANDRES & PREUSS 1983). On the lower Moselle, BIBUS (1983a) has found a similar magnitude of post-Pliocene displacement. At the northern margin of the mountains, the Pliocene valley floors are warped or faulted downwards. In the low-lying area of the Cologne Embayment they form a thick sediment series beneath the Quaternary gravels (BRUNNACKER & BOENIGK 1983; fig.2 and fig.3). Their correlation with the Pliocene gravels of the middle and upper Rhine regions was established by BOENIGK (1978) on the basis of their lithology and heavy minerals.

3.3 Examples of the Quaternary geomorphological development in the Rhenish Slate Mountains

In central Europe, the Tertiary was undoubtedly a period of prevailing planation surface development and the Quaternary a period of prevailing valley formation. Valleys were also formed in the early Tertiary but the changeover from the formation of shallow, broad-floored valleys to the incision of narrow valleys did not, in general, take place until the second half of the Quaternary. The remnants of former valley floors of the Rhine and of its major tributaries provide an example of these phases of valley formation and of the influence of climate and tectonics (fig.2 and fig.3).

The oldest Quaternary terraces shown in fig.3 were also formed in broad valleys. On the middle Rhine, the Moselle and the Lahn several of these terrace levels have been recognized above the younger narrow valley. They have been subdivided into six separate levels (ANDRES & SEWERING 1983, MÜLLER 1974, BIBUS & SEMMEL 1977, BIBUS 1980). Because of their width, these early Quaternary terraces are termed the Main Terraces (Hauptterrassen). On the lower Nahe, GÖRG (1984) has distinguished 12 early Pleistocene terraces, the result probably of more differentiated tectonic movements at the southern margin of the Hunsrück. The composition of the fluvial deposits changed from the Pliocene to the early Quaternary. The Pliocene gravels consist entirely of white quartz pebbles but the gravels of the Quaternary Main Terraces have many different lithological components and are multicoloured. Only very easily weathered materials are absent from these gravels. The Main Terrace gravels also contain a much greater variety of heavy minerals than the Pliocene deposits (BIBUS & SEMMEL 1977, NEGENDANK 1978). Pediment-like transition slopes often occur between the Pliocene terraces and the oldest Quaternary levels.

Below the Main Terraces, the valleys of the Rhenish Slate Mountains are usually narrow. The Middle terraces (Mittelterrassen) occur on the valley slopes, in many cases only as bedrock terraces and flatter slope segments so that their correlation is difficult, particularly in the narrow valleys of the upper middle Rhine between Bingen and Koblenz and of the lower Lahn. The Middle Terraces in the lower middle Rhine valley, (downstream from Koblenz) do, however, usually have deposits of sediments that are several meters thick (BIBUS 1980, BRUNNACKER & BOENIGK 1983). The number of identified Middle Terraces varies in different segments of

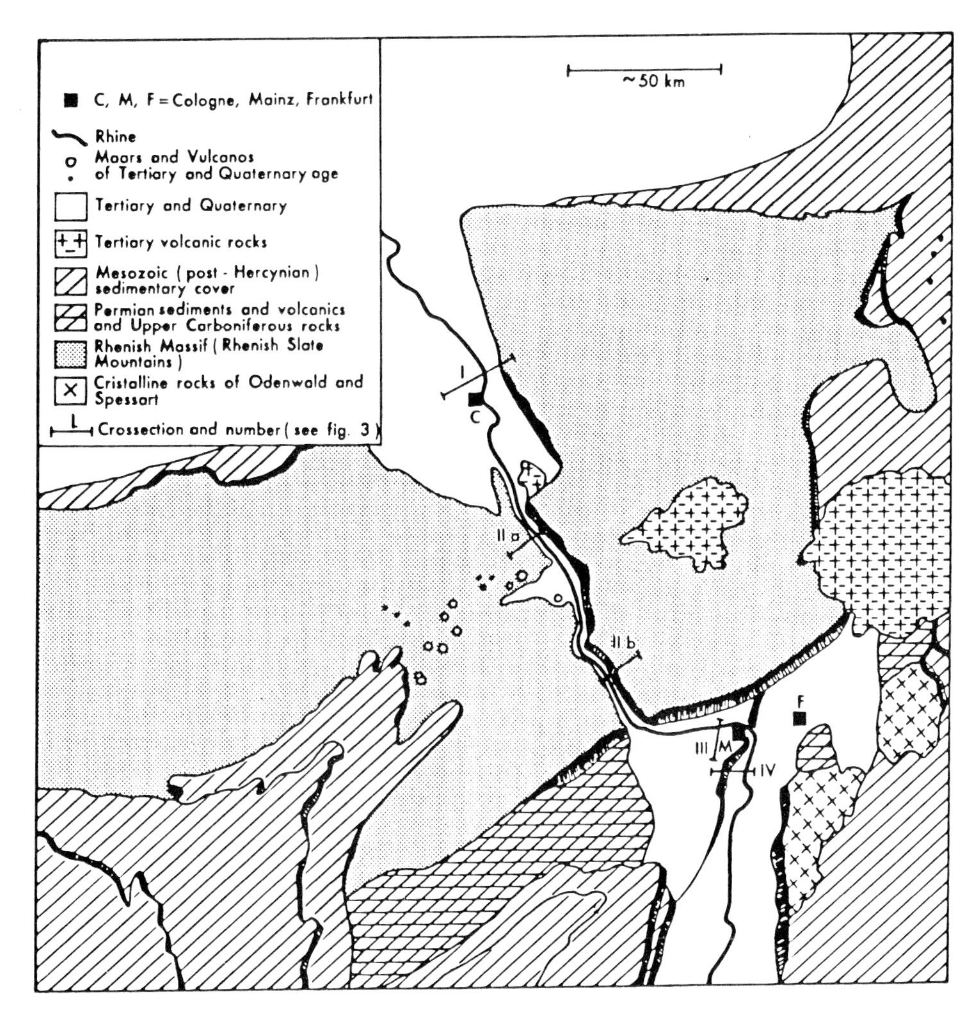

Fig. 2: *The geological structure of the Rhenish Slate Mountains and the position of the cross-sections in the Rhine Valley shown in fig.3.*

the valley between three and six, depending on their state of preservation. Most of the larger river valleys have, in addition, two Low Terraces (Niederterrassen) near the present valley floor (fig.3).

There are a total of 11 to 12 Quaternary terraces on the middle Rhine and the lower Lahn and 9 to 10 on the lower Moselle. The incision of the narrow younger valley into the level of the lowest broad Main Terrace occurred at the same time in all major valleys.

Along the upper courses of the Lahn and the Moselle there are fewer terraces (HEINE 1970, LIPPS 1985, M.J.MÜLLER 1976). Also, in the smaller tributary valleys, terraces occur mostly in the lower stretches close to the trunk valleys. In the up valley direction, first the Middle Terraces and farther up

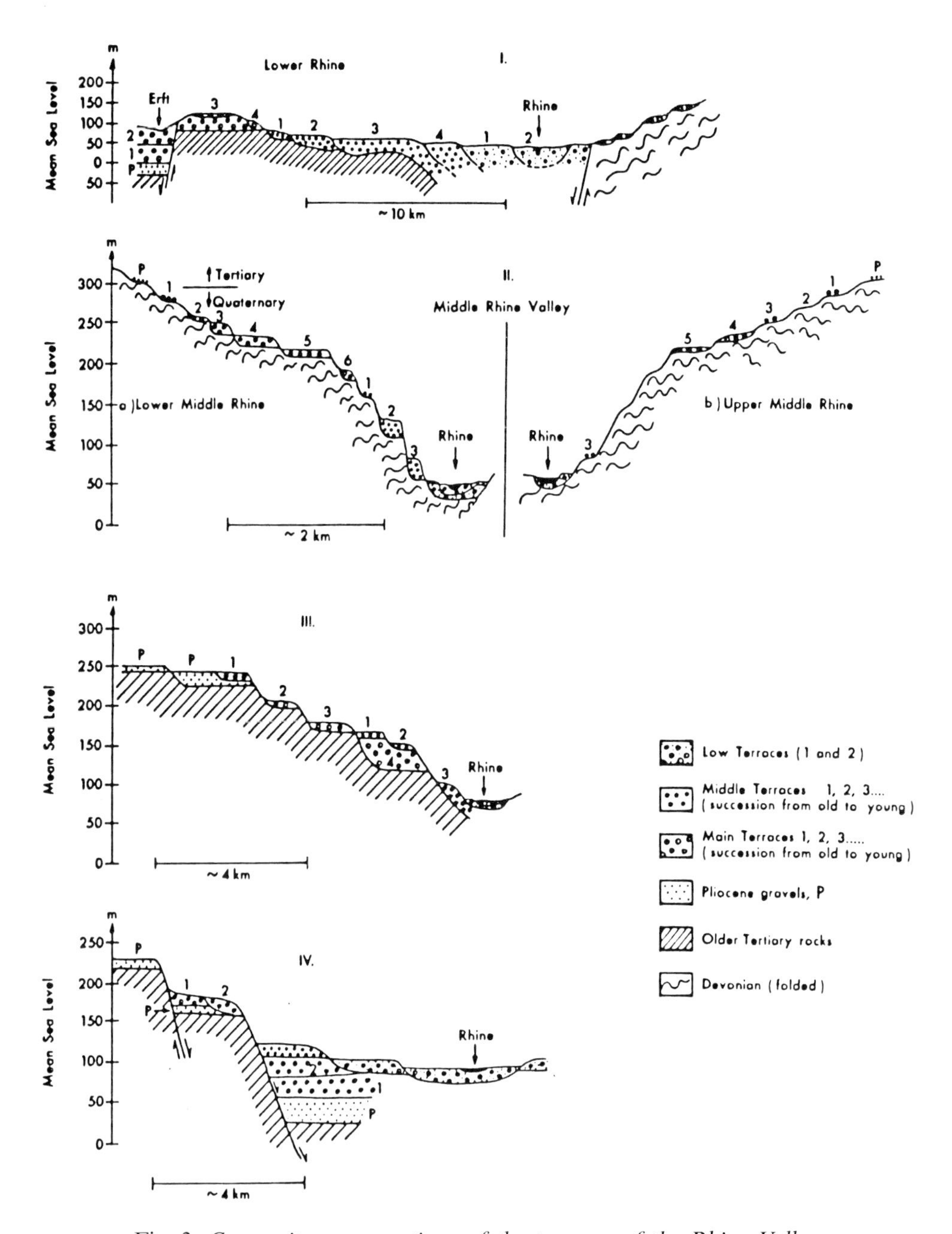

Fig. 3: *Composite cross-sections of the terraces of the Rhine Valley.*

I = Cologne Embayment, IIa = lower middle Rhine valley, IIb = upper middle Rhine valley, III = Rheingau between Mainz and Bingen, IV = northern Rhine Rift Valley.
(Based on BIBUS 1980, BIBUS & SEMMEL 1977, BRUNNACKER 1978, KANDLER 1971, SEMMEL 1983a, ANDRES unpublished)

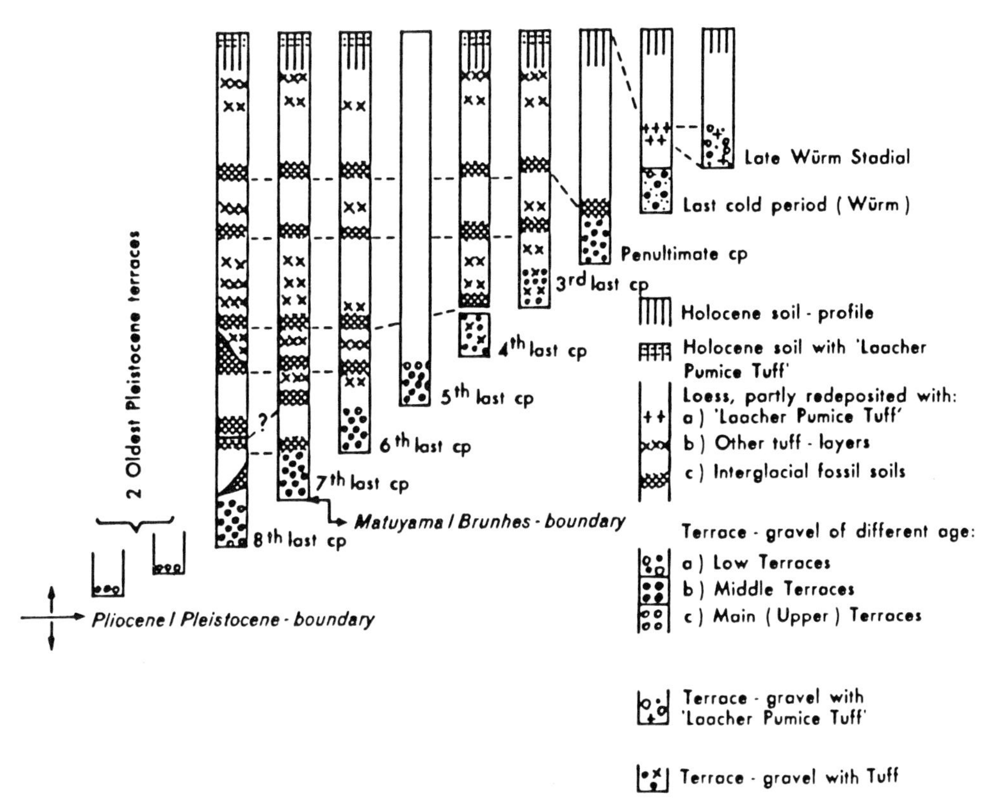

Fig. 4: *The terraces of the middle Rhine valley and their sedimentary cover dated by fossil soils, tuff layers and paleomagnetic remnants (based on BIBUS 1980, simplified).*

the Main Terraces merge into the valley floor. It can be assumed, therefore, that the shallow trough-shaped upper segments of these small valleys are little changed early Pleistocene or, in some cases, Tertiary relic forms. The typical distribution of terraces along these valleys indicates that the trunk streams controlled the processes of incision and accumulation.

The tendency towards valley deepening predominated in the entire central German uplands throughout the second half of the Quaternary. Uplift of the upland or subsidence of the marginal areas must, therefore, have occurred in order for the gradients to have been preserved. The total amount of valley deepening since the end of the Tertiary and consequently the probable amount of uplift, is about 200 m on the middle Rhine, the lower Lahn and the lower Moselle; on the smaller Rhine tributaries, the amount of deepening decreases up valley.

The typical sequence of terraces present in all major valleys shows that accumulation and vertical erosion alternated repeatedly during the general valley deepening. Lithological and biological evidence indicates that most of the terrace gravels accumulated during the Pleistocene cold periods (STEINMÜLLER 1982). The alternations between accumulation and incision must, therefore, be ascribed, primarily, to climatic causes.

A direct influence of the eustatic sea level variations on the erosion and accumulation by the rivers of the Rhenish Slate Mountains (ROHDENBURG 1968) appears not to have been possible because there was an area of continuous Quaternary sedimentation between the areas of erosion and the sea. In addition, the gradient of the Rhine, which extended into the North Sea Basin during the marine regression, was too low to cause headward erosion in the upper stretches of the rivers.

As a result of recent research, especially the investigation of the sediment covers of the terraces and the paleosols contained in them, of paleomagnetism and of volcanic tuff in the deposits, the dating of the terraces and their sediments has made considerable progress (fig.4).

The two Lower Terrace levels were developed during the last cold period. Paleomagnetic evidence shows that the deposits of the seventh-oldest cold period in the Main Terrace complex were accumulated at the time of the Matuyama/Brunhes magnetic reversal, that is, at about 700,000 years B.P. (BIBUS & SEMMEL 1977, SEMMEL 1983). Each of the subsequent cold periods is represented by a terrace and the mean duration of a cold period/warm period alternation was in the order of 100,000 years. During a period of more than a million years prior to 700,000 B.P., there was less than 100 m of valley deepening. On the other hand, the present narrow valley below the Main Terraces formed within a period of 500,000 years and the maximum incision was about 150 m (fig.3).

It has been suggested that the accelerated incision was caused by the greater severity of the cold periods (HEINE 1971). The presence of more intensively weathered reddish paleosol horizons on the older terrace deposits supports this interpretation (SEMMEL & BIBUS 1977, BIBUS 1980). Tectonic influences were, however, of at least equal importance. The Main Terraces have been warped on the margins of the Rhenish Slate Mountains and on the margins of the tectonic basins within the mountains (Limburg Basin, Neuwied Basin). The younger terraces are not deformed. The Main Terraces have been displaced vertically by about 35 m in the Neuwied Basin relative to the middle Rhine valley (BIBUS 1983) and by about 20 m in the lower Lahn valley relative to the Limburg Basin (ANDRES & SEWERING 1983). The displacement of the Main Terrace sediments within the Rhenish Slate Mountains relative to Main Terrace sediments of identical age in the adjacent areas are even greater. Such displacements can only be the result of intensified tectonic activity. Present-day seismic evidence indicates that these movements are continuing (AHORNER 1983). The renewal of volcanic activity in the western Eifel, the eastern Eifel and the Neuwied Basin is another indication of increased tectonic activity (LIPPOLT 1983, VAN DEN BOGAARD & SCHMINCKE 1988, ILLIES & FUCHS 1983).

Volcanic landforms such as cinder cones, stratovolcanoes and maars, and also covers of lava and volcanic ash, resulted from the most recent volcanism in the Rhenish Slate Mountains. The last major eruption created Lake

Laach (Laacher See) NW of Koblenz and spread a cover of pumice several meters thick on the adjacent landscapes. The Laach pumice, ejected 11,000 years ago, occurs in late Pleistocene sediments in nearly all parts of central Europe. There are also other, usually basic, tephra layers that form important index horizons in the Quaternary deposits (FRECHEN & LIPPOLT 1965, VAN DEN BOGAARD & SCHMINCKE 1988).

Maars do not indicate a final phase of the volcanic activity but are formed at the same time as cinder cones and lava volcanoes; which of these three types of eruption takes place depends on the differing influence of ground water on the rising magma (phreatomagmatism, BÜCHEL & LORENZ 1982). The series of volcanic events during the Quaternary suggests that at the present time volcanism in the Rhenish Slate Mountains is dormant but not extinct.

Quaternary relief development in the basins within the Rhenish Slate Mountains has not differed fundamentally from that in the upland areas. Although the terraces and the present valley floors in the Neuwied Basin and the Limburg Basin are much broader than elsewhere and their sediment covers, including the periglacial loess cover, are usually much thicker and more widely distributed, the very broad, shallow early Pleistocene valleys in the basins have also undergone a similar phase of intensified vertical erosion in the younger Quaternary. The basins were uplifted at this time but to a lesser extent than the upland areas and, therefore, subsided relative to the uplands (ANDRES & SEWERING 1983, BIBUS 1983).

The terraces of the Rhine descend steeply northward in the direction of the Lower Rhine Embayment and converge (BRUNNACKER & BOENIGK 1983, SCHNÜTGEN 1974, BIBUS 1980 and 1983). The complicated tectonic movements have caused the gravels of the older Main Terraces to be placed as a sedimentary sequence on top of one another. The younger Main Terraces and the Middle and Lower Terraces are, however, arranged in a normal terrace sequence (fig.3). This indicates, in addition to an eastward shift, an erosional lowering of the course of the Rhine in the later phases, although by a much smaller amount than occurred in the Rhenish Slate Mountains. Farther north the entire sequence of late Tertiary and Quaternary sediments lies below the present land surface (ZAGWIJN 1974).

Although not all the sediments and terrace levels in the Rhenish Slate Mountains and the adjacent areas have been correlated (SEMMEL 1984), considerable progress has been made in the dating and correlation of Quaternary profiles in recent years (BRUNNACKER & BOENIGK 1983, BIBUS 1980 and 1983, BIBUS & SEMMEL 1977. The Quaternary stratigraphy has also provided a chronological base for prehistorical research (VAN DEN BOGAARD & SCHMINCKE 1988).

The Pliocene and early Pleistocene terraces (Main Terraces) in the major river valleys retain nearly constant absolute elevations while the Middle and Lower Terraces have longitudinal gradients that are very similar to the gradients of the present river. Tilting cannot be the cause since the older terraces have a constant elevation in different valleys, including some in which the rivers flow in opposite directions, such as the Lahn and the Moselle. The Main Terraces cannot have had low gradients at the time of their formation because of the coarse grain sizes of their gravels.

QUITZOW (1974) has suggested that the more deeply excavated lower valley stretches have been subject to greater isostatic ulift as compensation for the greater removal of material. SEMMEL (1984) has pointed out that in this case the younger terraces would also have been affected. Possibly, however, isostatic compensation since the formation of the younger terraces has not been sufficient for a deformation of their levels to be apparent.

There are spatial and temporal variations of the vertical crustal movements in the Rhenish Slate Mountains. At the present time the uplift rate is highest in the Ardennes and the Eifel, owing perhaps to the presence of a mass of relatively lower density and higher temperature in the upper mantle (FUCHS et al. 1983). An uplift tendency at the margins of the Neuwied Basin has also been detected. (MÄLZER, HEIN & ZIPPELT 1983). Since the youngest volcanic activity, which has been related to the uplift-causing diapir in the upper mantle, took place on the western margin of the Neuwied Basin, the wider environment of the Basin has probably also been affected subsequent to the formation of the Main Terraces by a phase of uplift the centre of which lay west of the Rhine. The lack of gradient on the Main Terrace levels in the valleys of the middle Rhine, the Lahn, the Moselle and other Rhine tributaries could also have resulted from this phase of tectonic uplift.

4 Other areas of the Central Uplands

4.1 The Harz

The geologic structure and the Tertiary and Quaternary geomorphological history of the Harz differs from the Rhenish Slate Mountains in two essential respects. First, a granite batholith has intruded into the Variscan fold complex and now forms the highest elevation in the Harz, the Brocken (1142 m). Secondly, as a consequence of its greater height and exposed position, there was a local Pleistocene glaciation in the Harz (DUPHORN 1968). Landforms and material from the last glaciation are present, particularly in the Oker valley. The area was probably also glaciated during earlier cold periods.

The old peneplains of the Harz truncate the Variscan folds and occur at several levels in the form of a piedmont benchland (Rumpftreppe). A younger Harz plateau surface (Harzhochfläche) is thought to be the equivalent of the Trogflächen of the Rhenish Slate Mountains. The Pliocene phases of the present drainage net are apparent from the denudation surfaces. During the Miocene, the Harz did not stand much above its foreland. Uplift occurred mainly in the late Tertiary and in the Quaternary on the northern edge along a sharply delimited fault and in the south and west along a broad monoclinal upwarp. Valley development was affected by the alternation of glacial and interglacial conditions. The transport of material by the streams took place mainly in the cold periods, as shown by the morphological and stratigraphical relationship between the large alluvial fans and the low terraces on the valleys at the northern margin of the Harz. No reliable dates have been established for the early Quaternary terraces and denudation levels.

4.2 The Lower Saxon Upland

The upland of Lower Saxony projects for some distance into the North German Lowland. The area consists of several distinct mountain systems of Mesozoic rocks (fig.1). These rocks were deformed during the Saxonic orogeny more strongly than in other areas of the Mesozoic cover, resulting in synclinal, anticlinal and monoclinal structures. Salt tectonics have also been important. The area drains northward by way of the Weser and the Leine. The strike of the beds is predominantly Hercynian (NW-SE). The Rhenish orientation (NNE-SSW) also occurs frequently and is particularly apparent in the Leine Rift Valley near Göttingen, a structural continuation of the Rhine Rift Valley and of the Hessian Depression, and in the Weser valley.

Hogbacks supported by resistant Mesozoic sedimentary rocks are the predominant landform. Summit heights range from 250 m in the north to more than 500 m in the south. Where the strata are less deformed, cuestas and extensive tablelands occur. Basins and major valley zones separate the individual mountain systems. The basic structural pattern originated during the Cretaceous period. Remnants of Tertiary marine sediments indicate that the phase of intensified denudation and of hogback and cuesta formation began after the late Oligocene (BRUNOTTE & GARLEFF 1979).

The extent to which the pattern of structural landforms that dominates at the present time in the Lower Saxon Uplands existed when denudation was initiated, has been discussed extensively (see SPÖNEMANN in this volume). It is generally thought that the relief was relatively low until the late Tertiary but that the structural landforms on resistant rocks were present, although not prominent. The relief became more pronounced during the Pliocene and, particularly, during the Pleistocene. The relief depended greatly on local conditions, such as the distance from the trunk streams which formed local base levels, and upon the amount of local uplift. In some areas salt tectonics were also important. For example, solutional removal of salt underground has caused subsidence locally. Erosion deepened the major valleys by 50 to 120 m during the Quaternary (BRUNOTTE & GARLEFF 1979). Pediments developed at several levels on less resistant rocks in the basins.

4.3 The Hessian Upland and the Hessian Depression

The Hessian Upland consists mainly of subhorizontal beds of Bunter Sandstone (lower Triassic). Muschelkalk (middle Triassic) also occurs in graben-like depressions and in some areas Tertiary volcanic rocks are present. The Hessian Depression, a zone of subsidence, extends through the entire area from SSW to NNE. In the north, it is connected to the Leine Rift Valley a continuation of the Rhine Rift Valley. The Mesozoic sedimentary rocks subsided in the depression and were covered by marine and limnic Tertiary deposits. Basaltic lava subsequently covered the Tertiary deposits in many areas. In the western Vogelsberg, very thick lava flows filled the depression and the area was further modified later by uplifts.

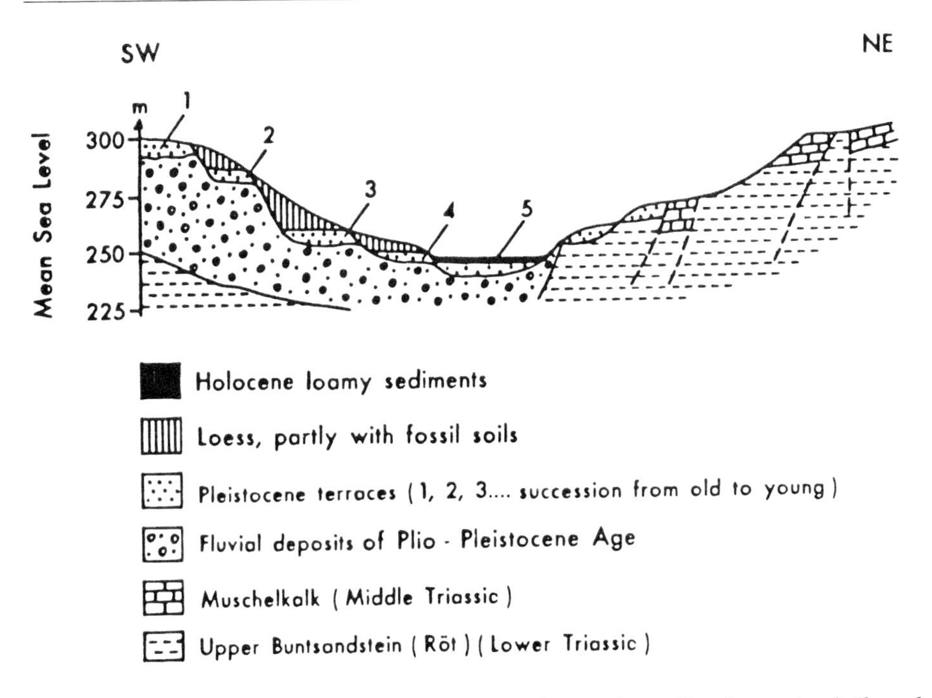

Fig. 5: *Cross-section of the Haune Valley in the northern Hessian upland (based on SEMMEL 1984, simplified).*

In the transition area from the Rhenish Slate Mountains to the Hessian Upland, peneplains extend from the Variscan basement rocks into the unfolded rocks of the Mesozoic cover. There are also cuestas and denudation surfaces, adjusted to the rock structure, that indicate a structure-dependent older phase of form development (SEMMEL 1966). The old denudation surface was dissected primarily in the Quaternary. In some valleys, Pliocene gravels occur at very low elevations, probably because of local subsidence due to solutional removal of salt deposits underground (fig.5).

The volcanic landforms in the Hessian uplands are for the most part small volcanic necks. At the margin of the Rhön they appear as a group of dome-shaped mountains. In the central area of the Rhön, extensive sheets of basalt have covered a landscape of low relief.

The Vogelsberg is a more or less circular basaltic upland with a central summit nearly 800 m above sea level. The slopes are almost uniform on all sides and the drainage is radial. The Vogelsberg is made up of a series of basaltic lava flows with a cumulative thickness of more than 300 m which covered a low-relief Miocene landscape. The area was uplifted subsequently. Borings have revealed underlying beds of Bunter Sandstone at more than 400 m above sea level below the basalt. The Tertiary deep weathering probably persisted longer in the basaltic areas than in the areas of the Variscan basement rocks or the Mesozoic rocks. Lateritic soils were formed as late as the Pliocene (BIBUS 1983). The intensive weathering on the Miocene basalt suggests that the high denudation surfaces on the Vogelsberg were being formed in the Pliocene, in contrast to other upland areas where the planation tendency had ended by this time. Valley formation took place, however, only in the Quaternary.

The Hessian depression, which is divided along part of its length into an eastern and a western branch, has a sequence of basins and subsidence zones. The subsidence zones generally contain early Tertiary marine and limnic sediments, a large part of which had been removed by fluvial erosion during the late Tertiary and early Quaternary. Much less fluvial downcutting occurred in these areas during the middle and late Pleistocene than in the adjacent uplands. Gentle valley forms and thick loess deposits are characteristic for this area. Pliocene lignite deposits overlain by Quaternary sediments indicate that only in the Horloff Graben, a branch of the Wetterau at the western margin of the Vogelsberg, did subsidence continue into the Quaternary.

4.4 The North Palatinate Upland and the Rhine-Hessen Tableland

The North Palatinate Upland (Saar-Nahe Bergland) differs from the Rhenish Slate Mountains primarily because of its geological structure. The two areas are separated by the southern marginal fault of the Hunsrück (AHORNER & MURAWSKI 1975; figs 1 and 2) which is also the boundary between the Rhenohercynian Zone and the Saxothuringian Zone of the former Variscan mountains. A deep synclinal trough within the Variscan mountains was filled with thick sediments of greatly varying grain sizes during the upper Carboniferous and the lower Permian (MURAWSKI et al. 1983). In the lower Permian major volcanic eruptions resulted in large porphyry massifs such as the Donnersberg and also lava flows (melaphyr) that were interbedded with the sedimentary series.

The landform development was similar to that in the Rhenish Slate Mountains. The oldest landform generation is a peneplain at about 600 m that was formed until the Eocene. Another denudation surface and a few areas of the trough levels (Trogfläche) can be distinguished at lower elevations (LIEDTKE 1969). The development of the trough levels was preceded by local sedimentary infilling (ZÖLLER 1985).

Valley development began in the Pliocene and attained its maximum in the late Pleistocene. ZÖLLER (1985) has assumed that the Quaternary uplift in this area varied greatly. Many of the valleys were superimposed, some by incision into covering rocks of low resistance, others by incision into local sedimentary fills (LIEDTKE 1969). The shape of the valley cross profiles is greatly influenced by rock resistance. Along the upper Nahe and its tributaries, narrow water gap valleys have developed in the resistant porphyries and melaphyrs, in contrast to the Permian sediments where there are broad valleys.

The Rhine-Hessen Tableland lies in the area between the Saar- Nahe Upland and the northern Rhine Rift Valley. The tableland does not belong to the Central Uplands but to the subsidence zone of the upper Rhine. It is geologically identical to the Mainz Basin into which the sea penetrated in the early Tertiary and deposited clayey, marly and calcareous sediments. Denudation has occurred in this area since the upper Miocene. Valleys subdivide the nearly horizontally bedded marine and brackish sediments into several plateaus which lie at about 250 m. The late Tertiary and Quaternary landform development in this area has been discussed in section 3 (figs 2 and 3).

References

AHLBURG, J. (1915): Über das Tertiär und das Diluvium im Flußgebiet der Lahn. Jb. preuß. geol. Landesanstalt 1914, **36**, 269–373.

AHORNER, L. (1983): Historical seismicity and present-day microearthquake activity of the Rhenish Massif, Central Europe. In: FUCHS, K. et al. (eds), Plateau Uplift, Springer Verlag, Berlin, Heidelberg, 198–221.

ANDRES, W. (1967): Morphologische Untersuchungen im Limburger Becken und in der Idsteiner Senke. Rhein-Main. Forsch. **61**, 88 pp.

ANDRES, W., BIBUS, E. & SEMMEL, A. (1974): Tertiäre Formenelemente in der Idsteiner Senke und im Eppsteiner Horst (Taunus). Z. Geomorph., N.F. **18**, 339–349.

ANDRES, W. & PREUSS, J. (1983): Erläuterungen zur Geomorphologischen Karte 1:25,000 der Bundesrepublik Deutschland, GMK 25 Blatt 11, 6013 Bingen, Berlin, 69 pp.

ANDRES, W. & SEWERING, H. (1983): The Lower Pleistocene Terraces of the Lahn River between Diez (Limburg Basin) and Laurenburg (Lower Lahn). In: FUCHS, K. et al. (eds), Plateau Uplift, Springer Verlag, Berlin, Heidelberg, 93–97.

BIBUS, E. (1973): Untersuchungen zur jungtertiären Flächenbildung, Verwitterung und Klimaentwicklung im südöstlichen Taunus und in der Wetterau. Erdkunde **27**, 16–26.

BIBUS, E. (1974): Das Quartärprofil im Braunkohlentagebau Heuchelheim (Wetterau) und seine vulkanischen Einschaltungen. Notizbl. hess. L.-Amt Bodenforsch. **102**, 159–167.

BIBUS, E. (1980): Zur Relief-, Boden- und Sedimententwicklung am unteren Mittelrhein. Frankf. geowiss. Arb., D1, 296 pp.

BIBUS, E. (1983): Distribution and dimension of young tectonics in the Neuwied Basin and the Lower Middle Rhine. In: FUCHS, K. et al. (eds), Plateau Uplift, Springer Verlag, Berlin, Heidelberg, 55–61.

BIBUS,E. & SEMMEL, A. (1977): Über die Auswirkung quartärer Tektonik auf die altpleistozänen Mittelrhein-Terrassen. CATENA **4**, 385–408.

BIRKENHAUER, J. (1971): Verharren und Änderung der Hauptabdachung am Rheindurchbruch bei Bingen und im Gebiet der Idsteiner Querfurche, Westdeutschland. Z. Geomorph., N.F., Suppl. Bd. **12**, 73–106.

BIRKENHAUER, J. (1972): Modelle der Rumpfflächenbildung und die Frage ihrer Übertragbarkeit auf die deutschen Mittelgebirge am Beispiel des Rheinischen Schiefergebirges. Z. Geomorph. N.F., Suppl. Bd. **14**, 39–53.

BOENIGK, W. (1978): Zur Ausbildung und Entstehung der jungtertiären Sedimente in der Niederrheinischen Bucht. Kölner geogr. Arb. **36**, 59–68.

BOENIGK, W. (1982): Der Einfluß des Rheingrabensystems auf die Flußgeschichte des Rheins. Z. Geomorph., N. F., Suppl. Bd. **42**, 167–175.

van den BOGAARD, P. & SCHMINCKE, H.-U. (1988): Aschenlagen als quartäre Zeitmarken in Mitteleuropa. Die Geowissenschaften 6. Jg., H. **3**, 75-84.

BREMER, H. (1959): Flußerosion an der oberen Weser. Göttinger geogr. Abh. **22**, 192 pp.

BREMER, H. (1978): Zur tertiären Reliefgenese der Eifel. Kölner geogr. Arb. **36**, 195–225.

BRUNNACKER, H. & BOENIGK, W. (1983): The Rhine Valley between the Neuwied Basin and the Lower Rhenisch Embayment. In: FUCHS, K. et al. (eds), Plateau Uplift. Springer Verlag, Berlin, Heidelberg, 62–72.

BRUNOTTE, E. & GARLEFF, K. (1979): Geomorphologische Gefügemuster des Niedersächsischen Berglands in Abhängigkeit von Tektonik, Halokinese, Resistenzverhältnissen und Abflußsystemen. In: Gefügemuster der Erdoberfläche. Festschr. z. 42. Dt. Geogr. Tag, Göttingen, 21–42.

BÜCHEL, G. & LORENZ, V. (1982): Zum Alter des Maarvulkanismus in der Westeifel. N. Jb. Geol. Paläont. Abh. **163**, 1-22.

BÜDEL, J. (1963): Klima-genetische Geomorphologie. Geogr. Rdsch. **15**, 269–285.

DUPHORN, K. (1968): Ist der Oberharz im Pleistozän vergletschert gewesen? Eiszeitalter und Gegenwart **19**, 164–174.

FRECHEN, J. & LIPPOLT, H.J. (1965): Kalium-Argon-Daten zum Alter des Laacher Vulkanismus, der Rheinterrassen und der Eiszeiten. Eiszeitalter und Gegenwart **16**, 5-30.

FUCHS, K., von GEHLEN, K., MÄLZER, H., MURAWSKI, H. & SEMMEL, A., (eds) (1983): Plateau Uplift, the Rhenish Shield - a case history, Springer Verlag, Berlin, Heidelberg, 411 pp.

GLATTHAAR, D. & LIEDTKE, H. (1984): Die tertiäre Reliefentwicklung zwischen Sieg und Lahn. Ber. z. dt. Landeskunde **58**, 129–146.

GÖRG, L. (1984): Das System pleistozäner Terrassen im Unteren Nahetal zwischen Bingen und Bad Kreuznach. Marburger geogr. Schriften **94**, 194 pp.

HEINE, K. (1970): Fluß- und Talgeschichte im Raum Marburg. Bonner geogr. Abh. **42**, 195 pp.

HEINE, K. (1971): Über die Ursachen der Vertikalabstände der Talgenerationen am Mittelrhein. Decheniana **123**, 307–318.

HÜSER,K. (1973): Die tertiärmorphologische Erforschung des Rheinischen Schiefergebirges. Karlsruher geogr. Hefte **5**, 135 pp.

ILLIES, J.H. & FUCHS, K: (1983): Plateau Uplift of the Rhenish Massif - introductory remarks. In: FUCHS, K. et al. (eds) Plateau Uplift, Springer Verlag, Berlin, Heidelberg, 1-8.

JESSEN, O. (1938): Tertiärklima und Mittelgebirgsmorphologie. Z. Ges. Erdkunde., Berlin, Jg. 1938, 36–49.

KAISER, K. (1961): Gliederung und Formenschatz des Pliozäns und Quartärs am Mittel- und Niederrhein sowie in den angrenzenden Niederlanden unter besonderer Berücksichtigung der Rheinterrassen. Festschr. Dt. Geogr. Tag. Köln, 236–278.

KANDLER, O. (1971): Die pleistozänen Flußterrassen im Rheingau und im nördlichen Rheinhessen. Mainzer Naturw. Arch. **9**, 5–25.

LIEDTKE, H. (1969): Grundzüge und Probleme der Entwicklung der Oberflächenformen des Saarlandes und seiner Umgebung. Forsch. dt. Landeskunde **183**, 63 pp.

LIEDTKE, H. (1984): Geomorphological mapping in the Federal Republic of Germany at scales 1:25000 and 1:100000. Bochumer geogr. Arbeiten **44**, 67–73.

LIPPOLT, H.J. (1983): Distribution of volcanic activity in space and time. In: FUCHS, K. et al. (eds) Plateau Uplift, Springer Verlag, Berlin, Heidelberg, 112–120.

LIPPS, S. (1985): Relief- und Sedimententwicklung an der Mittellahn. Marburger geogr. Schriften **98**, 100 pp.

LÖHNERTZ, W. (1978): Zur Altersstellung der tiefliegenden fluviatilen Tertiärablagerungen der SE-Eifel (Rheinisches Schiefergebirge). N. Jb. Geol. Paläont. **156**, 179–206.

LOUIS, H: (1953): Über die ältere Formentwicklung im Rheinischen Schiefergebirge, insbesondere im Moselgebiet. Münchener geogr. Hefte **2**, 97 pp.

MÄLZER, H., HEIN, G. & ZIPPELT, K. (1983): Height changes in the Rhenisch Massiv: Determination and analysis. In: FUCHS, K. et al. (eds) Plateau Uplift. Springer Verlag, Berlin, Heidelberg, 164–176.

MÜLLER, K.H. (1974): Zur Morphologie der plio-pleistozänen Terrassen im Rheinischen Schiefergebirge am Beispiel der Unterlahn. Ber. dt. Landeskunde **48**, 61–80.

MÜLLER, M.J. (1976): Untersuchungen zur pleistozänen Entwicklungsgeschichte des Trierer Moseltals und der Wittlicher Senke. Forsch. dt. Landeskunde **207**, 185 pp.

MURAWSKI, H. et al. (1983): Regional tectonic setting and geological structure of the Rhenish Massif. In: FUCHS, K. et al. (eds), Plateau Uplift. Springer Verlag, Berlin, Heidelberg, 9–38.

NEGENDANK, J. (1978): Zur Känozoischen Geschichte von Eifel und Hunsrück. Forsch. dt. Landeskunde **211**, 90 pp.

PANZER,W. (1967): Einige Grundfragen der Formenentwicklung im Rheinischen Schiefergebirge und ihre Erforschung. Die Mittelrheinlande, Festschr. 26. Deutsch Geogr. Tag, Bad Godesberg, 1–15.

PENCK, W. (1924): Die morphologische Analyse. Stuttgart, 283 pp.

QUITZOW, H.W. (1974): Das Rheintal und seine Entstehung. Conf. Soc. Geol. Belg., Liège, 53–104.

RATHJENS, C. (1977): Beobachtungen und Überlegungen zur älteren geomorphologischen Entwicklung des Hunsrücks und des Nahegebietes. Mannh. geogr. Arb. **1**, 259–276.

ROHDENBURG, H. (1968): Zur Deutung der quartären Taleintiefung in Mitteleuropa. Die Erde **99**, 297–304.

SCHMIDT, K.-H. (1975): Geomorphologische Untersuchungen in Karstgebieten des Bergisch-Sauerländischen Gebirges. Bochumer geogr. Arb. **22**, 157 pp.

SCHNÜTGEN, A. (1974): Die Hauptterrassen am linken Niederrhein aufgrund der Schotterpetrographie. Forsch. Ber NRW, 2399, 150 pp.

SEMMEL, A. (1966): Zur Entstehung von Flächen und Schichtstufen im nördlichen Rhönvorland. Dt. Geogr. Tag Bochum, Tagungsber. u. wiss. Abh., 340–350.

SEMMEL, A. (1983): The early Pleistocene terraces of the Upper Middle Rhine and its southern foreland. Questions concerning their tectonic interpretation. In: FUCHS, K. et al. (eds), Plateau Uplift. Springer Verlag, Berlin, Heidelberg, 49–54.

SEMMEL, A. (1983a): Die plio-pleistozänen Deckschichten im Steinbruch Mainz-Weisenau. Geol. Jb. Hessen **111**, 219–233.

SEMMEL, A. (1984): Geomorphologie der Bundesrepublik Deutschland. Geogr. Zeitschr., Beihefte, Stuttgart, 192 pp.

SEMMEL, A. (1984a): Reliefentwicklung im Rheinischen Schiefergebirge — neue Befunde, neue Probleme. Zur präquartären Entwicklung. 44. Dt. Geogr. Tag Münster, Tagungsber. u. wiss. Abh., 71–74.

SONNE, V. (1970): Das nördliche Mainzer Becken im Alttertiär. Betrachtungen zur Paläoorographie, Paläogeographie und Tektonik. Oberrh. geol. Abh. **19**, 1–28.

STEINMÜLLER, A. (1982): Die Flußterrassen und die Klimastratigraphie des Pleistozäns. Peterm. Geogr. Mitt. **126**, 113–118.

WOLDSTEDT, P. & DUPHORN, K. (1974): Norddeutschland und angrenzende Gebiete im Eiszeitalter. Stuttgart, 500 pp.

ZAGWIJN, W.H. (1974): Bemerkungen zur stratigraphischen Gliederung der pliopleistozänen Schichten des niederländisch-deutschen Grenzgebietes zwischen Venlo und Brüggen. Z. dt. geol. Ges. **125**, 11–16.

ZÖLLER, L. (1985): Geomorphologische und quartärgeologische Untersuchungen im Hunsrück-Saar-Nahe-Raum. Forsch. dt. Landeskunde **225**, 240 pp.

Address of author:
Prof. Dr. Wolfgang Andres
Fachbereich Geographie der Philipps-Universität Marburg
Deutschhausstraße 10
D-3550 Marburg/Lahn

CATENA SUPPLEMENT 15 p. 45–67 Cremlingen 1989

ON THE GEOMORPHOLOGY OF THE SOUTH GERMAN SCARPLANDS

Hanna **Bremer**, Köln

1 Introduction

The south German scarplands lie between the rivers Main and Danube. The cuesta landscape in the centre of the region is surrounded by uplands and the entire area is made up of dissected old denudation surfaces, escarpments, basins and valleys. To the south are the Alps and the Alpine Foreland (see FISCHER in this volume). The highest elevations within the region are the Feldberg (1493 m) in the Black Forest and the Große Arber (1458 m) in the Bavarian Forest. The lowest areas are in the northern part of the Rhine Rift Valley (Rhine Graben) where the elevation is less than 100 m and in the southeast in the Danube valley where it is about 300 m.

Several basic concepts relating to cuestas (Schichtstufen; QUENSTEDT 1843, WAGNER 1930), piedmont benchlands (Rumpftreppen; W.PENCK 1924) and the climato-genetic explanation of landforms (BÜDEL 1957) have been, and are still being, discussed in this region.

ISSN 0722-0723
ISBN 3-923381-18-2

3-923381-18-4/89/5011851/US$ 2.00 + 0.25

2 Geological evolution and structural landforms

Precambrian shields, partly modified in the Variscan (Hercynian) orogeny, flanked the south German sedimentation basin in the Mesozoic era. The basin also extended northward into Hessen and Lower Saxony and was connected in the west with the Paris Basin. The oldest Mesozoic beds, the Bunter Sandstone (Buntsandstein, lower Triassic) and the Muschelkalk (middle Triassic) increase in thickness towards the north and northeast. The Keuper beds (upper Triassic) were, for the most part, deposited in very shallow seas so that they contain frequent changes of facies. The Jurassic, by contrast, decreases in thickness northwards. Its lithology varies little regionally, with the exception of the Malm (upper Jurassic) in which layered limestones and marls occur next to the massive limestones of reef stocks that are in general round, up to 150 m in height and 1 to 2 km in diameter. The thickness of the Mesozoic sedimentary cover (Deckgebirge) ranges from 1200 m in areas of synsedimentary upwarps to 2000 m in the basins.

Uplift and denudation began in the lower Cretaceous. Evidence for this occurs in the Franconian Jura where the resistant limestones of the reef stocks formed domal hills which were sub-

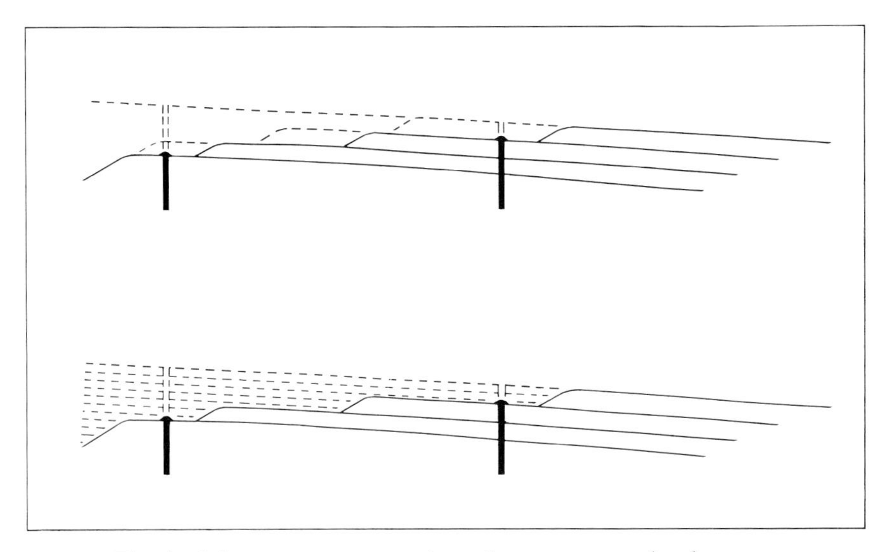

Fig. 1: *Schematic representation of cuesta scarp development.*

(a) Scarp retreat (based on SCHMITTHENNER 1954), (b) Scarps also develop with areal downwearing. Debris of overlying strata can be found in volcanic necks.

sequently buried under marine Cretaceous sands and must, therefore, have existed before the Cretaceous transgression. Other domal hills composed of limestone and sand were formed in the early Tertiary (PFEFFER 1986).

The sedimentary rocks of the Mesozoic in south Germany consist, therefore, of resistant limestones and sandstones interbedded with shales and marls, locally also salt and gypsum. The major scarps are formed by the Bunter Sandstone, by the Keuper sandstones and by the Jurassic limestones. The beds dip at an angle of 1° to 2° to the east, southeast and south so that the lithological and structural conditions for the development of a cuesta landscape are present. Dip slopes are 10 to 60 km wide, undulating slightly but often also quite flat. cuesta scarps have a relative height of 100 m or more and are dissected by valleys that have a wide variety of forms.

It is usually suggested that cuesta landscapes develop as a result of rock-controlled denudation (fig. 1) and that the scarps form after the removal of the weaker rocks. Landslides, undercutting, incision by obsequent valleys etc. are thought to cause their retreat. However, in south Germany, this explanation seems to be unlikely for the following reasons:

1. The dip slope coincides only rarely with a bedding plane of the resistant rock. Usually the resistant rock is truncated near the crest of the lower scarp and the weak rock is truncated in the foreland of the next higher scarp.

2. Along its horizontal extent the crest of the scarp lies in different stratigraphic horizons of the resistant rock (SPÄTH 1973, ZIENERT 1986).

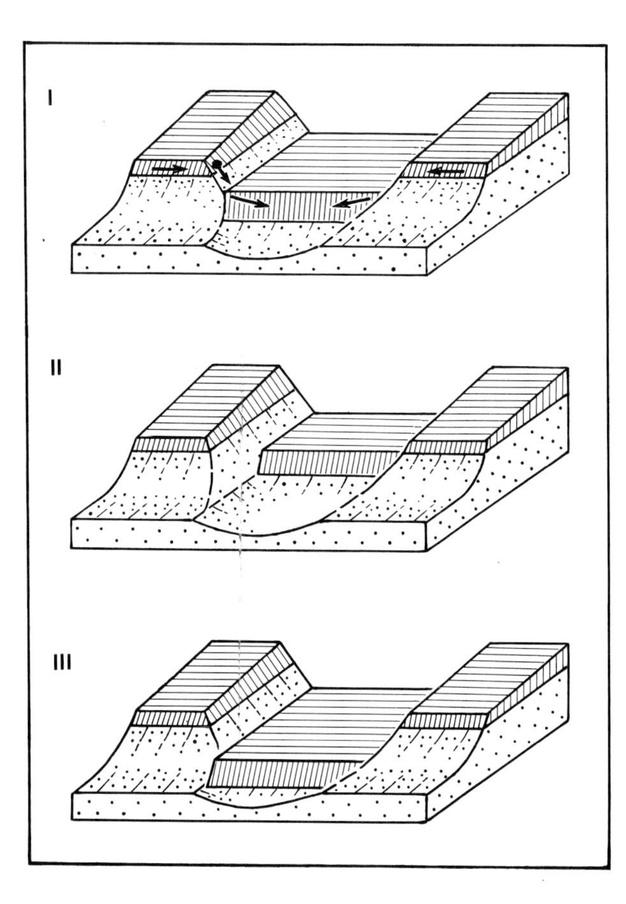

Fig. 2: *Schematic representation of a tectonic graben in the cuesta landscape.*

(a) In the case of very young subsidence the scarp in the graben would be very low.
(b) In the case of scarp retreat controlled by spring and seepage sapping, the scarp in the graben would retreat more because groundwater converges there and is, therefore, more plentiful.
(c) In reality, scarps frequently project in grabens. This is explained by areal downwearing, because the resistant caprock in the graben is reached later than that on the flanking uplands and, therefore, has a protecting effect.

3. The scarps vary frequently in height and form along their horizontal extent. Over large distances such variations might be explained by tectonics, denudation of the resistant rock or changes of facies but the variations in these scarps also occur over short distances. In addition the scarps are often sharply accentuated in the vicinity of rivers, become less distinct towards the divides and in some areas may be absent altogether.

4. Many faults have had no effect upon the configuration of the cuesta scarps. The same is true for upwarps (BÜDEL 1957). Salients of the scarps and residual outliers are frequently associated with tectonic low positions (SCHMITTHENNER 1954). This association is more easily explained by areal downwearing (fig.2).

On the basis of these observations it seems probable that the scarps developed from planation surfaces similar to those present in the adjacent crystalline upland areas and that both climatically and tectonically controlled processes have been of importance in the development of the cuesta landscape.

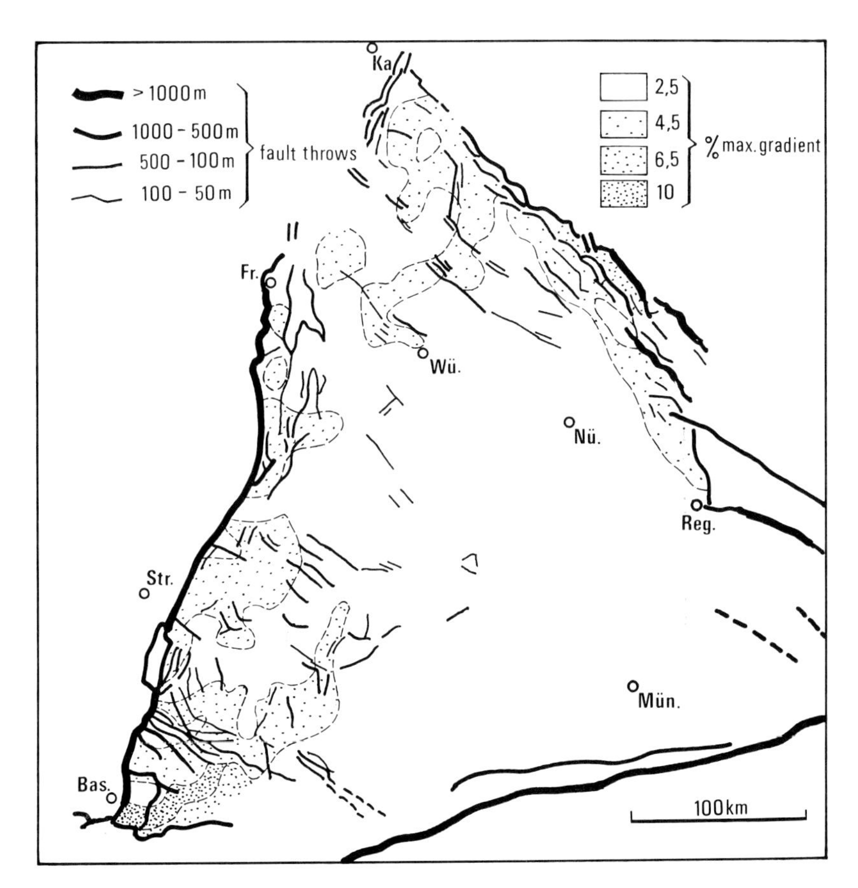

Fig. 3: *Tectonic limitation of the south German crustal block according to CARLÉ (1955). Small throws with less than 50 m have been omitted. The "Gefälle" refers to the dip of the strata in %.*

3 Tectonic development of south Germany and the spatial pattern of cuestas

The south German crustal block (CARLÉ 1955) is triangular in shape (fig.3). Its western and northwestern border is the Rhine Rift Valley and the Hessian Depression and its eastern border the east Bavarian marginal fault zone on the edge of the Bohemian Massif. In the south, the area is bordered by the marginal overthrust of the Alps. In addition to the main faults that surround the crustal block, there are numerous smaller faults with throws of, in general, less than 100 m. The strikes are varied but there are three orientation maxima: the SSW (Rhenish), WNW-NW (Hercynian) and the ENE (Swabian) directions. Only the main faults have fault scarps, usually interrupted by valleys.

The dip of the strata in the south German crustal block is particularly low towards the southeast so that the widths of surface exposure of the strata varies.

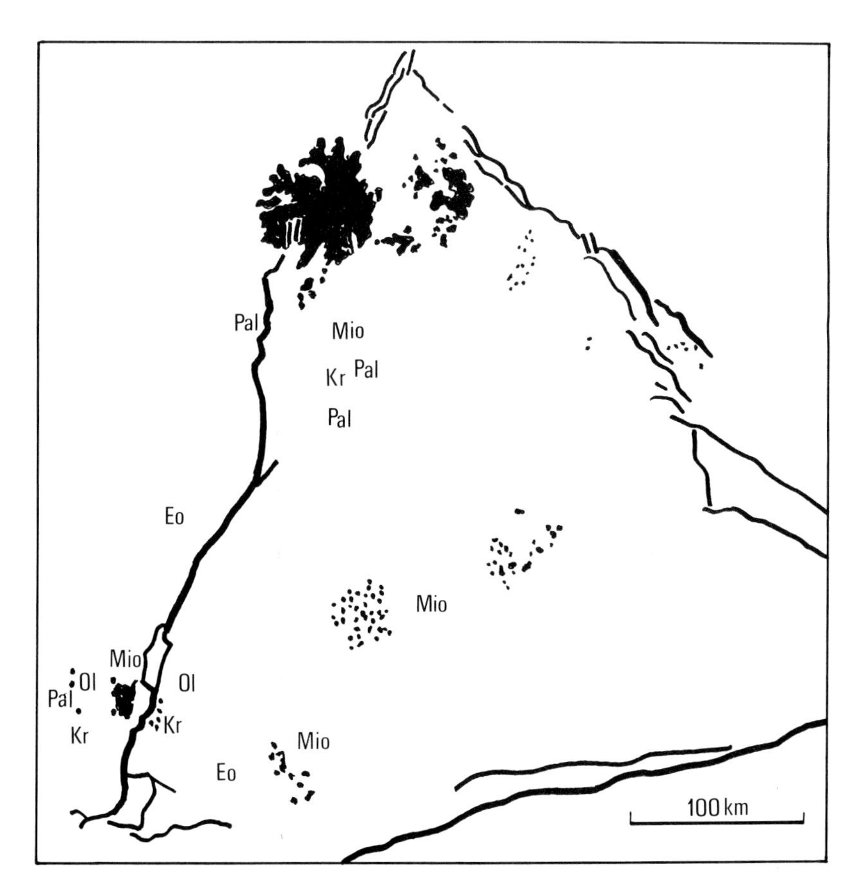

Fig. 4: *Volcanic areas and single volcanoes in south Germany and adjacent areas to the north (based on CARLÉ 1955, GEYER & GWINNER 1986).*

Along the upwarped margins of the block, particularly in the Black Forest, the Odenwald/Spessart and along the Bohemian Massif, the angles of the dips are greater, generally attaining about 3° but locally up to 6°. There are low hills and planation surfaces in these areas but no hogbacks.

4 Volcanoes

The largest volcanic areas of Germany, the Vogelsberg and the Rhön, lie on the margin of the south German crustal block. The eruptions began in the middle Miocene (Helvetian) and continued until the lower Pliocene. There is also a large volcanic area in the southwest of the area (fig.4). The Kaiserstuhl in the southern Rhine Rift Valley was active 18-13 million years B.P. Subsequently the volcanic activity shifted eastward to the Hegau at the western end of Lake Constance, where the last eruption probably occurred 9 million years B.P.. The volcanism accompanied the Tertiary diastrophism in these areas. Elsewhere volcanic necks occur on zones of fissures.

Examples are the Katzenbuckel in the eastern Odenwald and the Heldburg dike series in the northeast of the region.

The volcanic necks are a factor in the determination of the landscape evolution. Dogger (middle Jurassic) xenoliths occur in the Paleocene volcano of the Katzenbuckel and Malm (upper Jurassic) xenoliths in the Oligocene volcanics of the northeast. The Jurassic beds must, therefore, have extended this far north at the time of volcanic activity. It has been suggested that the Jurassic scarp formed soon after the uplift of the south German crustal block from the Cretaceous sea in the west and northwest and that it has retreated since then by up to 150 km. The same spatial pattern could, however, have been produced by areal lowering.

5 The Ries

Another feature of importance is the nearly circular depression of the Nördlinger Ries which was caused by a very large meteorite fall in the middle Miocene (Badenian/Sarmatian). The cuesta scarp of the Jura existed at that time in the area because debris masses from the impact have been found at the foot of the scarp (R in fig.9). The debris also dammed a valley that ran from north to south and caused the formation of the Rezat-Altmühl lake in which sands, clays and freshwater limestones (Hydrobia Limestone, L in fig. 9) were deposited. These deposits lie in part at the same, or at a slightly higher, elevation as the present Jura scarp foreland. Debris masses also occur in the valleys that dissect the scarp. Some of the valleys had probably not yet reached their present depth at the time of the meteorite fall.

6 The river network and fluvial deposits

The northern and western areas of the scarplands drain to the Rhine, the southern and eastern areas drain to the Danube (fig. 5). Stream density is highest on the shales and marls of the Keuper and also on the crystalline marginal uplands. The lowest stream density occurs on the Swabian and Franconian Jura which are deeply karstified. Karst development is less, and the stream density higher, on the Muschelkalk. In the Rhine Rift Valley, the higher sandy surfaces also have a very low stream density, particularly in the north and in the extreme south of the Rift Valley.

A conspicuous feature of the drainage pattern is the northward extension of the Danube watershed in the drainage basins of the Wörnitz and the Altmühl which rise near the Keuper scarp. Unlike the other rivers that rise near the scarp, they both flow southward into and through the Franconian Jura (fig. 9). The Wörnitz also crosses the Ries depression. Farther east, tributaries of the Altmühl follow old valley zones that have been developed by an earlier northward flowing drainage system. A similar phenomenon occurs in the Keuper scarp, where the Aisch and its largest tributary flow in broad funnel-shaped valleys into and through the scarp.

The main divide between the Danube and the Rhine systems is very low in some areas. Examples are the divide between the upper Tauber and the sources of the Wörnitz and the Altmühl, and that between the Red Main and the Pegnitz near Bayreuth. At, and in front of, the Keuper scarp, the rivers flow to the north, west, east and south although there is no major upland in this area.

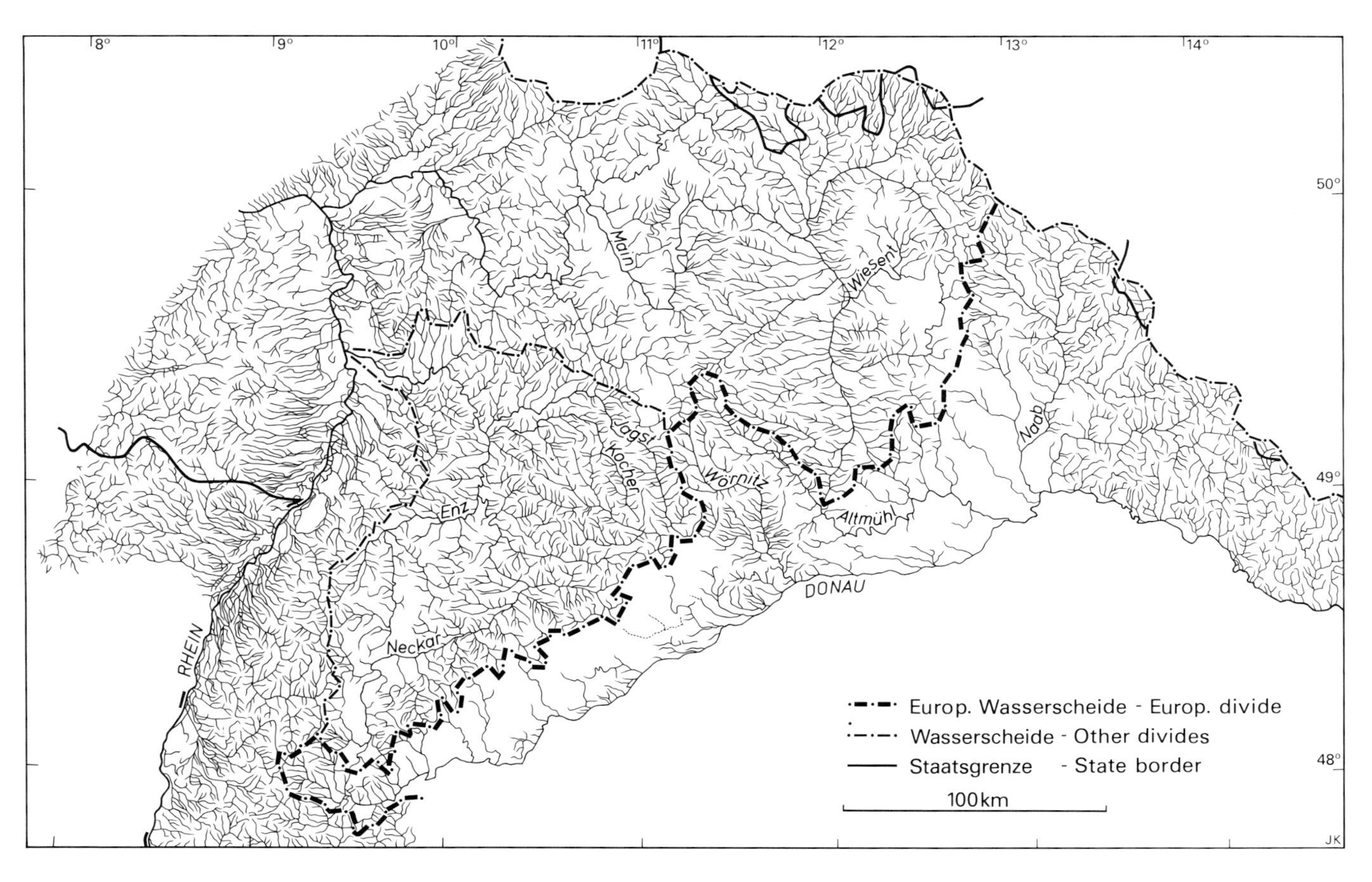

Fig. 5: *Pattern of streams of third and higher order.*
Note the lower densities on the Swabian Franconian/Jura and in the sandy fills of the Rhine Rift Valley. Particularly high densities occur on the clayey-marly lower Keuper. The Danube drainage basin extends far to the north in the Altmühl-Wörnitz area.

Some of the rivers flow into higher areas. It seems likely, therefore, that the stream pattern originated on a slightly upwarped old planation surface. Subsequent downwearing and also some captures (ZEESE 1972) have resulted in the present pattern so that, in general, the drainage network in this central part of the south German crustal block must be older than the Keuper scarp.

In the southern Black Forest, the streams follow the pattern of tectonic structures. In most other areas of the south German scarplands, such structural control is absent because of the great thickness of the regolith when incision began on a prior denudation surface. The general absence of structural control of the stream patterns suggests a relatively steady rate of weathering and denudation and also, therefore, a comparatively slow uplift during the establishment and incision of the stream network. Similar cases exist in the tropics (BREMER 1971, 1981).

The gradients of tributaries flowing to the Rhine are generally higher than those of the streams flowing to the Danube and the Rhine drainage basin is being enlarged. The Wutach, a Rhine tributary, captured a southern headwater of the Danube during the last glaciation. Other indications of drainage diversions occur in the conglomeratic deposits (Juranagelfluh) on the Swabian Jura. These conglomerates extend into the Alpine Foreland to the south where they are interbedded with the Upper Marine Molasse of the middle Miocene and the Upper Freshwater Molasse of the upper Miocene and contain pebbles of Muschelkalk from the north and of Bunter Sandstone from the Black Forest (SCHREINER 1965). East of the Ries, lydite-bearing Tertiary deposits indicate that a connection existed from the upper Main southward to the Alpine Foreland trough during the Tertiary. Along the present valley zone of the Naab, kaolinitic, sandy marly deposits partly interbedded with lignite indicate that there have been several phases of incision and filling (LOUIS 1984).

The lithologic composition of the gravels in the east and southwest of the south German crustal block show that they were transported from the north. However, with the exception of the conglomerates in the Hegau, it is not possible to reconstruct the transport routes precisely and thereby to obtain gradients from which tectonic displacements could be estimated. The age of these gravels and the patterns of reworking and redeposition are not well-known; most of the deposits are very small. There seem to be two types of facies in the pre-Quaternary gravels. The older gravels are usually well-rounded and frequently flattened. The grain size is varied and the long axis can attain 20–30 cm. The gravels also have a loamy-sandy, usually reddish, matrix. The gravel components include not only quartz and other highly resistant materials but also sandstones, despite the relatively great age of the gravels. There is no recognizable layering and little sorting. The gravels are probably the result of a special transport mechanism whereby the pebbles and cobbles were rounded in sandy river beds and transported later in large mudflows. The clay components were washed out during and after deposition. The transport of the gravel components over great distances, 100 km or more, probably took place in several stages and may have occurred at relatively low gradients. The younger pre-Quaternary gravels are well-sorted or are sandy deposits with individ-

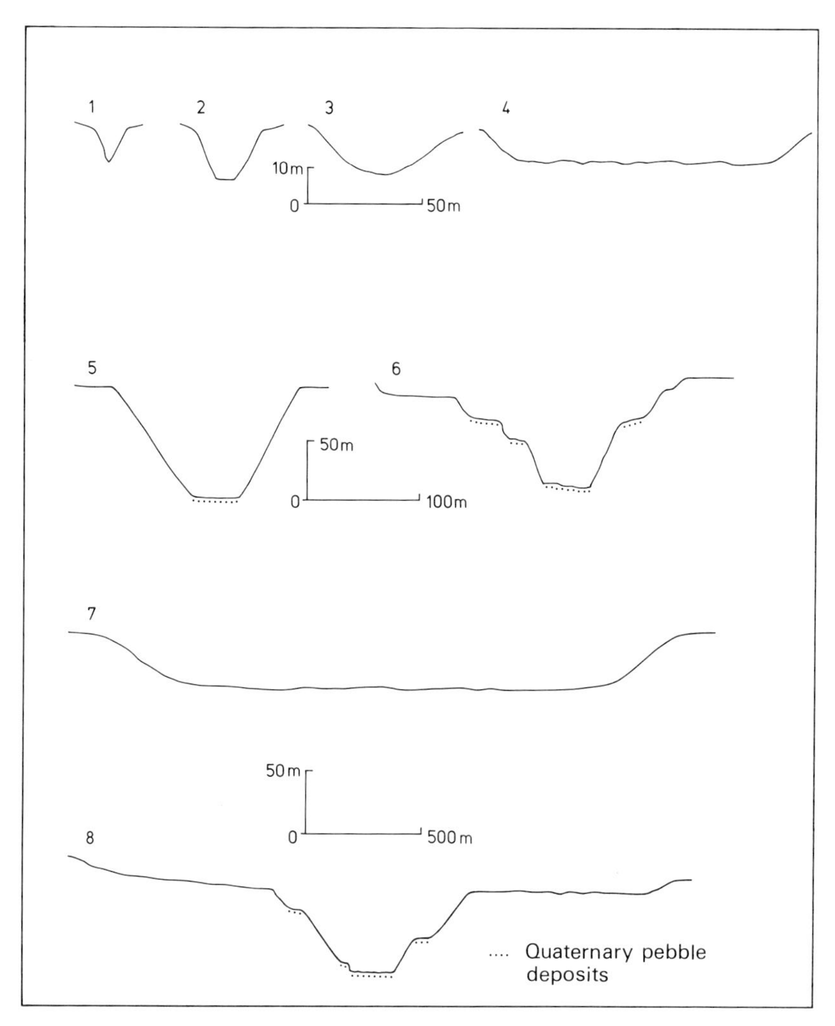

Fig. 6: *Valley cross-profiles in southern Germany.*
No.1–4 are different valley heads; No.3 and No.4 are possibly modified relic forms. No.5 is the cross-profile of a narrow valley, No.6 represents more or less normal valleys. No.7 is a planation band and No.8 a composite form of a Quaternary valley with terraces that has been incised into a planation band.

ual pebbles. These Quaternary deposits are fluvial in origin and contain easily weathered components. In general, they lie at lower elevations than the Tertiary gravels and are useful indicators of a younger relief generation.

7 Forms of valleys

Of particular significance is the pattern of narrow, in some cases gorge-like, valleys and wide valley zones (fig.6). In some wide valleys in-valley divides occur from which rivers flow in opposite directions (fig. 7). The causes of this

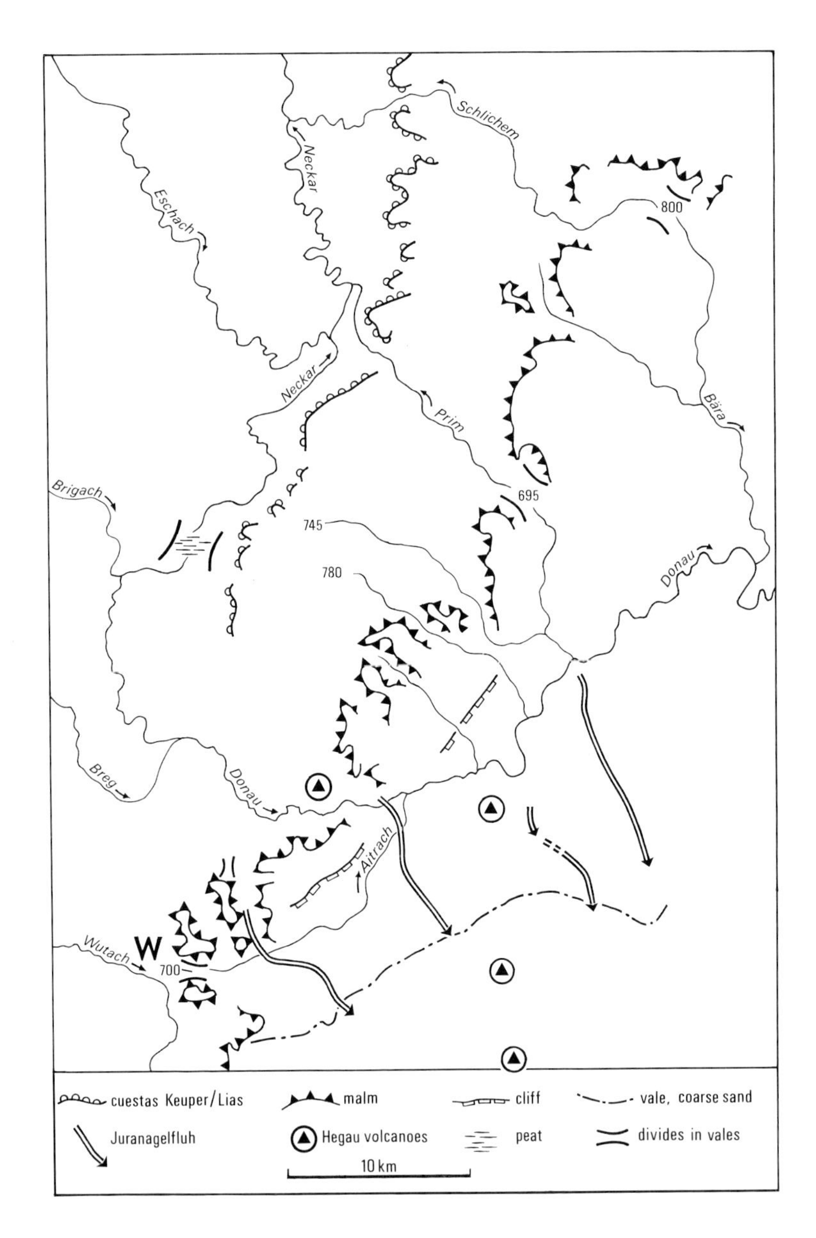

Fig. 7: *Planation bands in the western Swabian Jura. W = elbow of capture of a Danube headwater by the Wutach.*

pattern have not been investigated but it is not possible to ascribe it entirely to the lithology. In the Jura, for example, there are a number of gorges in the limestone plateau, such as the gorge of the Danube near Kelheim or that of the Wiesent. On the other hand, the Brenz valley in the Swabian Jura and several valleys in the Franconian Jura are 1 or 2 km wide. Narrow valleys occur in the Muschelkalk, particularly on the Neckar, but there are also a large number of wide valley stretches in tributaries of the Neckar and in the Tauber valley in this formation. Narrow and wide stretches are also present in the sandstones, although the contrast is not as great. It is probable that the narrow valley stretches developed as a result of more rapid incision, perhaps following more rapid uplift, and that the wide valley stretches have developed from planation bands (Flächenstreifen; fig.7). The upper parts of the cross profiles of these valley stretches are very wide and have slopes of only 1° to 5°. The lower slopes are, in general, between 10° and 20° although in narrow stretches they attain up to 40°. In limestone, they may be almost vertical. Terraces occur, only some of which are accumulation terraces of Quaternary gravels. In narrow valleys there are also structural benches associated with resistant beds. Along the Main and its southern tributary the Regnitz, there are sequences of Quaternary terraces but they are not as well-developed as the terraces of the Rhine (see ANDRES in this volume).

The valley heads usually have the form of shallow troughs (dells). On Muschelkalk some of these are several kilometers long. In the marginal uplands, especially on the crystalline and Bunter Sandstone, some V-shaped valleys change into box-shaped valleys farther downstream. However, an adjacent drainage basin may quite frequently have a very wide valley floor in the upper reach which, in some cases, is swampy. In the Bunter Sandstone area of the Palatinate there are valley widenings that can be termed intramontane basins which were probably formed in the Tertiary.

The composite valley forms of the middle and lower courses of the rivers are polygenetic. The wide, gentle upper slopes developed during the restricted planation period of the late Tertiary and the steeper lower slopes in the Quaternary, as shown by gravels and weathering residues. In some cases, the upper wide valleys have been lowered over their entire widths but the present-day rivers often do not correspond in size to these widths. There is also no evidence of much larger rivers having flowed in the valleys in the past. The wide valley floors are probably planation bands that have been lowered which would also account for the presence of in-valley divides. Corresponding forms have been described in the humid tropical parts of Sri Lanka (BREMER 1981). Planation embayments (Flächenbuchten) also belong to this group of landforms. These are broad funnel-shaped reentrants that occur particularly in the Keuper scarp but also in the Bunter Sandstone area of the Odenwald and on the Murg and Kinzig rivers in the Bunter Sandstone area of the Black Forest.

In the Swabian Jura, the wide valleys end in wind-gaps at the Jura scarp. WAGNER (1930) and others have suggested that this indicates that the scarp has retreated. Such a pattern can, however, develop by areal lowering. Wide, seemingly beheaded valleys at the margins of the intramontane basins occur, for example, in central Australia and Sri Lanka. Their origin is clearly due to areal lowering, not scarp retreat.

8 Products of weathering

Isolated remnants of weathering residues from a hot and humid period occur in all parts of south Germany, with the exception of the "Gäuflächen", agriculturally used denudational plains on the Muschelkalk where they are rare. Lateritic materials, soil relics, karst joint fillings and grus are all present. Apart from karst joint fillings and transported soil relics found in Tertiary sediments, dating of these residues is possible only in relation to the relief development. Research in this area has been initiated only recently.

Among the lateritic materials are the "Bohnerze", pisolithic ores, which were used formerly for iron production. The oldest lie on the highest planation surface remnants and are, therefore, early Tertiary in age or perhaps even Cretaceous. A pre-upper Cretaceous weathering horizon occurs in the Oberpfalz in northeastern Bavaria, where thick deposits of iron ore have developed as a result of weathering in situ. They were later covered by sediments. In the Oberpfalz Depression, iron indurations several meters thick developed in Dogger beds (middle Jurassic) close to the present land surface. The Mössbauer spectra indicate that they are probably early Tertiary or older (SPÄTH & MBESHERUBUSA 1982).

Soil relics in the limestone areas of the Swabian and Franconian Jura consist of red loams. In the Swabian Jura they occur at the highest elevations (ZEESE 1976/77). Flints have often been left in these residual loams after solution weathering of the limestone. At lower elevations there are ochre-coloured loams which are probably younger. In the Franconian Jura red loams and ochre loams also occur in which the quartz grains have been pitted by solution, indicating a considerable age, perhaps early Tertiary (BURGER 1988). Relic soils are particularly frequent on the Keuper sandstones. They appear to be the lower horizons of former red loams and are often gleyed and bleached. An exception are the flint loams on the Hohenlohe Plain that occur mainly on the Muschelkalk (MÜLLER 1958).

In the Swabian Jura and in the southern part of the Franconian Jura there are karst fissures (joint lapies) that usually reach depths of 12 to 15 m. These are filled with red or yellow loams and in some cases contain iron oxide indurations. DEHM (1961) has collected Tertiary fossils of small animals from these loams. The fissures were developed in the period from the middle Eocene to the Pleistocene at about the same level and often in proximity to one another (fig.8). This would indicate that the surface in which the fissures developed has not been lowered since the formation of the oldest fissures (KUPPELS 1981). The fissures occur in the "Kuppenalb", the part of the Jura that was not covered by the Miocene molasse sea, and on the "Flächenalb" which lies south of the Miocene cliff line. In the latter area, they were covered by molasse deposits but were later re-exhumed.

There are zones of grus on the granite of the Bavarian Forest that, in some areas, attain depths of 10 to 50 meters. They probably represent the roots of a very deep zone of weathering that developed in hot and humid climates (KUBINIOK 1988). In the Odenwald, grus also occurs on granite and in a few ar-

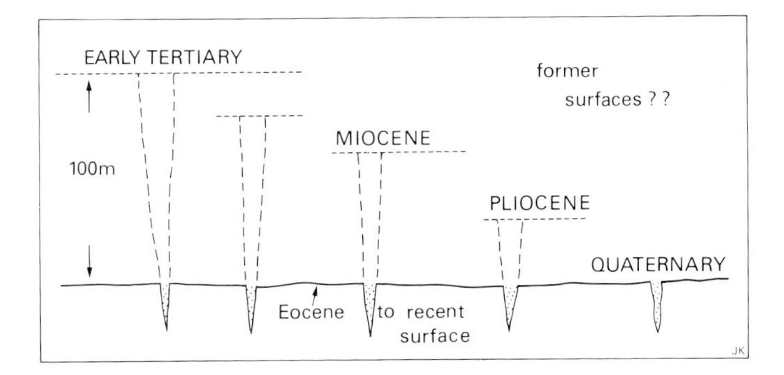

Fig. 8: *Schematic representation of the karst fissures in the Swabian Jura and the southern Franconian Jura.*
It is unlikely that the oldest, Eocene, fissures were much deeper originally and the younger ones less deep, that all were worn down to the depth of 12 m that is present now, and that the youngest fissures developed to this same depth without downwearing. It seems more probable that, regardless of age, each fissure was widened by solution to approximately the same depth and filled with a residual loam that prevented its further development. Each fissure contains fossils of only one age.

eas on old weathered zones of the crystalline rocks beneath the Bunter Sandstone. Here the basement rocks have, therefore, been weathered before the deposition of the early Mesozoic sediments. Occurrences of grus are, however, so widespread and are found in such varied positions that they cannot generally be ascribed to pre-Permian weathering or to pneumatolytic processes. In the Odenwald the grus is rarely deeper than 5 m and maximally 10 m deep. There appears to have been less grus development on the granite of the Black Forest. The grus profiles are often found on old denudation surfaces but also occur in broad valleys and intramontane basins. In the Franconian Jura there are grussed dolomites but it is not clear whether their weathering has been of a similar nature.

9 Loess

Loess deposited during the Pleistocene cold periods covers the lower denudations surfaces in south Germany. The upper limit of the deposits rises from an altitude of about 300 m in the northern Rhine Rift Valley to about 500 m southeast of Regensburg. These elevations can be interpreted as the altitudinal limit of the cold period tundra. The tundra vegetation trapped the wind-transported material that was blown from the seasonally dry river beds and the areas of frost-weathered debris above the tundra zone; winds from the west and southwest dominated. On the leeward slopes, the loesses of several cold perods have been deposited on top of each other, particularly in the region of the Main river. Older surfaces that date at least from the late Quaternary have been preserved in these areas. Many loess profiles have, however, been truncated, indicating that during the colder phases of the cold periods the loess was removed by deflation.

Solifluction sheets dating from the last cold period are widely distributed above the altitudinal limit of the loess. They are generally about one meter thick. At the base of the slope there may be 3 to 5 m thick accumulations of debris in a fine-grained matrix that are often interbedded with loess. The solifluction debris from the last cold period cannot always be distinguished from the debris of earlier periods. Frequently, however, there is a layer of redeposition, up to 20 cm thick, that dates from the younger tundra period (younger Dryas, late Würm glaciation). This characteristic layer has been first recognized and termed "Deckschutt" (cover debris) by SEMMEL (1964).

At lower elevations, loess is largely absent on late Pleistocene Low Terraces of the rivers so that the loess is of about the same age or older. On the very wide Low Terrace in the Rhine Rift Valley there are late glacial dune fields which have caused the tributaries from the east to flow along the eastern edge of the Rift Valley. The Neckar, for example, formerly joined the Rhine near Darmstadt. Such dune fields and sand sheets, which are usually covered with pine forest, also occur in the Pegnitz lowland near Nürnberg.

10 Regional example 1: Tectonics and relief development in the Rhine Rift Valley and its marginal uplands

The subsidence of the Rhine Rift Valley began in the middle Eocene. During the early Tertiary it was particularly intensive in the south of the Rift Valley but in the late Tertiary and Quaternary it took place primarily in the northeast and continues in this area at the present time. There are traces of marine influences in the south dating from the Oligocene and in the north from the Miocene.

In the north, the Tertiary sediments lie on Paleozoic rocks and near Ludwigshafen, on Bunter sandstone. Successively southwards they lie on the entire stratigraphic sequence of the Triassic and in the southernmost part of the Rift Valley on Jurassic rocks. The Mesozoic strata must, therefore, have been removed in the north by denudation before the subsidence began, indicating that the north has been uplifted earlier. The thickest early Tertiary deposits are, however, in the south of the area. If it is assumed, on the basis of the dynamics of tectonics whereby the subsidence of the Rift Valley would occur along the axis of maximum upwarp, that the marginal uplands were uplifted simultaneously with the subsidence of the adjacent parts of the Rift Valley, the Odenwald would then have to have been uplifted later than the Black Forest. However, this is contradicted by the greater overall denudation and the Paleozoic subsurface of the Tertiary deposits in the north. An explanation might be differing modes of relief development due to tectonics. Probably with early slow uplift areal downwearing was active but later more rapid uplift resulted in incision and a complete interruption of erosion at the edge of the higher planation relic surface (BÜDEL 1982).

West of the Rhine, subsidence was limited during the later phases of Rift Valley development so that the streams flowing to the Rhine from the west are incised. Plateaus and broad interfluves alternate with low-lying triangular alluvial fans. The higher areas are loess-covered, fertile, and used for agriculture. Along the foot of the fault scarp there are vineyards. The sandy alluvial fans are covered by forests.

In the northern part of the Rift Valley the Rhine river incised 8 to 10 m

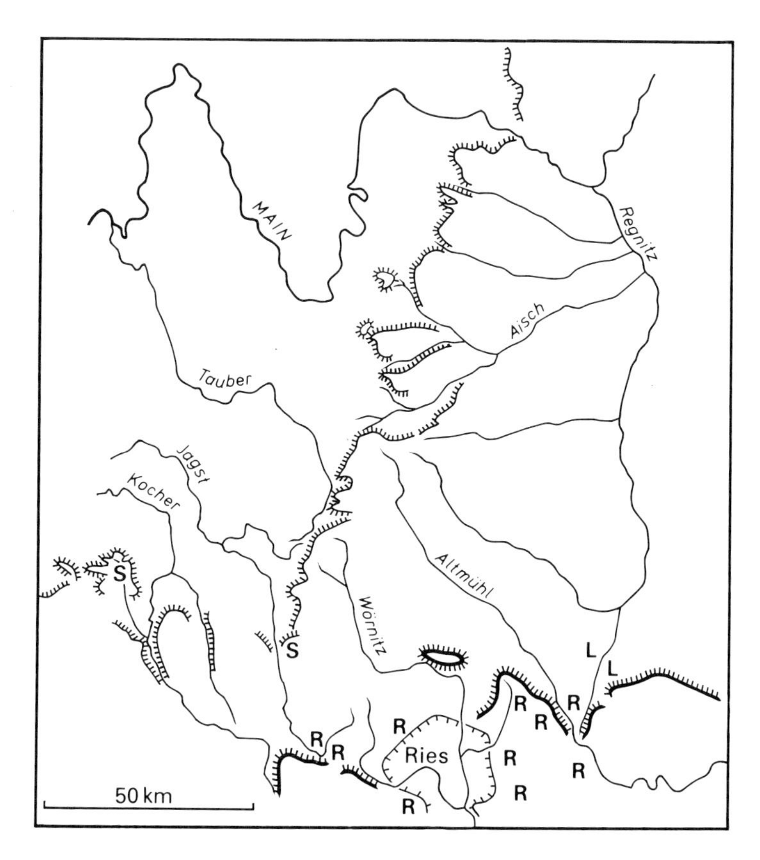

Fig. 9: *Keuper and Jura cuesta and drainage systems.*
S = slides, R = Ries debris, L = limestone (Tertiary).

into the Low Terrace, formed during the last Pleistocene cold period. In the south, by contrast, the river was braided and aggrading. Artificial straightening now causes it to flow in a single bed. The cutting off of meanders in the previously strongly meandering northern stretch of the Rhine has shortened its course and converted it into a major waterway. Since 1830, these modifications have caused a vertical incision of the channel of 10 m in the south and 1 to 3 m in the north. As a result, the water table in the floodplain has been lowered and it is now dry enough for field crops, although dams are necessary for flood protection.

11 Regional example 2: the Keuper scarp

The location of the Keuper scarp is independent of the drainage pattern (fig.9). A river network exists that was inherited from a pre-existing peneplain. In the Hassberge north of the Main, the major rivers flow parallel to the cuesta scarp. Some of their tributaries originate at the scarp. Those valleys that follow the dip of the beds are usually very wide and have extended headward for a considerable distance. In some areas they lower the scarp itself (SPÄTH 1973). The Steigerwald to the south is separated from the Hassberge by a broad planation pass through which the Main flows at the present time. In the Steigerwald, the rivers flow into the scarp in wide valleys that maintain their width over great distances. Farther south in the Frankenhöhe the large rivers rise on top of the scarp. The upper courses of many of the valleys of the rivers are wide and can be termed intramontane basins, although the widening took place in weak Keuper beds which tend to expedite erosion and denudation.

The pattern of rivers relative to the scarp indicates that the former is older. The scarp has been affected by the rivers only locally. It has been formed by planation processes. If it is assumed that it developed in different places at different times, the varying plan forms and profile forms of the scarp can be explained. In the north, in the Hassberge, the scarp is relatively straight with few large planation embayments and passes, the result of planation of the lower surface. Steep scarp slopes are irregularly distributed. In the central section, the planation embayments are smaller, the scarp crest is frequently bevelled and, in some cases, the scarp slope is uniformly inclined, indicating a restricted planation tendency at the lower level. In the area between the Jagst and Kocher rivers and farther to the west, the scarp is well-developed in some segments and has large embayments and wide valleys but the dip slope is mainly an area of hills. Locally recent landslips indicate that the scarp has been modified in the Quaternary. As a result, the ground plan of the scarp is very irregular in some areas although the scarp has receded overall by only a few kilometers from its general regional alignment.

There are indications that the Keuper cuesta was developed from a higher denudation surface predominantly by areal downwearing. Evidence for this denudation surface is provided by the truncation surface on which relic soils occur and by the presence of a Cretaceous cover on the neighbouring Franconian Jura. The apparent bevelling of the scarp crest would then be the remnant of an extended slope (Streckhang) which initially formed during downwearing and was later steepened and cut off by the denudational retreat of the slope foot.

12 Deformations of the Mesozoic beds

In south Germany, the Muschelkalk and the Keuper are the most widespread formations at and below the land surface. For this reason the elevation of the contact between them can be used as an indicator of tectonic deformations (fig. 10; CARLÉ 1955). Usually the synclines and anticlines are not apparent in the relief. Resistant and weak strata are usually truncated by a structure-independent denudation surface.

A comparison of the Muschelkalk/Keuper contact with the present land surface shows that while the stratigraphic boundary lies at similar elevations in the Odenwald, up to 1400 m, and in the Black Forest, up to 1700 m, the maximum surface elevations today are about 500 to 600 m in the former and nearly 1500 m in the latter. In the Odenwald, therefore, about 800 m has been removed by denudation between the Muschelkalk/Keuper boundary and the oldest denudation surface remnants

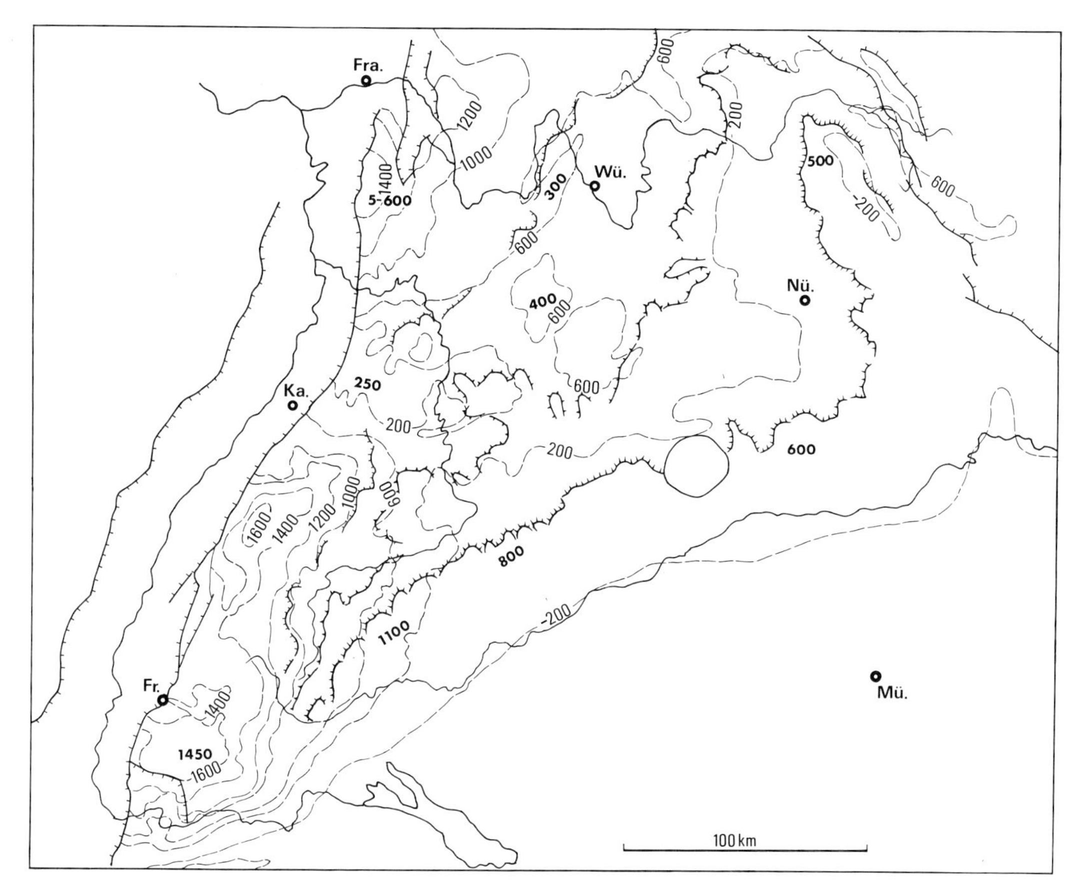

Fig. 10: *Amounts of uplift (strike contours of the Muschelkalk/Keuper stratigraphic boundary based on CARLÉ, simplified) and present elevation of the highest surfaces in the different areas.*

but in the Black Forest, only 200 m to 300 m. Keuper and Jurassic beds were present above the boundary, as is shown by Jurassic xenoliths in the volcanic neck of the Katzenbuckel. The Keuper decreases in thickness southward, the Jurassic northward, together they are about 1000 m thick. This means that in the Odenwald approximately 1800 m and in the Black Forest approximately 1200 m have been removed. Between both areas lies the tectonic depression of the Kraichgau where the same method of estimation indicates a removal of only 900 to 1000 m.

In the central area of the Franconian cuesta region (Franconian shield) and in the Thüngersheim anticline, north of Würzburg, the stratigraphic boundary of the Muschelkalk and Keuper has been upwarped by up to 500 to 600 m; in the area of the Swabian Jura, it lies in the north at +100 m and in the south at -200 m elevation. Corresponding values in the broadly synclinal Franconian Jura are +100 m at the margins and -

200 m along the central axis. The present land surface in the northern Franconian Gäuland (Thüngersheim anticline) lies at about 300 m and in the southern Franconian Gäuland at about 400 m. Denudation surface remnants in the northern Franconian Jura occur at about 500 m elevation. In the vicinity of the Ries, the surface remnants lie at about 600 m; they rise westward in the Swabian Jura to 800 to 1100 m. Therefore, in the Thüngersheim anticline the amount of denudational lowering is about 1200 m, in the southern Franconian Gäuland, about 1000 to 1100 m, in the Franconian Jura 300 to 500 m and in the western Swabian Jura only about 100 m.

Even though these estimates may have a possible error of 100 to 300 m because of unknown thickness variations, the differences are nevertheless significant. Their spatial pattern cannot be explained by the length of denudation. They are the result of tropoid relief development which is characterized by divergent weathering and denudation (BREMER 1971, 1981). Exposed bedrock surfaces are being lowered very slowly. Therefore, uplifted old denudation surfaces can be preserved for a very long time, particularly limestone surfaces that have been subject to deep karst development. Their further downwearing occurs mainly as a result of restricted planation, that is, by the lowering of planation bands.

13 Age of epirogenic movements and of denudation

The Molasse basin north of the Alps was the baselevel for the drainage southward. From the upper Cretaceous to the upper Miocene this basin formed a large sea in which the prevailing transport directions of sediments varied between a predominantly eastward transport and, in the later Tertiary, a westward transport.

With few exceptions the Tertiary deposits consist of sands and clays, in part because of the weathering in a hot and humid climate but also because fine-grained material was transported into the seas from land areas that had not yet been uplifted much. Uplift was relatively slow, as evidenced by the large number of remnants of old denudation surfaces at various levels. Alluvial fans and deltas were formed on the coasts of these seas and the gravel components of these deposits show which rocks had been eroded and indicate the nature of the local relief. The total amount of denuded material in the Tertiary was considerable despite the low elevations because the planation processes produced large quantities of material.

The paleogeographic evidence indicates that denudation probably started in the Cretaceous period. In the Franconian Jura, the upper Cretaceous sea transgressed onto a denudation surface (BÜDEL 1957). Recently a karst development at depth has been discovered below the Alpine Foreland which must have been initiated before the subsidence of the Jurassic beds and was perhaps intensified later. The Jurasssic limestones must, therefore, have been above sea level. In the Franconian Jura the age of these karst forms is indicated by fills of upper Cretaceous sediments in karst systems that attain depths of up to 200 m.

In addition, valleys, probably lowered planation bands, occur with deposits of Ries debris in the Franconian Jura. In the Naab depression, early Miocene valley systems must have been at least 100 m deep (LOUIS 1984). Further west,

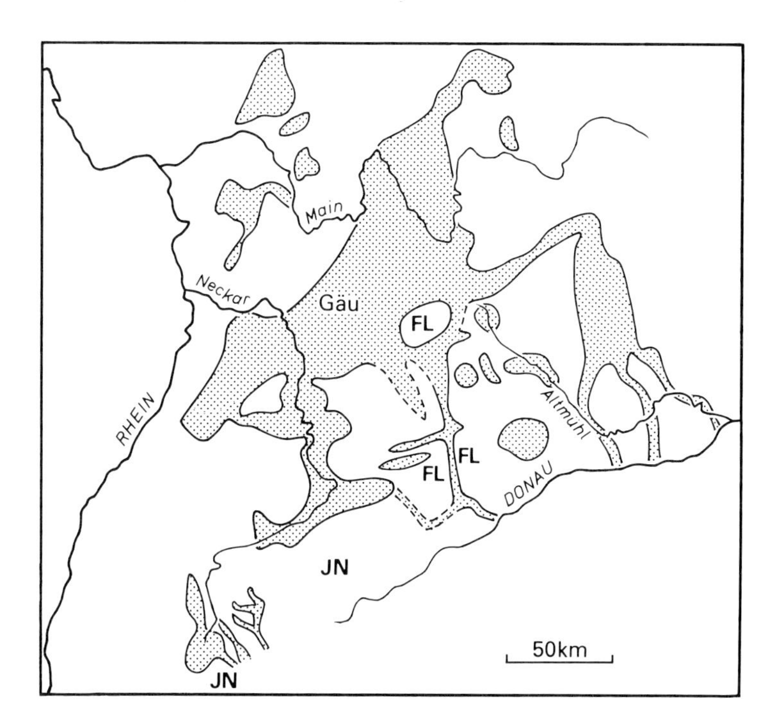

Fig. 11: *Distribution of the lowest denudation surface, probably dating from the last planation phase which was especially extensive in the Muschelkalk areas where, as a consequence, cuestas are largely absent. The denudation surface areas become narrower southward and change to planation bands in the Swabian Jura (fig.7).*
JN = Juranagelfluh, FL = Flint loam.

KLEBER (1987) has derived Quaternary tectonic movements. In the Rhön, pre-Miocene valleys have been filled to depths of 150 m (MENSCHING 1957).

In the lower Miocene (Burdigalian), the Molasse sea eroded a conspicuous cliff in the Swabian Jura which now lies at 500 m above sea level in the eastern Jura and at almost 900 m in the western Jura (GLASER 1964). The cliff was uplifted after the middle Miocene but younger deposits cannot be correlated accurately enough to date tectonic movements.

The old denudation surfaces of the Franconian Jura and of the eastern Swabian Jura both lie at 500 to 600 m above sea level so that it can be assumed that this old relief has been uplifted en bloc. The old surfaces of the Odenwald, the Spessart and the Keuper scarp lands have similar elevations and could, therefore, be interpreted as an indication of an extensive initial denudation surface. A major uplift of this old surface has taken place in the Black Forest, in the western Swabian Jura and also in the eastern marginal uplands.

An extensive younger surface, probably of Miocene age, is formed by the loess-covered Gäu surface on the Muschelkalk (fig. 11). The relatively low-lying levels in the marginal uplands (often planation bands, small planation embayments or intramontane plains) and similar forms in the cuesta areas of the Keuper and Jura also belong to this younger planation phase.

Additional evidence of epirogenic deformation can only be found outside the area of the scarplands in the Alps, where strong uplift occurred in the early Tertiary and again in the Pliocene. It is possible that plate tectonics related to the Alpine orogeny and the foreland trough also affected the area further north, including the cuesta region. A relationship between the two uplift phases and the development of the denudation levels is possible.

14 Tectonically and climatically controlled landform development

Since denudation began in the Cretaceous, the present landscape has developed over a very long period. Temporally and spatially differentiated tectonics have been the cause, in part, of the pattern of exposure of the different beds and their elevations. The simplest explanation would be that the earlier and greater the uplift, the greater the denudation. However, the difference between the northern and southern parts of the Rhine Rift Valley, the contrast between the Black Forest and the Odenwald and between the Swabian Jura and the Gäuland indicate that other influences were also present. If there are geomorphological processes of the kind active in the tropics at the present time, a rapid uplift can lead to rapid incision of streams, to the divergence of weathering and denudation, to the preservation of old denudation surfaces and to an overall high relief. A slow uplift, by contrast, results in a steady areal downwearing and a greater total amount of denudational removal. This could apply to the northern part of the south German crustal block. The weathering relics, the river network and the valley forms support this interpretation but further investigation and mapping are needed.

There seem to have been two phases of uplift and, correspondingly, two denudation surfaces. The phases of uplift probably lasted for long periods of time. Tertiary sedimentation has been relatively continuous and is more differentiated by crustal movements in the subsidence area than by variations in the sediment supply. There are, however, subordinate planation levels within the major denudation surfaces, which are probably the result of slow rates of uplift (BREMER 1981). Apart from fault tectonics, uplift probably did not vary spatially more than 200 m, with the exception of the western Swabian Jura, the marginal uplands in the east and northeast and the Black Forest. In the southern Franconian Jura the presence of Tertiary gravels, sands and limestones suggests that more differentiated tectonic movements have occurred.

The processes of landform development have changed over time. Because of the similarity of the higher relief elements in south Germany to relief elements found in parts of the tropics, it has been suggested that those in south Germany were developed in the Tertiary in a hot and humid climate which lasted until the middle Tertiary. By the late Tertiary, there was a pronounced cooling trend although if the summers remained hot these morphogenetic processes could have continued, at least seasonally.

The planation bands, small planation embayments and wide valleys can be termed forms of restricted planation (BREMER 1981). They can occur as a result of uplift, of a change to a subhumid climate or perhaps of a change to a climate with cooler winters. The decisive climatic change took place, however, at the beginning of the Quaternary about 2.5 million years ago. After a transitional period with broad terraces, narrow valleys with gravel deposits of resistant and non- resistant pebbles were first formed in the Günz cold period.

15 Geomorphological processes

Based on the general pattern of relief development in south Germany, conclusions can be drawn about the processes that have shaped the major landform types in the region and particularly whether the cuesta landscapes are the result of areal downwearing or of scarp retreat. In addition to points already discussed, scarp retreat would appear to have been limited for the following reasons.

1. In the area in which the Swabian/Franconian Jura cuesta changes its strike from northeast to north, a scarp was present before the Ries event; the distribution of Ries debris and of Miocene limestones show that this scarp has retreated by an insignificant distance since then.

2. The scarp has also not retreated in the valleys of the rivers that flow into it; it can be assumed, therefore, that the embayments of the cuesta scarp in other places where the rivers flow into the scarp are also old forms.

3. Since the conglomerates in the western part of the Swabian Jura contain Muschelkalk pebbles, a continuous scarp cannot have existed in that area. Either the scarp was interrupted by broad planation bands through which the transport took place or the scarp was formed only later.

4. In the area of Kocher and Jagst, early Pleistocene or perhaps late Pliocene sands extend to the foot of a 200 m high scarp (ZEESE 1976/77).

5. The middle Jurassic rock (Dogger) was originally grey (BRUNNACKER 1978); its brown colour is thought to be due to subsequent weathering. Along the scarp only this "brown Jura" can be observed so that the scarp must be very old wherever this rock is present in the scarp slope.

Recent landslips which have occurred in the marly Feuerletten of the lower Jurassic and in the Knollenmergel of the Keuper indicate that scarp retreat is continuing. The landslips are widespread in the area north of the central Swabian Jura. It is difficult to determine whether small obsequent valleys are being widened and deepened at the present time. The springs which supply water to these valleys are up dip contact springs or fault springs. Their shifting is controlled by the geological conditions.

The number and size of residual outliers may be possible indicators of scarp retreat but they have not yet been evaluated. There are relatively few outliers and no outlier generations although these might be expected with a more or less uniform rate of scarp retreat. The varying form of the cuesta scarps and the wide, and in places undercut, bevel of the scarp crest have also not yet been evaluated.

Measurements of recent processes have been limited primarily to the dissolved and suspended load of streams. The suspended load has certainly increased considerably because of the high density of settlements. The dissolved load also seems to have increased. Extrapolations of the measurements yield such high values that they do not correspond to the natural conditions. The wide distribution of periglacial covers, on which the postglacial soils have developed, shows that there has been little change in the last 10,000 years. Also, most valley floors have changed relatively little. A series of postglacial terraces such as those that have been identified in the valley floor of the Main (SCHIRMER 1983) is not of great significance. Investigations show that in some valley heads minor changes of form can be observed. The landforms cannot, however, be explained by an extrapolation of the present-day processes.

Acknowledgements

I am very grateful to my colleagues Karl-Heinz Pfeffer, Heinz Späth and Reinhard Zeese with whom I have been able to discuss problems related to this paper.

References

BAYERISCHES GEOLOGISCHES LANDESAMT (1964): Erläuterungen zur Geologischen Karte von Bayern 1:500,000. München.

BLUME, H. (1971): Probleme der Schichtstufenlandschaft. - Erträge der Forschung **5**, Darmstadt.

BREMER, H. (1971): Flüsse, Flächen- und Stufenbildung in den feuchten Tropen. Würzburger geogr. Arb. **35**.

BREMER, H. (1981): Reliefformen und reliefbildende Prozesse in Sri Lanka. In: BREMER et al. (1981), 7–183.

BREMER, H., SCHNÜTGEN, A. & SPÄTH, H. (1981): Zur Morphogenese in den feuchten Tropen. Verwitterung und Reliefbildung am Beispiel von Sri Lanka. Relief, Boden und Paläoklima 1., Stuttgart.

BRUNNACKER, K. (1978): Boden. In: Das Mainprojekt (1978). Schriftenreihe d. Bayr. LA f. Wasserwirtschaft. H. **7**, 21–23.

BÜDEL, J. (1957): Grundzüge der klimamorphologischen Entwicklung Frankens. Würzburger geogr. Arb. **4/5**, 5–46.

BÜDEL, J. (1982): Climatic Geomorphology. (Translated FISCHER & BUSCHE), Stuttgart.

BURGER, D. (1988): Exkursion der Kommission für Geomorphologie in der Bayerischen Akademie am 11. und 12. Juni, 1988.

CARLÉ, W. (1955): Bau und Entwicklung der südwestdeutschen Großscholle. Beih. Geol. Jb. 16.

DEHM, R. (1961): Spaltenfüllungen als Lagerstätten fossiler Landwirbeltiere. Mitt. Bayr. Staatss. Pal. u. Hist Geol. **1**, 58–72.

GEYER, O. & GWINNER, M. (1986): Geologie von Baden-Württemberg, Stuttgart.

GLASER, U. (1944): Die Miozäne Strandzone am Südsaum der Schwäbischen Alb.. Würzburger geogr. Arb. **11**.

KLEBER, A. (1987): Die jungquartäre und ältest-quartäre Entwicklung von Flächen und Tälern im nördlichen Vorland der Südlichen Frankenalb. Bayreuther Geowiss. Arb. **10**.

KUBINIOK, J. (1988): Kristallinvergrusung an Beispielen aus Südostaustralien und Mittelgebirgen. Kölner geogr. Arb. **48**.

KUPPELS, I. (1981): Die Karstspalten der Schwäbischen Alb als Leitformen für die Morphogenese. Kölner geogr. Arb **39**.

LOUIS, H. (1984): Zur Reliefentwicklung der Oberpfalz. Relief, Boden Paläoklima **3**. Stuttgart.

MENSCHING, H. (1957): Geomorphologie der Hohen Rhön und ihres südlichen Vorlandes. Würzburger geogr. Arb. **4/5**, 47–88.

MÜLLER, S. (1958): Feuersteinlehme und Streuschuttdecken in Ostwürttemburg. Jh. geol. LA in Baden-Württemberg **3**, 241–262.

PENCK, W. (1924): Die morphologische Analyse. Geogr. Abh. 2,2. Stuttgart.

PFEFFER, K.-H., (1986): Die Karstgebiete der nördlichen Frankenalb zwischen Pegnitz und Vils. Z. Geomorph., Suppl. Bd. **59**, 67–85.

QUENSTEDT, F.A., (1843): Das Flözgebirge Württembergs. Mit besonderer Rücksicht auf den Jura. Tübingen.

SCHIRMER, W. (1983): Symposium "Franken": Ergebnisse zur holozänen Talentwicklung und Ausblick. Geol. Jb. **71**, 355–370.

SCHMITTHENNER, H. (1954): Die Regeln der Morphologischen Gestaltung im Schichtstufenland. Pet. geogr. Mitt. **98**, 3–10.

SCHREINER, A. (1965): Die Juranagelfluh im Hegau. Jh. geol. LA Baden-Württemberg **7**, 303–354.

SEMMEL, A. (1964): Junge Schuttdecken in hessischen Mittelgebirgen. Notizbl. hess. LS Bodenforsch. **92**. 275–285.

SPÄTH, H. (1973): Morphologie und morphologische Probleme in den Hassbergen und im Coburger Land. Würzburger geogr. Arb. **39**.

SPÄTH, H. & MBESHERUBUSA, F. (1982): Die Datierung von Eisenanreicherungen mit Hilfe des Mößbauer-Effektes. Dargestellt an Sedimenten des unteren Mittelrheins und seiner Umrahmung sowie der nördlichen Oberpfalz. Z. Geomorph., Suppl. Bd. **43**, 203–213.

WAGNER, G.G. (1930): Erd- und Landschaftsgeschichte mit besonderer Berücksichtigung Süddeutschlands. Öhringen.

ZEESE, R. (1972): Die Talentwicklung von Kocher und Jagst im Keuperbergland. Flußgeschichte als Beitrag zur Deutung der Schichtstufenmorphogenese. Tübinger geogr. Studien **49**.

ZEESE, R. (1976/77): Die Oberflächenform der Region Ostwürttemberg. Raumordnungsbericht Bd. 2: Regionalverband Ostwürttemberg, Naturraum. Schwäbisch-Gmünd, 7–35.

ZIENERT, A. (1986): Grundzüge der Großformenentwicklung Südwestdeutschlands zwischen Oberrheinebene und Alpenvorland. Heidelberg.

Address of author:
Prof. Dr. Hanna Bremer
Geographisches Institut der Universität Köln
Albertus-Magnus-Platz
D-5000 Köln 41

DAN H. YAALON (ED.)

ARIDIC SOILS and GEOMORPHIC PROCESSES

SELECTED PAPERS of the INTERNATIONAL CONFERENCE
of the INTERNATIONAL SOCIETY of SOIL SCIENCE
Jerusalem, Israel, March 29 – April 4, 1981

CATENA SUPPLEMENT 1, 1982

Price: DM 95,–

ISSN 0722–0723 / ISBN 3–923381–00–X

This CATENA SUPPLEMENT comprises 12 selected papers presented at the International Conference on Aridic Soils – Properties, Genesis and Management – held at Kiryat Anavim near Jerusalem, March 29 – April 4, 1981. The conference was sponsored by the Israel Society of Soil Science within the framework of activities of the International Society of Soil Science. Abstracts of papers and posters, and a tour guidebook which provides a review of the arid landscapes in Israel and a detailed record of its soil characteristics and properties (DAN et al. 1981) were published. Some 49 invited and contributed papers and 23 posters covering a wide range of subjects were presented at the conference sessions, followed by seven days of field excursions.

The present collection of 12 papers ranges from introductory general reviews to a number of detailed, process oriented, regional and local studies, related to the distribution of aridic soils and duricrusts in landscapes of three continents. It is followed by three papers on modelling and laboratory studies of geomorphic processes significant in aridic landscapes. It is rounded up by a methodological study of landform–vegetation relationships and a regional study of desertification. Additional papers, related to soil genesis in aridic regions, are being published in a special issue of the journal GEODERMA.

D.H. Yaalon
Editor

CATENA SUPPLEMENT 15 p. 69–83 Cremlingen 1989

THE LANDFORMS OF THE GERMAN ALPS AND THE ALPINE FORELAND

Klaus **Fischer**, Augsburg

1 Introduction

The two major landform units in the southernmost part of Germany, the Alps and the Alpine Foreland, contain several morphological and structural zones. The German Alps can be divided into the High Limestone Alps and the Marginal Alps, and the latter can be further subdivided into the Marginal Limestone Alps, the Flysch zone and the Helveticum zone. The Alpine Foreland is composed of a zone of former foreland glaciation, in which there is a wide range of glacial landforms, a zone of gravel deposits in front of the terminal moraines and a zone of "Tertiary Hills", that is, hills in Tertiary deposits, that was not covered by the Pleistocene glaciers.

2 The German Alps

2.1 Geology and structural landforms

The Alps are characterized by a northward-vergent structure of folds, imbricated thrust wedges and nappes or overthrust sheets. In the German Alps, the Flysch zone overlies the Helveticum zone and is in turn overlain by the three nappes (Allgäu Nappe, Lechtal Nappe and Inntal/Berchtesgaden Nappe) of the Limestone Alps.

Most landforms are adapted to the structure and lithology. For example, the front of the Allgäu Nappe, which is made up of resistant calcareous rocks, rises steeply above the forested lower mountains of the weak flysch zone along the entire margin of the German Alps, from the vicinity of Oberstdorf in the west to the area of Berchtesgaden in the east.

In addition to the effects of the nappes and folds, the influence of structure is also apparent from the adjustment of erosion to the lines and zones of faulting. Some of the major transversal valleys and passes have developed on Upper Eocene to Middle Miocene diagonal, normal or transcurrent faults. Examples are the transversal valleys of the Ammer and Loisach rivers near Garmisch-Partenkirchen.

The development of crests, arêtes, and peaks and of slope forms is influenced primarily by the strike and dip of the strata. In large areas of the northern Limestone Alps the beds dip southward. In addition, the baselevel of erosion for most of these areas lies to the north. Consequently, the majority of the rock faces and steep slopes is oriented northward. Because of the frequent change of facies and because of the complicated tectonic structure, rocks of very differ-

ISSN 0722-0723
ISBN 3-923381-18-2

3-923381-18-4/89/5011851/US$ 2.00 + 0.25

ent resistance lie next to, or on top of, one another, and there is, as a result, a great variety of forms over short distances. On limestones and dolomites, high rock walls, steep slopes, sharp isolated peaks, arêtes and broad summit massifs occur. Clayey, marly rocks, by contrast, tend to have less steep slopes, gently inclined high pastures and large hollows. Rocks of high resistance in synclines have, in many cases, resulted in an inversion of relief (fig. 3). Structure and lithologic variations are of great importance in the present landform pattern of the Alps, especially at the chorological (landscape) and the local scale.

The major sections of the High Limestone Alps are, from west to east,

1. the Allgäu High Alps,
2. the Lechtal Alps,
3. the Wetterstein Mountains and the Mieminger Mountains,
4. the Karwendel Mountains and the Rofan Mountains,
5. the Loferer and Leoganger "Steinberge" and
6. the Berchtesgaden Alps.

Over wide areas the Limestone Alps lie above 2000 m and also, therefore, above the tree line. They have a high internal relief and there are indications of past glaciation. Glacial erosion and physical weathering have combined to shape their "Alpine landforms".

The High Limestone Alps are composed primarily of massive carbonaceous sedimentary rocks of the Alpine Triassic. In the west, the Main Dolomite (Norian) predominates, in the Wetterstein and Karwendel Mountains, the Wetterstein Limestone (Ladinian) and in the Steinberge and the Berchtesgaden Alps, the Dachstein Limestone (Norian), each with thicknesses of about 1000 m.

The Main Dolomite in the Allgäu and Lechtal Alps is characterized by sharp peaks and rugged arêtes. It has a breccious structure which causes ravines to form rapidly in rock walls and on slopes and for large accumulations of talus to occur at the base of the slopes and in the valleys. The shape of the glacial troughs has been destroyed, bedrock steps and bedrock floors in valleys covered and the cirques filled with debris. The Main Dolomite produces more debris than any other rock in the Limestone Alps. Much of its waste disintegrates into "Gries" (grus). Another morphological component are the steep Allgäu Grass Mountains which are covered with vegetation to their summits. They are made up of Jurassic radiolarites and Aptychia beds.

In the Mieminger and Wetterstein Mountains and the Karwendel, the rock walls and nearly horizontal arêtes are composed of Wetterstein Limestone. Because of the slow rate of weathering and the karstification of the limestone, stepped cirques and glacial troughs have been better preserved here than in the Allgäu. In the Rofan Mountains and Kaiser Mountains, the light grey Wetterstein Limestone also forms sharp peaks, and where the beds dip steeply, bedding planes have been widened into chimneys and ravines, leaving pinnacles and towers between them.

The landform pattern in the Berchtesgaden Alps is very different. Instead of mountain chains, massifs of thick-bedded horizontal Dachstein Limestone overlying breccious Ramsau dolomite are surrounded on all sides by valleys. The near-horizontal bedding of these large tectonic units, lying on a little-disturbed basement, has resulted in the development of a gentle relief in the summit zone of the massifs which has been preserved because, largely, of the karstification of the Dachstein Limestone. Lapies, sinkhole fields and cave systems all occur in the limestone. This paleo-relief of, presumably, Middle to Upper Miocene age has a local relief of up to 1000 m. It is an old upland of hills and isolated mountain groups which contrasts with the very long, more or less uniformly steep outer slopes of these mountain massifs.

Only small remnants of the old relief have been preserved in the northern Limestone Alps. Examples are the Zugspitzplatt and the Osterfelder in the Wetterstein. Others occur in the Mieminger Mountains, in the Lechtal Alps and in the Allgäu. The paleo-relief remnants are independent of the lithology; their surface extends across various rock types. Their gentle hill forms, termed "Rax landscape" after the type location in Austria, have evolved from a still older relief that was formed when the major east Alpine longitudinal valleys did not exist. On these relic forms "Augensteine", rounded smoothly polished pebbles and sand grains, occur individually or as deposits at greatly varying heights within the old relief, in most cases probably after reworking and repeated deposition. It is not clear whether the "Rax landscape" represents a genetically unified system that has been separated into a mosaic of blocks that lie at different altitudes because of later differentiated uplifts, or whether it has been developed at different levels because of progressive lowering of the baselevel.

Below these denudation surface systems there are flattened slope segments along the valleys which are, probably, remnants of old valley floors. Successive "valley generations" have been ascribed by some authors exclusively to successive phases of uplift and corresponding incision. The climato-genetic interpretation also acknowledges the effects of uplift but assumes a more important cause to be a change from a humid, or at least sub-humid, tropical climate to a subtropical and finally to a mid-latitude climate, a sequence which is thought to have led from a gentle relief to one with deeply incised V-shaped valleys between which remnants of older relief generations have been preserved. Neither the first nor the second interpretation is without contradiction.

Between the High Limestone Alps and the Marginal Limestone Alps there are long stretches of low terrain and basins which are partly aligned with fault zones (fig.1). The Marginal Limestone Alps have a maximum height of about 2000 m and summits that are rocky but not sharply pointed. Their structure is more complicated than that of the High Limestone Alps, particularly towards the northern margin where the strata are very deformed. In many areas they dip steeply, are intensively folded and traversed by a large number of faults. The variety of rock types and of facies has resulted in a wide range of landforms. Near the margin of the Alps on the resistant Wetterstein limestones, several high rock faces and summits, such as the Wendelstein and the Hochstaufen, provide good views of the surrounding landscapes. The Main Dolomite and its debris slopes are characterized by forested mountains and debris-filled ravines and valley floors (Griese). Examples are the longitudinal valleys of the Ammer and the Isar. On the limestones of the upper Rhät, tower-shaped rock bastions and

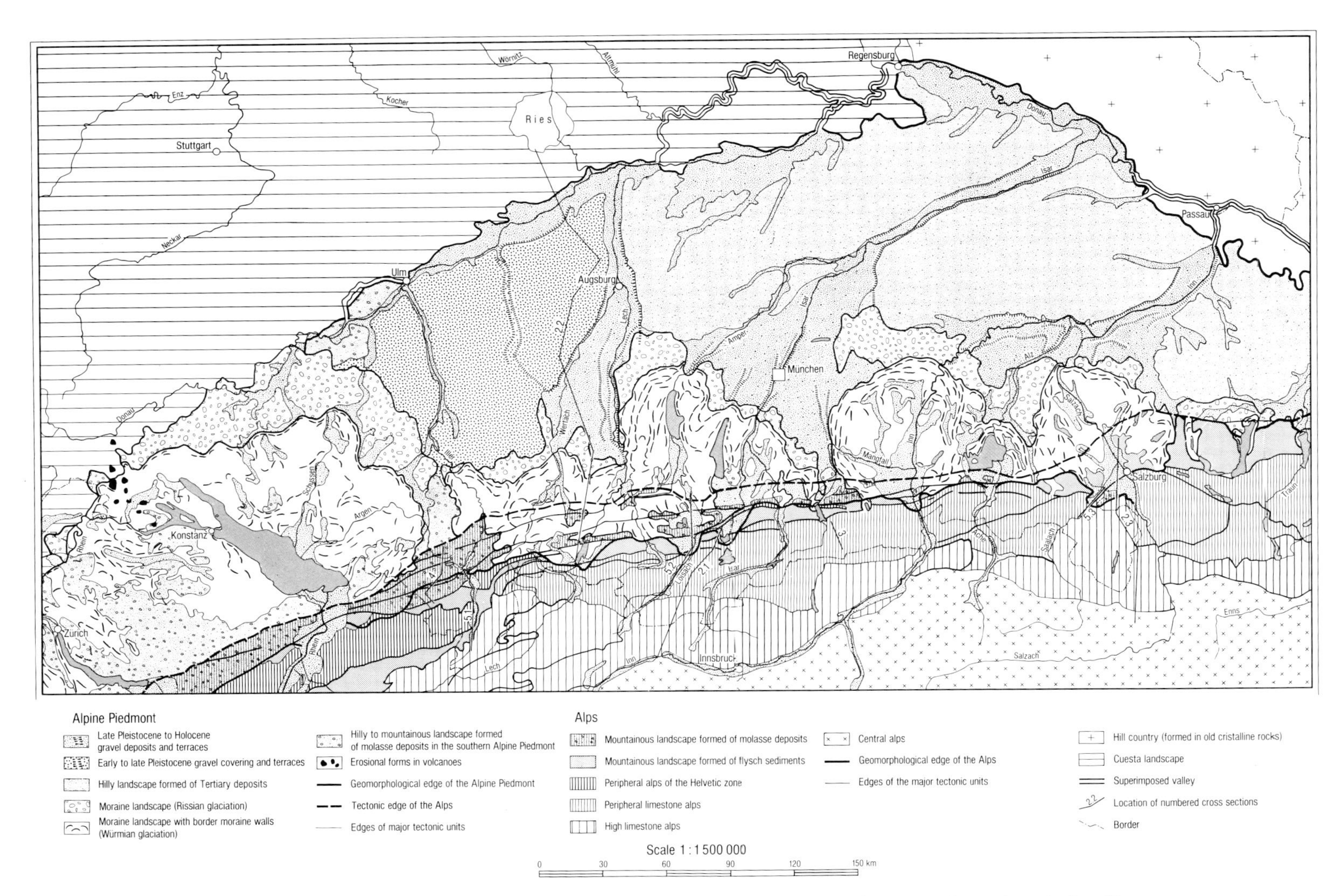

Fig. 1: *Geomorphological overview of the Alpine Foreland (Alpine Piedmont) and the Alps.*

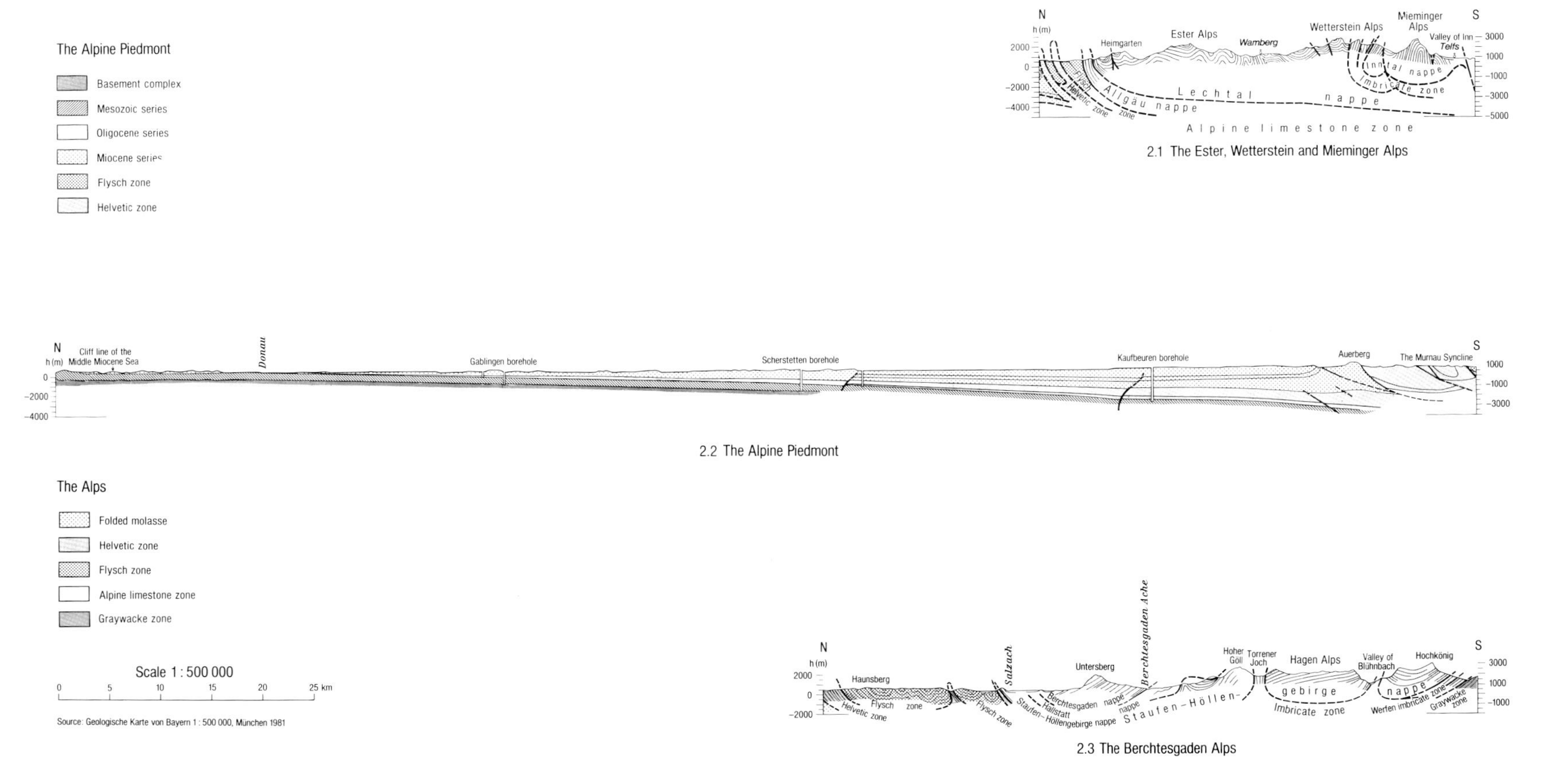

Fig. 2: *N-S cross-section through the Alpine Piedmont and the Alps (to scale).*

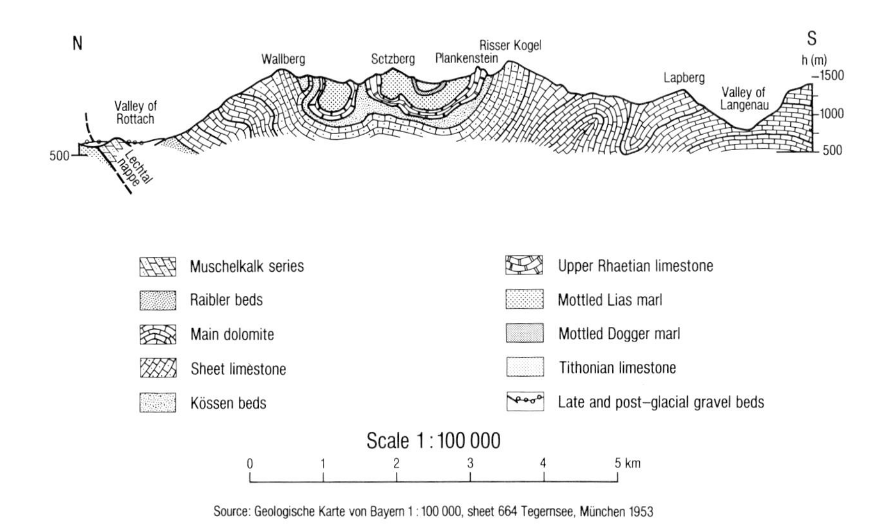

Fig. 3: *N-S cross-section through the "Synclinorium" south of the Lake Tegernsee (to scale).*

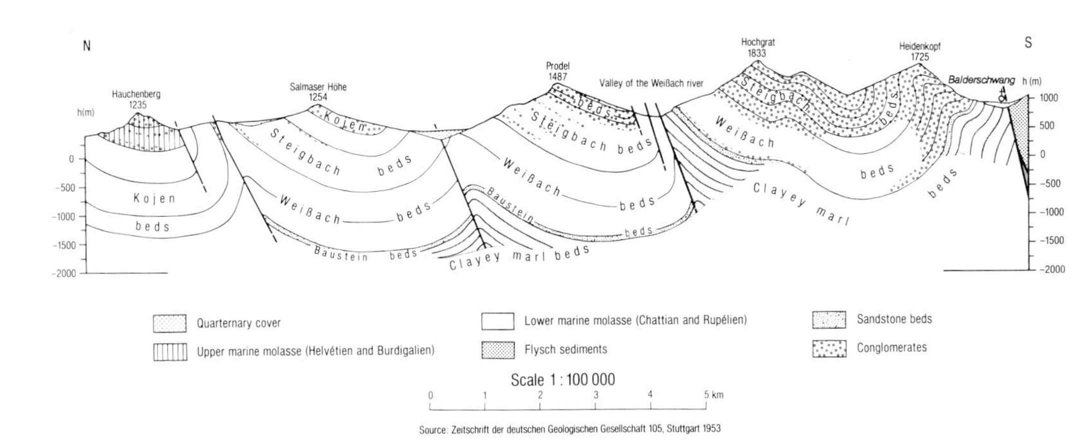

Fig. 4: *N-S cross-section through the folded molasse of the West Allgäu (to scale).*

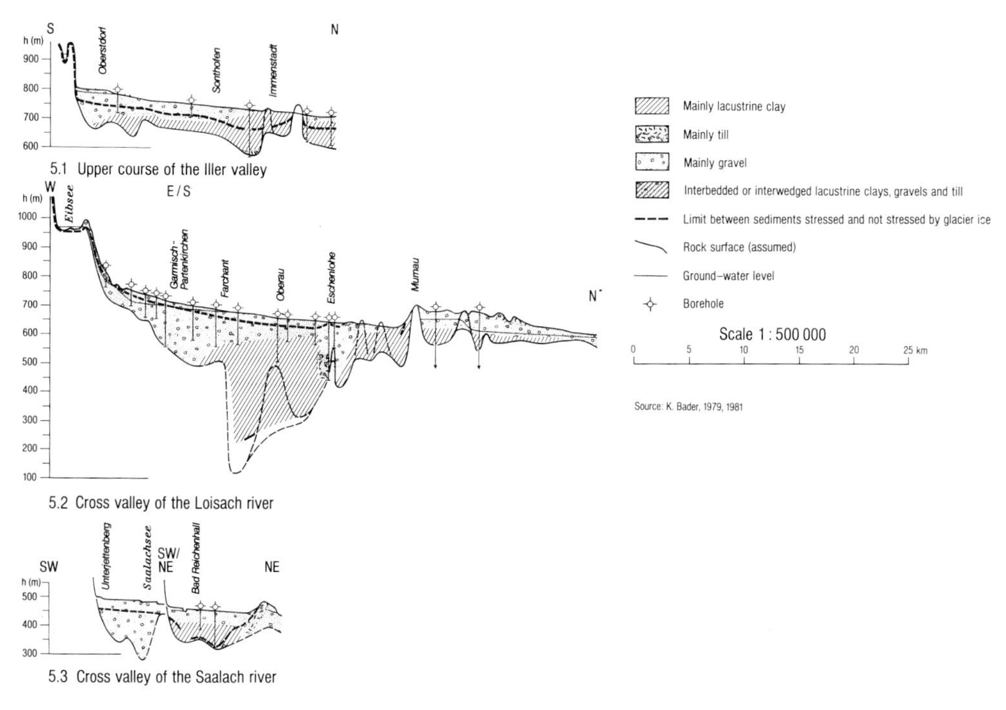

Fig. 5: *Longitudinal section of over-deepened valleys (Vertical scale 25 times exaggerated).*

small rock faces have developed in the Ammergau, near the Schliersee and the Tegernsee. The weaker Partnach beds, Raibler Beds, Kössener Beds, Lias Marls and Neocomian rocks have been eroded more strongly to form zones of lower elevation.

Densely-spaced valleys, which are connected by in-valley divides and low passes, subdivide the Marginal Limestone Alps into many small mountain groups and individual mountains. Since there are a large number of captures, the divides have shifted considerable distances. There are few paleoform remnants because the interconnected valleys are closely-spaced. It is also very difficult to distinguish between lithologically and structurally caused benches and river terraces in the valleys.

The Flysch and Helveticum zones lie beyond the northern boundary of the Marginal Limestone Alps. East of the Iller river the zones are only a few kilometers wide. In some sections they have been completely overrun by the Alpine limestone nappes. The Flysch consists of an internally imbricated sequence of beds of weak rock. It was formed from the sediments of turbidity currents in a narrow deep sea trough of the east Alpine geosyncline. The Flysch is easily weathered and denuded; its landforms are rounded ridges with moderate slope angles and relatively low summits. No natural rock outcrops occur on the slopes. Because of the danger of landslides on the clayey, marly impervious rocks, the Flysch is kept largely under forest. A characteristic feature of the region is the large number of narrowly incised V-shaped gullies and small valleys, termed Tobel.

West of the Iller, Flysch and Helveticum form a single area of upland. Greatly varying rock resistance has resulted in a variety of landforms, particularly on the Helveticum. The Schratten limestone, a thick-bedded pure limestone, is the most important summit-forming rock. It also forms rock walls. Where it lies horizontally, high plateaus with karst landforms have developed, such as the Gottesacker Plateau and the high rock faces of the Hoher Ifen and the Gottesacker Walls. On the Helveticum, anticlines form the high ground and the valley zones coincide with synclines. The axial planes of the folds dip to the south so that north-facing slopes tend to be steep and subdivided by ledges and south-facing slopes are often almost parallel to the bedding. Many summits in the Flysch zone are made up of the conglomeratic Reiselberg Sandstone; examples are the "horns" in the Allgäu Marginal Alps.

2.2 The effects of glaciation and periglacial processes

The typical high mountain landforms of the Alps are largely the result of the glaciations during the Pleistocene. Slopes that were above the glaciers were worn back by intensive weathering so that ridges became arêtes, summits became horns because of the retreat of cirque walls and the valley heads were transformed into semi-circular trough heads. In the valleys, irregular long profiles developed and tributary valleys became hanging valleys.

At the height of glaciation, the glaciers moved along transversal pathways (the transversal valley of the Alpine Rhine, the Mittenwald Gap and the transversal valleys of the Inn and the Salzach) from the northern longitudinal valley zone of the Alps into the Northern Limestone Alps, bringing with them central Alpine glacial drift. The glaciers bifurcated at transfluence or diffluence passes and joined with the local glaciers to form a network of ice streams. In the mountains only the summits and highest arêtes remained uncovered. On the northern margin of the Alps, there was an apron of ice fans or foreland glaciers. The ice was more than 1000 m thick in the Northern Limestone Alps and the snow line at 1200 m to 1300 m above sea level. Erosion by the glaciers straightened, widened and deepened the valleys. Large basins developed in many areas but there were few well-defined glacial troughs. Examples are the valley zone of Königssee-Obersee-Fischunkel in the Berchtesgaden Alps and the Rein Valley in the Wetterstein Mountains. Intensive glacial erosion is apparent from the large amount of over-deepening in valley stretches upstream of transversal rock barriers, particularly in the Marginal Limestone Alps. The largest known over-deepening (500 m) occurred in the transversal Loisach valley between Farchant and Oberau (fig.4). The distribution of the over-deepened valleys depends not only on the erodibility of the bedrock and the pre-glacial valley form but also on the location of the valley relative to the ice flow, for example, at confluences. Ice-pressured sediments have been found which indicate that the over-deepening by glaciers occurred before the Würm glaciation. The erosive effect of the Würm glaciation was comparatively limited and confined mainly to the removal of loose materials. This corresponds, for example with evidence that the Würm glacier in the Alpine Inn valley advanced rapidly for a short period and did little geomorphological work. Also, there are few older interglacial deposits (breccias, gravels) in the Alps.

As the glaciers back-wasted, the filling-in of the over-deepened valleys began. Delta gravels and lake clays indicate the existence of late glacial to postglacial lakes in the valleys. The longitudinal Isar valley is an example. There are lakes in some basins at the present time, including the 200 m deep Königssee and the Tegernsee. Streams in ice-free side valleys were dammed by the trunk valley glaciers and deposited gravels termed Verbauungsschotter (ice-blocked gravel deposits).

Rock barriers, steps in the long profiles and at the exits of hanging valleys have been dissected frequently by saw-cut gorges below which alluvial and mudflow fans were deposited that became important for the location of settlements. Landslides and rockfalls were also important in late glacial and postglacial slope development. In addition, in almost all areas of the mountains, there are local moraines which can be correlated with a snowline 300 m to 900 m below the present snowline. The present glaciation of the Northern Limestone Alps in Germany is insignificant and limited to the Zugspitze, the Watzmann and the Hochkalter. It covers an area of about 1 km^2.

At the present time, the dominating morphological processes below the treeline are slow mass movement and stream work. Above the treeline, the most important processes are frost weathering, solifluction and gravitational mass movements. Some of the latter, such as rockfalls, avalanches and mudflows, have high momentary transport rates. In addition, there are karst processes on the limestones which began in the geological past.

3 The Alpine Foreland

The geomorphological boundary and the tectonic boundary between the Alps and the foreland are not identical. A.PENCK (1894) defined the boundary as the steep marginal slopes of the mountain range which intersect at a noticeable angle with the more gently sloping low relief of the foreland. The tectonic boundary of the Alps lies partly to the north, partly to the south of this geomorphological boundary. In the areas of the broader valley exits to the foreland, the geomorphological boundary lies further to the south. West of the Iller it crosses into the zone of the subalpine Molasse which elsewhere belongs to the Alpine Foreland to the north. South of the line Immenstadt-Oberstaufen-Bregenz parts of the folded Molasse form mountain chains with arêtes and hogbacks of high Alpine character that are caused by the imbricated structure, the greater uplift and the thick conglomeratic beds of the Molasse in this area.

3.1 The geological evolution

The origin and development of the Alpine Foreland is closely linked to that of the Alps. It was formed as the pre-Alpine depression simultaneously with the uplift of the Alps in the Upper Eocene and was filled with the denudational and erosional debris, the Molasse, transported out of the rising mountain region. The more than 3000 m of Molasse consists, therefore, of sediments that can be correlated with the phases of uplift and dissection of the mountains.

Close to the Alps, the Molasse is separated from the underlying rocks by a shear zone. It has been compressed in a series of broad, east-west striking synclines, partly imbricated and pushed northward across the autochthonous or unfolded Molasse. Structurally the folded (or subalpine) Molasse belongs to the Alps. In the west of the zone there are four synclines, and in the east, south of the Chiemsee, there is only one. Small anticlines, modified by steep thrust planes, are also present in the synclines and apparent in the morphology (fig.5). An example is the Murnau syncline southwest of Penzberg, the axis of which rises eastward. Conglomeratic series on the flanks of the synclines support ridges and individual peaks.

To the north is the autochthonous unfolded Molasse which includes clays, marls, sands and gravels of varying composition and origin and frequent vertical and horizontal changes of facies. Over much of the area the Molasse is covered by Pleistocene and Holocene sediments.

The Alpine Foreland was first uplifted in the Pliocene at a rate that was higher in the west than in the east. Evidence for this is provided by the elevations of the cliffed former coastline of the Burdigalian Molasse sea in the Swabian Jura. The drainage system of the Danube developed on the surface that descended to the east in the middle Pliocene in a landscape of low relief in which there were isolated Molasse mountains to the south. The divide between the Rhone and the Danube systems was in the southward extension of the southern Black Forest. The Alpine Rhine was tributary to the Danube until the Middle Pleistocene (TILLMANS et al. 1983). The Wutach drainage area on the southeastern margin of the Black Forest was captured by the Rhine in the late Pleistocenc. It was thought that the Aare drained to the Danube in the Pliocene, but new sediment-petrographical evidence suggests that this may not be the case.

3.2 Glaciations of the Alpine Foreland

During the Pleistocene, large parts of the Alpine Foreland were repeatedly covered by foreland glaciers which spread, fanlike, from the exits of the large transversal valleys of the Alps and eroded trunk ice lobe basins (Stammbecken) and branch basins (Zweigbecken) that extended radially outward. On the rims of these over-deepended glacial basins are several rows of hummocky young moraines and less well-defined older moraines. The multiple glaciation has been analyzed in detail by A. PENCK and E. BRÜCKNER (1901-1909) in "Die Alpen im Eiszeitalter" (The Alps in the Ice Age). The main results of their research are still valid although modified in detail by more recent investigations. PENCK and BRÜCKNER distinguished four main Alpine glaciations and named them, from oldest to youngest, Günz, Mindel, Riss and Würm, after four small rivers in the Alpine Foreland.

Individual foreland glaciers are named after the river valleys from which they have flowed into the foreland. Because of its large Alpine tributary area, the Rhine Glacier was the largest foreland glacier. It probably had its greatest extent during the Mindel glaciation, at least in its eastern part. There are only a few greatly modified remnants from this glacial. The northern part of the trunk ice lobe basin of the Rhine Glacier is occupied by Lake Constance (Bodensee), the largest lake in the Federal Republic of Germany. Because of the depth of the lake, it has been suggested that tectonic processes were involved in the development of the lake basin. However, fluvial incision and glacial over-deepening seem to have been the decisive factors (SCHREINER 1979). In the Hegau, at the western end of the lake, several massive volcanic necks of Upper Miocene basalt and phonolite stand above the moraine landscape of the Rhine Glacier.

Further east, the Iller-Wertach-Lech Glacier probably also reached its maximum extent in the Mindel glacial (see also HABBE & RÖGNER in this volume). Its tributary area lay almost entirely in the Limestone Alps and it was, therefore, considerably smaller than the Rhine Glacier. The moraines are composed almost exclusively of limestone material. Its contact with the Loisach-Isar Glacier to the east is difficult to determine morphographically although sediment studies have made it possible to separate the two glaciers in the vicinity of Schongau on the Lech river (PIEHLER 1974).

The Loisach-Isar Glacier has also eroded several over-deepened branch basins of varying sizes. The Ammersee and the Starnberger See, remnants of formerly larger lakes, occupy two of the basins; a lake in the Wolfratshausen basin disappeared in the Holocene (JERZ 1979). Even during the period of the maximum extent of the Loisach-Isar Glacier, an area of ice-free terrain existed between it and the Inn-Chiemsee glacier to the east.

In the basin of the Inn-Chiemsee Glacier, which also seems to have to attained its maximum extent in the Mindel glaciation, thick varve clays near Rosenheim indicate the former existence of a 10 km long lake (Rosenheimer See). Sedimentation and vertical erosion by the Inn river into the moraines near Wasserburg caused the lake to become dry. The Chiemsee, remnant of a larger lake that originally extended into the Alps, occupies the second trunk basin of this large forland glacier. Several rows of end moraines outline the lobes of the contiguous glaciated area.

The easternmost foreland glacier in Germany, the Salzach Glacier, may have had its greatest extent in the Günz Glaciation. Seven branch basins, most of which have lakes, radiate from its trunk basin. The trunk basin also contained a late glacial lake 30 km in length, which extended south and southwestward to Hallein and Bad Reichenhall. The Traun glacier to the east in Austria formed the eastern limit of Pleistocene foreland glaciation. It had three branches and reached no further than the northern edge of the Alps.

The landforms and materials of the Würm glaciation cover the largest part of the the area that was covered by foreland glaciers. Landforms dating from this glaciation are relatively unaltered. The end moraines have irregular, hummocky surfaces; the glacial forms associated with the trunk and branch basins, including moraine ridges, drumlins and drumlin fields, eskers, kames and dead-ice features such as kettles, have been changed little by denudation processes. The weathering depth in the unconsolidated material is only about 0.5 m.

The branch basins also contain landforms related to the retreat of the glaciers. The retreat moraines consist of successive ridges suggesting that the retreat stages were characterized by readvances that remained within the maximum ice limit. It is not known how far the glaciers melted back before any readvance occurred. In the basins of the Würm foreland glaciers moraines from three to four readvances have been found (TROLL 1924). At each stage of readvance a new system of outwash channels and gravel deposits was established that now form dry valley floors or contain underfit streams. In some valleys, sharply accentuated late glacial meltwater terraces exist that are linked to the moraines of the individual retreat stages. These valleys have been termed Trompetentälchen (small trumpet valleys) because of their funnel-shaped widening from the moraine outward (TROLL 1926). A centripetal drainage system developed in the over-deepened trunk and branch basins after the ice had retreated from the area. Examples are the Schussen and Argen rivers in the area of the Rhine glacier and the Vils river in the area of the Lech glacier.

The older moraine landscapes have gently undulating surfaces on which there are neither lakes nor peat bogs. Periglacial denudation processes and the widespread deposition of cover sediments, such as loess and aeolian sands, have reduced the local relief. The surface materials are deeply weathered, in places to a depth of several meters.

3.3 Glaciofluvial landforms

Beyond the young and the older moraines there are large gravel surfaces that correspond to the outwash plains (sandurs) of north Germany. The material was deposited by meltwaters and is partly interbedded with moraine material. In general, the glacio-fluvial cobbles, pebbles and sands were deposited in each of the glaciations at progressively lower levels, so that usually each younger gravel deposit accumulated on a valley floor that had been incised into the level of the next older deposit. The gravel terraces of the last two glaciations are particularly well-preserved: the High Terrace (Hochterrasse) of the Riss glaciation and the Low Terrace (Niederterrasse) of the Würm glaciation.

In the area of the Iller and Lech rivers, older remnants have been preserved on the major divides and interfluves in the form of cover gravel plateaus (Deckenschotterplatten). A clear linkage with terminal moraines has, however, only been established for the Low Terrace gravels, the High Terrace gravels and the lower (younger) cover gravels. The highest and oldest cover gravels indicate that there were cold periods prior to the Günz glaciation. They have been termed the Danube Cold Period and the Biber Cold Period. Research to determine the number of cold periods and of gravel deposits is in progress.

An area of about 1800 square kilometers known as the Munich Inclined Plain (Münchener Schiefe Ebene) differs fundamentally from the other gravel areas. In this area the oldest glacio-fluvial gravels are not at the highest elevation but are buried underneath the younger deposits in a normal depositional sequence in which the youngest deposits are on top. The gravels, are of different ages and are, for the most part, consolidated conglomerates separated by thick weathered zones that extend down into the underlying conglomerates, along vertical joints or pipes. These gravels are exposed in the Isar valley south of Munich. SCHAEFER (1968) has shown that the uplift of the Landshut Swell northeast of Munich was the probable cause of the reversal of the vertical sequence of the gravel deposits.

3.4 The Tertiary Hills

The Tertiary Hills extend to the north of the gravel plains and older moraines and to the east of the Lech river. They are underlain by the weak sedimentary rocks of the Upper Freshwater Molasse and the Upper Marine Molasse. The hills are denudational landforms formed by stream incision and periglacial denudation, mainly during the Pleistocene. The low resistance of the rocks facilitated these processes. The area is dissected by a dense network of small valleys and hollows. The slopes are generally gentle and the maximum relief is about 50 m. Only east of the line Ortenburg-Simbach on the Inn river where there are consolidated sands and gravels of the Upper Freshwater Molasse and marls of the Upper Marine Molasse are the forms less rounded. The Tertiary Hills are characterized by asymmetrical valley cross profiles and a large number of dells. BÜDEL (1944) and others have suggested that the valley asymmetry is the result of snow accumulation by prevailing winds on slopes exposed to the east. The snow would have kept the soil saturated in the summer for a long time causing intensive gelisolifluction and slope flattening and a high rate of waste supply to the streams. The streams would, therefore, have pushed laterally against their eastern banks and the west-facing slopes would have been undercut and thereby steepened. Asymmetry is, however, not limited to valleys that are oriented transversal to the prevailing winds. Thermal differences in exposure have also been suggested as a more probable, primary factor and asymmetrical snow cover to be of secondary importance (KARRASCH 1970).

Solifluction combined with linear downwearing by converging meltwater has caused the development of the dells. In spite of the periodically high concentration of runoff along their axes, a fixed channel could not be formed because of the high supply rate of solifluction debris from the sides. The solifluction smoothed irregularities on the side slopes and shaped the shallow trough form of the dells. Unlike stream valleys, curving dells have symmetrical cross profiles except where there is asymmetry due to differences of lithology or of exposure.

There are two types of valley in the Tertiary Hills: the broad, steep-sided but flat-floored valleys, with terraces, of the Pleistocene meltwater rivers that flowed across the area and the, for the most part, meandering, shallow, trough-shaped valleys in which the autochthonous rivers flow. The contrast results from differences in the intensity of stream work and of slope denudation. The number of terraces in the former meltwater valleys varies depending on local conditions. The lower Isar valley, for example, has only a low terrace along its left side; this is related to conditions on the Munich Inclined Plain. Along the lower Lech, the Aindlingen Terrace series is one of the most fully subdivided Quaternary landform areas of the Alpine Foreland. Nine different gravel deposits have been distinguished, beginning possibly with the Biber Cold Period, and including deposits from the Danube Cold Period to the Würm glaciation (TILLMANS et al. 1983).

During the glacial maxima when vegetation was absent, silt and fine sands were blown out of the wide gravelly valley floors and from areas of weathered Flinz (fine-grained Molasse sediments). On the leeward, east or north-facing slopes, this material was deposited as loess. It is often a component in solifluctive material. The calcium carbonate contents of 30% in the loess in the Tertiary Hills is relatively high. The thickness of the loess exceeds 10 m in some areas.

Scattered remnants indicate that the Tertiary deposits were formerly present as a continuous cover at higher elevations on the southern margin of the Swabian Jura, on the Franconian Jura and on the Bohemian Massif (cf. Geologische Karte von Bayern 1:500,000, 1981). The former existence of this cover is the reason for the gorges of the Danube downstream of Sigmaringen where the river leaves the Swabian Jura and flows alternately in broad valleys within the Alpine Foreland and in narrow gorges within the resistant rocks in marginal areas of the Swabian Jura, the Franconian Jura and in the Austrian part of the Bohemian Massif. Examples of the gorges and narrow V-shaped valley stretches are the Neuburg Gap, the Weltenburg Narrows and the course of the Danube from these Narrows to Regensburg. There are, in addition, old valley stretches that have been abandoned by the Danube such as the Schmiech-Blau valley zone between Ehingen on the Danube, Blaubeuren and Ulm, the Wellheim valley zone and the lower Altmühl valley downstream from Dollnstein.

This pattern of incised valley stretches exists because the predecessor of the Danube in the Upper Pliocene was flowing on a higher land surface of little-consolidated Tertiary deposits that covered more resistant older rocks. During a new phase of downcutting in the late Upper Pliocene and early Pleistocene the river incised into the resistant rocks and became fixed in its valley along some stretches. Subsequently, the Tertiary sediments were removed. These water gaps (Durchbruchstäler) are the result of superimposition and constitute the local baselevels for the erosion of the broad, open valley stretches that lie up river from them, the Donauried, the Donaumoos, including the area of Ingolstadt/Neuburg a.d. Donau, and the Dungau downstream from Regensburg. They also belong to the Alpine Foreland and extend across the Danube northward. The Jurassic limestone spurs of Neuburg and of Kelheim are parts of the Jura and the Sauwald, downstream from Passau, is a part of the crystalline Bavarian Forest that lies south of the Danube. The Danube is, therefore, only an approximate northern boundary of the Alpine Foreland.

References

BADER, K. & JERZ, H. (1978): Die glaziale Übertiefung im Iller und Alpseetal (Oberes Allgäu). Geologisches Jahrbuch, **A 45**, Hannover, 25–45.

BADER, K. (1981): Die glaziale Übertiefung im Saalachgletschergebiet zwischen Inzell und Königssee. Eiszeitalter und Gegenwart **31**, Hannover, 37–52.

BAYERISCHES GEOLOGISCHES LANDESAMT (1981): Erläuterungen zur Geologischen Karte 1:500,000, 3rd Edition, München, 168 pp.

BÜDEL, J. (1944): Die morphologischen Wirkungen des Eiszeitklimas im gletscherfreien Gebiet. Geologische Rundschau **34**, Stuttgart, 482–519.

FISCHER, K: (1976): Das Formenbild der Allgäuer Alpen. Mitteilungen Verband Deutscher Höhlen- und Karstforscher, Jg. 22, München, 43–48.

FLIRI, F., HILSCHER, H. & MARKGRAF, V. (1971): Weitere Untersuchungen zur Chronologie der alpinen Vereisung (Baumkirchen, Inntal, Nordtirol). Zeitschr. Gletscherkunde und Glazialgeol. **7**, Innsbruck, 5–24.

FRANK, H. (1979): Glazial übertiefte Täler im Bereich des Isar-Loisach-Gletschers. Eiszeitalter und Gegenwart **29**, Hannover, 77–79.

JERZ, H., STEPHAN, W., STREIT R. & WEINIG, H. (1975): Zur Geologie des Illar-Mindel-Gebietes. Geologica Bavaria **74**, München, 99–130.

JERZ, H. (1979): Das Wolfratshausener Becken, seine glaziale Anlage und Übertiefung, Eiszeitalter und Gegenwart **29**, Hannover, 63–69.

KARRASCH, H. (1970): Das Phänomen der klimabedingten Reliefasymmetrie in Mitteleuropa. Göttinger Geogr. Abh. Heft **56**, Göttingen, 299 pp.

PENCK, A. (1894): Morphologie der Erdoberfläche. 2 vols, Stuttgart.

PENCK, A. & BRÜCKNER, E. (1901–1909): Die Alpen im Eiszeitalter. Leipzig, 3 vols., 1199 pp.

PIEHLER, H. (1974): Die Entwicklung der Nahtstelle von Lech-, Loisach- und Ammergletscher vom Hoch- bis Spätglazial der letzen Vereisung. Münchener Geographische Abhandlungen. München. 105 pp.

SCHAEFER, I. (1968): Münchener Ebene und Isartal. Mitteilungen Geogr. Gesellsch. München **53**, München, 175–203.

SCHREINER, A. (1979): Zur Entstehung des Bodenseebeckens. Eiszeitalter und Gegenwart **29**, Hannover, 71–76.

SEILER, K.-P. (1979): Glazial übertiefte Talabschnitte in den Bayerischen Alpen. Eiszeitalter und Gegenwart **29**, Hannover, 35–48.

SEMMEL, A. (1974): Geomorphologie der Bundesrepublik Deutschland 4th Ed., Steiner, Stuttgart, 192 pp.

TILLMANS, W., BRUNNACKER, K. & LÖSCHER, M. (1983): Erläuterungen zur Geologischen Übersichtskarte der Aindlinger Terrassentreppe zwischen Lech und Donau. Geologica Bavaria **85**, München, 31 pp.

TROLL, K. (1924): Der diluviale Inn-Chiemsee-Gletscher. Forschungen zur deutschen Landeskunde, vol. **23**, Heft 1, Stuttgart, 121 pp.

TROLL, K. (1926): Die jungglazialen Schotterfluren im Umkreis der deutschen Alpen. Forschungen zur deutschen Landes- und Volkskunde, vol. 24, Heft 4, Stuttgart, 158–256.

Address of author:
Prof. Dr. Klaus Fischer
Lehrstuhl für Physische Geographie
Universität Augsburg
Universitätsstraße 10
D-8900 Augusburg
Federal Republic of Germany

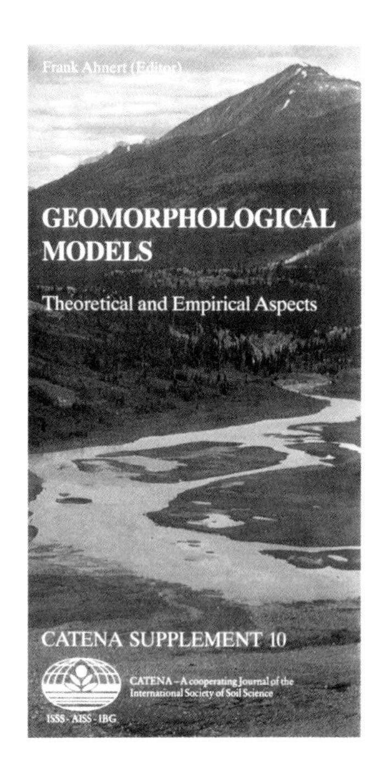

Frank Ahnert (Editor):

GEOMORPHOLOGICAL MODELS

Theoretical and Empirical Aspects

CATENA SUPPLEMENT 10, 1987

Price DM 149,– / US $88.–

ISSN 0722-0723 / ISBN 3-923381-10-7

CONTENTS

CATENA SUPPLEMENT 15 p. 85–94 Cremlingen 1989

GLACIAL AND PERIGLACIAL MORPHOLOGY OF THE LÜNEBURG HEATH

Jürgen **Hagedorn**, Göttingen

1 Introduction

The Lüneburg Heath in Lower Saxony lies between the valleys of the Elbe in the north and the Aller in the south. In the east it is bordered by the Jeetzel Basin, which extends into the German Democratic Republic, and in the west by the lowlands of the Este and Wümme. It is primarily a region of loamy, bouldery moraines and of glaciofluvial sediments. Few elevations exceed 130 m above sea level; the highest summit is the Wilseder Berg (169 m). The valley floors and basins are generally at about 50 m above sea level.

The morainic hills on the heath are part of a continuous alignment that extends NW to SE from the Harburger Berge, south of Hamburg to the Wilseder Berg, the Lüß Plateau, and the Wierener Berge in the Federal Republic and the Letzlingen Heath and the Fläming, east of Magdeburg, in the GDR (fig.1). Southwest of this zone are extensive, gently sloping glaciofluvial outwash plains, such as the Munster Sandur and the Sprakensehl Sandur. On the more steeply sloping northeastern side, there are terminal moraines and basins, such as the Uelzen Basin in which widespread deposits of boulder clay occur. The terminal moraines are usually push moraines that consist of glaciofluvial materials. The entire landform association is the most extensive and most typical example of a marginal area of glaciation in the northwest German Lowland. The ice marginal proglacial spillway (Urstromtal) that drained the area is the valley of the present Aller river.

The landforms date from the Saale glacial. This glacial has been subdivided into the older Drenthe stadial and the younger Warthe stadial (WOLDSTEDT 1929; LIEDTKE in this volume for table of dates and nomenclature of glaciations). The forms have been modified, particularly during the Weichesl period by later denudation and erosion.

The line of hills at the eastern margin of the Lüneburg Heath has been interpreted as a second terminal moraine ridge of the Warthe stadial that links up with the outer terminal moraines in the south. HÖVERMANN (1956) has suggested that the Warthe advance in north Germany consisted of two lobes, one that came from the direction of the present North Sea and reached the western margin of the Lüneburg Heath and the other that came from the direction of the present Baltic Sea.

Until recently the stratigraphic subdivision of the sediments has been derived from the spatial sequence of land-

ISSN 0722-0723
ISBN 3-923381-18-2

3-923381-18-4/89/5011851/US$ 2.00 + 0.25

forms. A direct stratigraphic analysis of the ground moraine materials indicates, however, that the stratigraphical dating of the terminal moraines, the reconstruction of the proglacial drainage systems and also the dating of the later periglacial modification of the old moraine area must be reexamined.

2 Ice margins and ground moraines of the Saale glacial and their morphological and stratigraphical significance

The relationships between morphologically identifiable ice margin positions and the boundaries of lithological moraine components were investigated by MILTHERS (1934) and HESEMANN (1939). RICHTER (1958) used the relative amount of the east Fennoscandian Rapakivi rocks in the crystalline stony components of the moraines as the criterion to distinguish the different ice advances of the Saale glacial. In the Lüneburg Heath it was found that the area east of the main ice margin position of the Warthe stadial had a much larger share of Rapakivi material than the areas west of it. However, the Rapakivi share of the outwash sediments and of the terminal moraines associated with this ice margin, which were previously also believed to belong to the Warthe stadial, corresponded to the Rapakivi share of the Drenthe terminal moraines that lay farther to the west. RICHTER concluded from this evidence that the outwash sediments were fluvial sediments of the Drenthe stage and that the Warthe ice margin in this area coincided with an earlier Drenthe ice margin. The Drenthe terminal moraine possibly also halted the Warthe advance.

LÜTTIG (1958), GROETZNER (1972), DUPHORN et al. (1973) and the Geological Map 1:200,000 confirmed these conclusions. The Saale glacial was subdivided into an older Drenthe stadial with moraines that contain a predominance of material from southern and central Sweden, a younger Drenthe stadial, the moraines of which have a similar composition but also a large amount of upper Cretaceous material and some Silurian limestones, and the Warthe stadial in which east Baltic components predominate (MEYER 1983a). The typical composition of the Warthe moraines has also been found in western Lower Saxony above the older Drenthe boulder loam but has been interpreted in this case as a special facies of these older deposits (MEYER 1983a).

The ground moraine stratigraphy is based on the assumption that each stadial is an ice advance from the source area of those lithologic components that predominate in the moraine, and that this advance was preceded by a major backwasting of the ice of the previous advance during an interstadial.

The directions of the inland ice that are indicated from the composition of the moraines have also been supplemented by measurements of the axial orientations of the morainic boulders (EHLERS & STEPHAN 1983). However, not all ice advances can be correlated with definite positions of the ice margin. The inland ice of the Warthe stadial, for example, seems to have moved very slowly and had little morphological effect; it did not create terminal moraines and was stopped by the older Drenthe terminal moraines. Also, the Warthe ground moraine has little morphological significance since it overlies the older land surface composed of Drenthe sed-

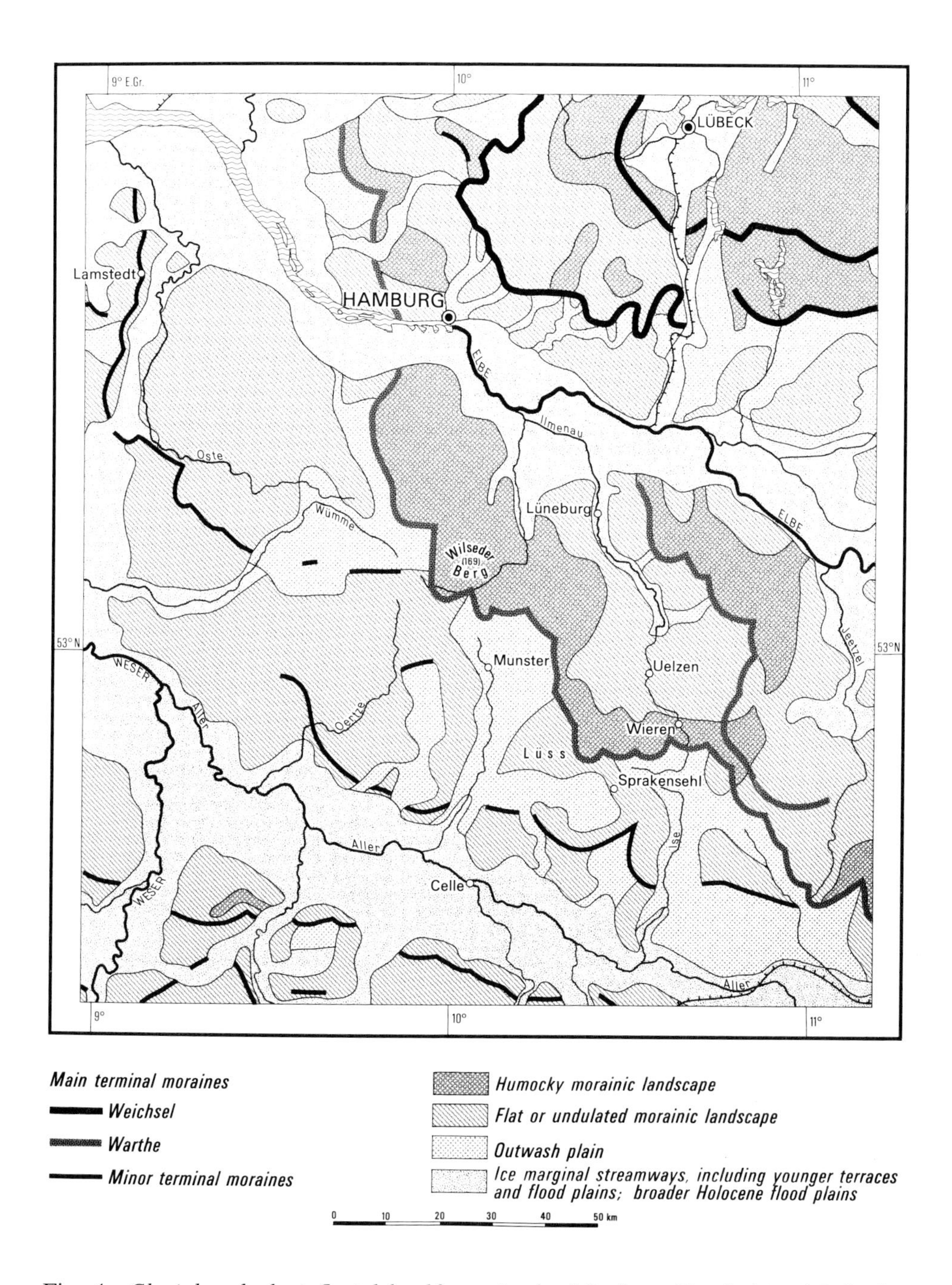

Fig. 1: *Glacial and glaciofluvial landforms in the Lüneburg Heath (simplified after LIEDTKE 1981).*

iments only as a thin cover. Moreover, the most accentuated ice-marginal landforms of the northwest German lowland are, in general, not connected with stadial, that is, with stratigraphically relevant, ice margin positions.

GRIMMEL (1971, 1973) suggested that the relief of the northeastern Lüneburg Heath, and of the north German Lowland in general, was formed primarily by subglacial erosion and sedimentation, including action by subglacial meltwaters, and that the relief was largely preserved during the melting of the ice. End moraines, especially push moraines are, therefore, rare, and the urstromtäler are not proglacial valleys along adjacent ice margins but preferred zones of subglacial erosion by ice and meltwaters. Subsequently, GRIMMEL (1976) also suggested that the terminal moraines were caused by subglacial push effects and were not, therefore, terminal moraines.

The glacial-morphological features of the Lüneburg Heath would appear, therefore, to be unrelated to a climatic-stratigraphical subdivision of the Saale glacial. A reassessment of the origin of the terminal moraines, sandurs, ice lobe basins and other relief forms in the old moraine areas, especially the more pronounced landforms, is, therefore, necessary. Research into the moraine stratigraphy of the Saale glacial by MARCUSSEN (1978), EHLERS (1981, 1983a, 1983b), STEPHAN (1980) and others has shown that in the moraines there is typically an early dominance of middle Swedish and south Swedish components but that later east Baltic components are present (WOLDSTEDT & DUPHORN 1974). This sequence occurs twice in the Saale ground moraines in Lower Saxony and also in sediments of Weichsel age. SCHUDDEBEURS & ZANDSTRA (1983) have found it in Saale ground moraines in the Netherlands.

WOLDSTEDT & DUPHORN (1974) suggest that the ice in the west Scandinavian centre of glaciation is likely to have formed rapidly at the beginning of a glacial period and to have advanced into north Germany. As the build up of the ice on the leeward side of the Scandinavian mountains continued, the ice divide shifted eastward, causing a greater ice flow from east Fennoscandia. If this hypothesis is correct, a lower unit with middle Swedish and south Swedish components and an overlying upper unit with east Baltic components must be distinguished in the stratigraphy of the ground moraine. Also, therefore, the Warthe stadial in Lower Saxony would not begin, but end, with the occurrence of east Baltic erratics. The boulder loams with prevailing middle and south Swedish components, that have formerly been interpreted as material of the younger Drenthe stadial, would occur at the beginning of the Warthe stadial. The stratigraphy has also been subdivided on this basis in Schleswig-Holstein (PICARD 1960, STEPHAN 1980), and in the vicinity of Hamburg (EHLERS 1981, 1983b) where the ground moraine of the older Drenthe glacial is usually clearly separated from the overlying younger ground moraine by glaciofluvial sediments. In addition, a fossil soil, the "Treene" soil, which was formed in a warmer phase, occurs between these two units in Schleswig-Holstein (PICARD 1960, STREMME 1960, 1964).

If this boundary between the Drenthe and Warthe stadials is accepted, there is no longer a discrepancy between the subdivision of the Saale glacial on the basis of the stratigraphy of the moraine materials, the climatic-stratigraphic subdivision and also the morphologically recognizable location of the main ice margin. Moreover, the sandurs in front of the main ice margin position would then be related to the terminal moraines of the main ice margin and to the Warthe stadial. This does not preclude the occurrence of relief forms, such as the Harburger Berge (EHLERS 1978), created by earlier glaciations. The advances of the Warthe ice beyond the main ice margin as well as the creation of younger push moraines adjacent to old ice margin positions have to be analysed further (HAGEDORN 1964).

The stratigraphic differentiation of the moraines is based on their facies but since the same facies of the same stadial can occur in different areas at different times, the stratigraphical correlation across larger regions is difficult. A morphological analysis must, therefore, be included in an interpretation of the stratigraphy.

3 Glaciofluvial landforms and the drainage of the Warthe stadial

Based on the occurrence of boulder loam shown in areas of sandurs, as shown on older geological maps and mentioned in the accompanying explanatory text, GRIMMEL & SCHIPULL (1975) have concluded that the large sandurs, which are commonly thought to belong to the Warthe stadial, were also partly overrun by later ice advances and that, for example, the "Sprakensehl Sandur" is not a genuine sandur but a land surface formed by glacial erosion and shear processes. The overrunning of older sandurs by the Drenthe ice is implied by the Braunschweig (1974) sheet of the 1:200,000 Geological Map (MEYER 1983c). GROETZNER (1972) has also found boulder marl in the area of the Sprakensehl Sandur which he interprets, however, as older ground moraine remnants that have been planed off during the sandur formation. More recent geological mapping at 1:200,000 has not provided evidence to the contrary. A glaciofluvial surface that essentially was formed as a surface of deposition but is penetrated by denuded remnants of boulder loam has been recognized in the southern Uelzen Basin by HAGEDORN (1964). Such occurrences do not appear to be unusual.

In addition to the sandurs, the urstromtäler which now form the valleys of the Aller and of the lower Elbe are the most important glaciofluvial landforms in the area of the Lüneburg Heath. There are also several smaller valleys, some of which run from the inner edge of the moraines northward to the Elbe. The Aller urstromtal was thought to be the main drainage way for the Warthe stadial and the Elbe urstromtal the drainage way for the Weichsel glacial. Recent research into the lithology and stratigraphy of the sedimentary fills indicates a more complicated pattern of development. The Aller urstromtal is part of the Breslau-Magdeburg-Bremen urstromtal which accompanied the ice margin of the Warthe stadial. Sediments on the upper flanks of the urstromtal (DUPHORN 1971, MEYER 1983c) indicate that a broad valley zone existed that received the drainage from the ice margin in the north and from the central uplands in the south.

The sandur plains lie about 20 m above the floor of the Aller valley

(HÖVERMANN 1956). The northern tributaries of the Aller have also incised into the sandur and these valleys as well as the valley of the Aller have a low terrace above the Holocene valley floor which, dated by the underlying Eem sediments, was formed in the Weichsel glacial (MEYER 1983c). The Aller valley and its tributary valleys were, therefore, incised after the deposition of the sandurs and before the Eem interglacial, that is, near the end of the Warthe stadial. The incision could have been caused by an increase in the spatial concentration of runoff combined with a decrease in sediment loads. In the Weichsel glacial the accumulation of the low terrace sediments was the result of periglacial runoff conditions and higher sediment loads.

The valley of the lower Elbe was thought to have originated during the Weichsel glacial when it formed the urstromtal segment that drained the runoff from the ice margin in the north, and also the eastern urstromtäler of the Weichsel glacial, towards the North Sea. At the southern rim of the Elbe valley, however, the Saale ground moraine reaches from the hills of the Lüneburg Heath into the Elbe valley to a height of 40 m above sea level (SCHROEDER-LANZ 1964, MEYER 1983c). In addition, in the Hamburg area, valley sediments of late Saale age lie below the Weichsel valley sediments to depths of 30 m below sea level (GRUBE & EHLERS 1975). The predecessor of the lower Elbe valley must, therefore, have existed during the Saale glacial. Its development probably began during the Holstein interglacial because the marine Holstein ingression reached this low-lying area (GRAHLE 1936, GRIMMEL 1973, MEYER 1983c). The Elbe valley was, therefore, available as a drainage way during the retreat phases of the Warthe stadial.

In the Weichsel glacial a sedimentary fill of about 15 m thickness was deposited that became the low terrace after Holocene incision. Near Schnackenburg, its surface lies at 20 m above sea level and near Cuxhaven at 20 m below sea level, so that the elevation of its base ranges from 5 m to -35 m (MEYER 1983). Before the accumulation of these low terrace sediments, considerable downcutting into the glacigenous sediments of the Saale glacial took place, at least in part, towards the end of the Warthe stadial. It is not certain whether the sediments in the valley were also eroded in the early Weichsel glacial.

According to the maps of KUSTER & MEYER (1979) and of HINSCH (1979), the valleys at the base of the Quaternary cover ascribed to subglacial erosion by ice and meltwater at the beginning of the Elster glacial (see LIEDTKE in this volume), do not follow but cross the Elbe valley so that there is no relationship between the development of the Elbe valley and erosional processes during the advance of the Elster ice. The earliest possible development of the Elbe valley took place during the downwasting of the Elster ice sheet.

The existence of a prior Elbe valley under the inland ice of the Warthe stadial led GRIMMEL (1973) to assume a subglacial meltwater discharge towards the west in this valley. No direct evidence, such as sediments, exists for this. However, HAGEDORN (1965) found that in the Uelzen Basin the drainage, which had previously been directed southward to the Aller urstromtal, was reversed to the north after the retreat of the ice sheet of the Warthe stadial. After this drainage reversal, the surface of the Basin was apparently lowered by about 20 m by denudation.

The estimate of downwearing would be smaller if a pre-existing subglacial slope towards the Elbe valley is assumed which corresponded to the slope of the ground moraine cover and which was hydrographically usable, at the latest, by the beginning of the disintegration of the inland ice sheet. A condition for the establishment of an initially subglacial northward drainage would be the continuation of this drainage subglacially in the Elbe valley. Such subglacial drainage from the Lüneburg Heath to the Elbe valley and farther to the northwest may have developed during the period when the ice sheet of the Warthe stadial began to disintegrate. The valleys that drain the Lüneburg Heath towards the Elbe were probably established at that time. Subsequently, they were rapidly incised by headward erosion from the nearby low baselevel of the Elbe. In some cases they extended headward, beyond the terminal moraines and captured areas that earlier had drained towards the south.

4 The periglacial landform changes in the old moraine area

Young and old moraine areas are distinguished on the basis of landform differences. The young moraine areas have relatively unaltered terminal moraines, a weakly developed valley network and a large number of depressions with no surface outlet. The old moraine areas have a more mature valley system, fewer depressions with no outlets and a generally subdued relief, particularly in the area that was glaciated during the Drenthe stadial. The forms on the Lüneburg Heath are well-developed and for this reason the area was formerly assumed to have been glaciated during the last glacial. The valley network is, however, also well-developed and there are very few depressions that have no outlet. The density of these depressions has been used by GRIPP (1924) to delineate the maximum extent of the last glaciation and this limit is essentially still valid. The depressions were caused by the melting of dead ice. In the old moraine area they were subject to erosion on their rims and to infilling, particularly during the Weichsel glacial when the ice-free areas were affected by periglacial processes.

The most important periglacial landforms of the Weichsel glacial include the low terraces in the larger valleys and the closely-spaced and deeply incised dry valleys that occur particularly in the terminal moraine ridges and along the sides of the Elbe valley. There are also covers of solifluction debris and indications of aeolian activity such as dunes, aeolian sand sheets, sand loess deposits and deflational pavements. A few depressions without outlets occur, of which most are deflation hollows or collapse sinks above the margins of salt plugs and, in a few cases, remnants of pingoes. Only very rarely are such depressions glacial kettle holes from the Saale glacial (GARLEFF 1968, LADE 1980).

The dry valleys are the result of intensive periodic discharge of snow meltwaters on the bare impervious permafrost that has left scars of ice wedges in the soil. The systematic asymmetry of the dry valleys is also an indication of their periglacial origin. Their valley floors enter the trunk valley at the level of the low terrace of the latter, indicating that they were formed in the Weichsel glacial. Some dry valleys end at the foot of terminal moraines on a colluvial footslope. GARLEFF & LEONTARIS (1971) have shown that this type of footslope on the Wilseder Berg was formed as a result of deposition on peat from the Eem interglacial, a further indication that the dry valleys are periglacial features of the Weichsel glacial, although they may have been initiated during the retreat of the Warthe ice when large amounts of meltwater were flowing on surfaces not yet protected by vegetation. Morphological changes in the dry valleys have also occurred during the Holocene mainly as a result of the destruction of the forest (HAGEDORN 1964).

The research by GARLEFF & LEONTARIS (1971) shows that the incision of these dry valleys has contributed to the denudational lowering of the hilltops and to aggradation of the lower ground and also, therefore, to the general levelling of these areas by periglacial processes. Locally where the trunk streams have been able to remove the sediment load that arrives, aggradation is absent and the relief has been accentuated. The Elbe, however, was not able to remove all of the material delivered from the older moraine areas to the south, as shown by alluvial fans of Weichsel age in the Elbe valley (MEYER 1983c).

Throughout the old moraine area, so-called morainic cover sand (Geschiebedecksand) is widely distributed. This is a layer of unstratified, partially loamy sand, about 0.5 m thick, in which individual erratics occur either irregularly or concentrated in a stone layer just above the base of the sand layer. The morainic cover sand was thought to be a type of ground moraine and to indicate that the underlying sediments had been covered by inland ice (STOLLER 1914). They are, however, of periglacial origin. Older glacigenous and glaciofluvial sediments have been reworked by cryogenic processes and transported downslope by solifluction and slope wash. On the slopes the morainic cover sand is generally solifluctional. On plateau surfaces it also occurs in situ and has been formed from the underlying ground moraine (SCHRÖDER 1977) but only rarely has the entire ground moraine been reworked into morainic cover sand. Where, therefore, there was no solifluctional transport, the presence of morainic cover sands may indicate a former ground moraine cover but it cannot generally be used as evidence of a former inland ice cover.

The layer of stones in the cover sand represents an old surface. It resulted from the selective removal of fines as shown by ventifacts. Frost heaving of stones to the surface probably also was a contributing factor. Subsequently the stone layer was covered mainly by aeolian sedimentation, as indicated by the presence of sand loess, aeolian sand sheets and dunes.

The glacigenous landforms of the Lüneburg Heath have been greatly modified under the periglacial conditions of the Weichsel glacial by fluvial erosion and accumulation, as the dry valleys, depositional footslopes and low terraces show, and by solifluction, wash and aeolian processes. The action of these processes explains not only the more subdued relief of the old moraine area compared to the young moraines but also why there is a mosaic of ground moraines and glaciofluvial sediments of different ages rather than a continuous cover of ground moraine. The major relief features that were created during the Warthe stadial by the inland ice and its meltwaters have , however, been preserved.

References

DUPHORN, K. (1972): II. Natur des Landes. A. Geologie. In: Die Landkreise in Niedersachsen, **26**: Der Landkreis Gifhorn. Bremen-Horn, 22–30.

DUPHORN, K., GRUBE, F., MEYER, K.-D., STREIF, H. & VINKEN, R. (1973): State of research on the Quaternary of the Federal Republic of Germany. A. Area of the Scandinavian Glaciation. 1. Pleistocene and Holocene. Eiszeitalter und Gegenwart, **23/24**, 222–250.

EHLERS, J. (1981): Problems of the Saalian Stratigraphy in the Hamburg Area. Medd. Rijks Geol. Dienst **34, (5)**, 26–29.

EHLERS, J. (1983a): Different till types in North Germany and their origin. In: EVENSON, E.B. et al. (ed.): Tills and related deposits. Proceedings of the INQUA Symposia on the Genesis and Lithology of Quaternary Deposits. Rotterdam. 61–8O.

EHLERS, J. (1983b): The glacial history of north-west Germany. In: EHLERS, J. (ed.): Glacial deposits in north-west Europe, Rotterdam. 229–238.

EHLERS, J., MEYER, K.-D. & STEPHAN, H.-J. (1984): The pre-weichselian glaciation of north-west Europe. Quartern. Sc. Rev., **3**, 1–4O.

EHLERS, J. & STEPHAN, H.-J. (1983): Till fabric and ice movement. In: EHLERS, J. (ed.): Glacial deposits in northwest Europe. Rotterdam. 267–274.

GARLEFF, K. (1968): Geomorphologische Untersuchungen an geschlossenen Hohlformen ("Kaven") des Niedersächsischen Tieflandes. Göttinger geogr. Abh. **44**, Göttingen, 142 pp.

GARLEFF, K. & LEONTARIS, S.N. (1971): Jungquartäre Taleintiefung und Flächenbildung am Wilseder Berg (Lüneburger Heide). Eiszeitalter und Gegenwart **22**, 148–155.

GRAHLE, H.-O. (1936): Die Ablagerungen der Holstein-See (Marines Interglazial I), ihre Verbreitung, Fossilführung und Schichtenfolge in Schleswig-Holstein. Abh. preuß. geol. L.-A N.F. **172**, 110 pp.

GRIMMEL, E. (1971): Geomorphologische Untersuchungen in der nordöstlichen Lüneburger Heide. Hamburger geogr. Stud. **27**, Hamburg, 55 pp.

GRIMMEL, E. (1973): Überlegungen zur Morphogenese des Norddeutschen Flachlandes, dargestellt am Beispiel des unteren Elbtales. Eiszeitalter und Gegenwart **23/24**, 76–88.

GRIMMEL, E. (1976): Bemerkungen über Stauch-"End"moränen. Eiszeitalter und Gegenwart **27**, 69–74.

GRIMMEL E. & SCHIPULL, K. (1975): Der Sprakensehler Sander: Ein klassischer "Sander" der Lüneburger Heide? Mitt. geogr. Ges. Hamburg **63**, 171–181.

GRIPP, K. (1924): Über die äußerste Grenze der letzten Vereisung in Nordwest-Deutschland. Mitt. geogr. Ges. Hamburg **36**, 159–245.

GROETZNER, J.-P. (1972): Geschiebekundlich-stratigraphische Untersuchungen in Randgebieten der Suderburger Bucht (Uelzener Becken). Mitt.gol. Inst. TU Hannover **11**, 75 pp.

GRUBE, F. & EHLERS, J. (1975): Pleistozäne Flußsedimente im Hamburger Raum. Mitt. geol. paläontol. Inst Univ. Hamburg **44**. 353–382.

HAGEDORN, J. (1964): Geomorphologie des Uelzener Beckens. Göttinger geogr. Abh. **31**. Göttingen, 200 pp.

HAGEDORN, J. (1965): Die Umgestaltung des glazigenen Reliefs der norddeutschen Altmoränengebiete am Beispiel des Uelzener Beckens. Eiszeitalter und Gegenwart **16**, 116–120.

HESEMANN, J. (1939): Diluvialstratigraphische Geschiebeuntersuchungen zwischen Elbe und Rhein. Abh. naturw. Ver. Bremen **31**, 247–285.

HINSCH, W. (1979): Rinnen an der Basis des glaziären Pleistozäns in Schleswig-Holstein. Eiszeitalter und Gegenwart **29**, 173–178.

HÖVERMANN, J. (1956): Beiträge zum Problem der saaleeiszeitlichen Eisrandlagen in der Lüneburger Heide. Abh. braunschweig. wiss. Ges. **VIII**. 36–54.

KUSTER, H. & MEYER, K.-D. (1979): Glaziäre Rinnen im mittleren und nordöstlichen Niedersachsen. Eiszeitalter und Gegenwart **29**, 135–156.

LADE, U. (1980): Quartärmorphologische und -geologische Untersuchungen in der Bremervörder-Wesermünder Geest. Würzburger geogr. Arb. **50**, 173 pp.

LIEDTKE, H: (1981): Die nordischen Vereisungen in Mitteleuropa. Forsch. dt. Landeskunde **204**, 160 pp.

LÜTTIG, G. (1958): Methodische Fragen der Geschiebeforschung. Geol. Jb. **75**, 361–418.

LÜTTIG,G. (1968): Möglichkeiten der Endmoränen-Verknüpfung im Gebiet zwischen Aller und Elbe. Mitt. geol. Inst. TU Hannover **8**, 66–73.

LÜTTIG, G. & MEYER, K.-D. (1974): Geological history of the river Elbe, mainly of its lower course. Cent. Soc. Geol. Belgique. L. evolution quarternaire des bassins fluviaux de la mer du nord meridional, Liege, 1–19.

MARCUSSEN, I. (1978): Über die Verwendbarkeit von Geschieben in Grundmoränen als Hilfsmittel der Stratigraphie. Der Geschiebesammler **12 (2/3)**, 13–20.

MEYER, K.-D. (1983a): Indicator pebbles and stone count methods. In: EHLERS, J. (ed.): Glacial Deposits in northwest Europe. Rotterdam. 275–287.

MEYER, K.-D. (1983b): Saalian end moraines in Lower Saxony. In: EHLERS, J. (ed.) Glacial deposits in northwest Europe. Rotterdam. 335–342.

MEYER, K.-D. (1983c): Zur Anlage der Urstromtäler in Niedersachsen. Z. Geomorph. N.F. **27 (2)**, Berlin, 147–160.

MILTHERS, V. (1934): Die Verteilung skandinavischer Leitgeschiebe im Quartär von Westdeutschland. Abh. preuß. geol.L.-A. NF, **156**, 1–74.

PICARD, K. (1960): Zur Untergliederung der "Saalevereisung" im Westen Schleswig-Holsteins. Z. dt. geol. Ges. **112**, 316–325.

RICHTER, K. (1958): Geschiebegrenzen und Eisrandlagen in Niedersachsen. Geol. Jb. **76**, 223–234.

SCHRÖDER, E. (1970): Geomorphologische Untersuchungen im Hümmling. Göttinger geogr. Abh. **70**, 113 pp.

SCHROEDER-LANZ, H. (1964): Morphologie des Estetales. Ein Beitrag zur Morphogenese der Oberflächenformen im nördlichen Grenzgebiet zwischen Stader Geest und Lüneburger Heide. Hamburger geogr. Stud. **18**, 180 pp.

SCHUDDEBEURS, A.P. & ZANDSTRA, J.G. (1983): Indicator pebble counts in the Netherlands. In: EHLERS, J. (ed.): Glacial deposits in northwest Europe. Rotterdam. 357–360.

STEPHAN, H.-J. (1980): Glazialmorphologische Untersuchungen im südlichen Geestgebiet Dithmarschens. Schr. naturwiss. Ver. Schleswig-Holstein **50**, 1–36.

STEPHAN, H.-J. & EHLERS, J. (1983): North German till types. In: EHLERS, J. (ed.): Galcial Depostis in north-west Europe. Rotterdam. 239–247.

STOLLER, J. (1914): Der jungdiluviale Lüneburger Eisvorstoß. Jahresber. niedersächs. geol. Ver. **7**, 214–230.

STREMME, H. E. (1960): Bodenbildungen auf Geschiebelehmen verschiedenen Alters in Schleswig-Holstein. Z. dt. geol. Ges. **112**, 299–308.

STREMME, H.E. (1964): Die Warmzeiten vor und nach der Warthe-Eiszeit in ihren Bodenbildungen bei Böxlund (westl. Flensburg). N. Jb. Geol. Paläontol. Mh. **4**, 237–247.

WOLDSTEDT, P. (1929): Das Eiszeitalter. Grundlinien einer Geologie des Diluviums. Stuttgart, 406 pp.

WOLDSTEDT, P. & DUPHORN, K. (1974): Norddeutschland und angrenzende Gebiete im Eiszeitalter. Stuttgart, 500 pp.

Address of author:
Prof. Dr. Jürgen Hagedorn
Geographisches Institut der Universität Göttingen
Goldschmidtstraße 5
D-3400 Göttingen
Federal Republic of Germany

CATENA SUPPLEMENT 15 p. 95–106 Cremlingen 1989

LANDFORM DEVELOPMENT AND SOIL STRUCTURE OF THE NORTHERN FEDERAL REPUBLIC OF GERMANY THEIR ROLE IN GROUNDWATER RESOURCES MANAGEMENT

Otto **Fränzle**, Kiel

Summary

The chemical pollution of groundwater is a function of three factors: the physico-chemical properties of the polluting compounds, the specific structure of the strata through which the water infiltrates downward and the specific structure of the aquifer.

The spatially differentiated pollution risks are determined by means of geographical information systems, at the regional planning level on the basis of infiltration rates and of filtering and buffering effects of the soils in the unsaturated and the saturated zones, at the local scale with additional consideration of the hydraulic properties of the aquifer. If the concentration level of chemicals is low, the transfer of chemical compounds is decisively influenced by the properties of the soil and the substrate. If there are high concentrations, however, the reaction potentials of the compounds present are, under similar conditions of soil and groundwater budgets, of greater importance.

ISSN 0722-0723
ISBN 3-923381-18-2

3-923381-18-4/89/5011851/US$ 2.00 + 0.25

1 Quaternary sediments and related soils

1.1 Tills and meltwater sediments

Unlike the basins in the Netherlands and East Anglia, north Germany did not subside during the early Pleistocene. In most cases, therefore, the deposits of the Elsterian and younger glaciations lie directly on Pliocene or Miocene strata; in southern Lower Saxony they cover older Tertiary and Mesozoic strata.

The Elsterian ice sheet reached the margins of the central uplands but in most places its sediments are covered by younger deposits. The most striking features of this glaciation are the over-deepened valleys which were incised deeply into the underlying strata, in several places to more than 400 meters below sea level. These valleys were probably formed by subglacial meltwater erosion (EHLERS et al. 1984, GRUBE 1979, HINSCH 1979, KUSTER & MEYER 1979). Their fill of glaciolimnic sands is the most important aquifer in northern Germany (fig.2).

The Holsteinian interglacial, according to varve chronology by MÜLLER (1974) and MEYER (1974), lasted about

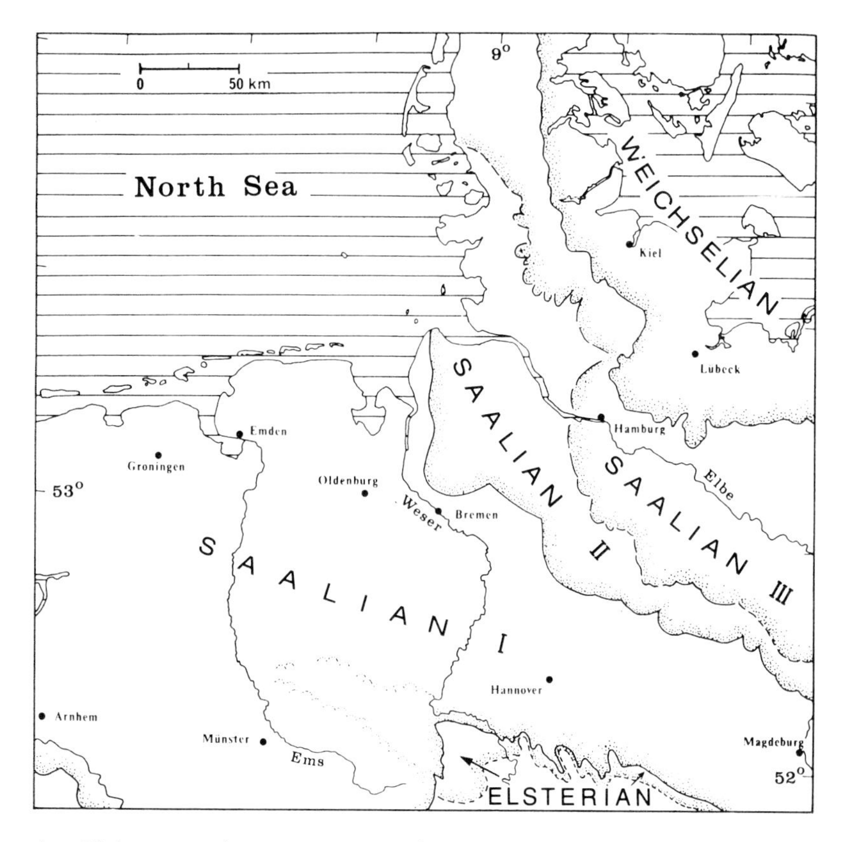

Fig. 1: *Pleistocene glaciations in northern Germany and the Netherlands (after EHLERS 1983, modified).*

16,000 years and was followed by the Saalian glaciation. Extensive meltwater deposits and three different tills were left by this glaciation. The tills are termed Older Saalian (I), Middle Saalian (II) and Younger Saalian (III). The extent of the inland ice sheets during the Saalian glaciation is shown on fig.1.

The Saalian moraines and outwash sediments owe their distinctive landforms to early and middle Weichselian periglacial solifluction, which varied regionally, and to the widespread slope wash by meltwater slope processes which resulted in extensive gently rolling planation surfaces (FRÄNZLE 1969, 1988a, GALBAS et al. 1980). With increasing pleniglacial aridity, aeolian activity also increased and a widespread layer of cover sand was deposited on top of the periglacially re-worked moraines and outwash plains. Intense cryoturbation accounts for the incorporation of the uppermost layers of the interglacially weathered tills and meltwater deposits into the sand deposits, forming the "morainic cover sand" (Geschiebedecksand). A second phase of aeolian activity in the late Weichselian left a largely stone-free sand cover of, generally, lesser extent but with dune fields deposited locally.

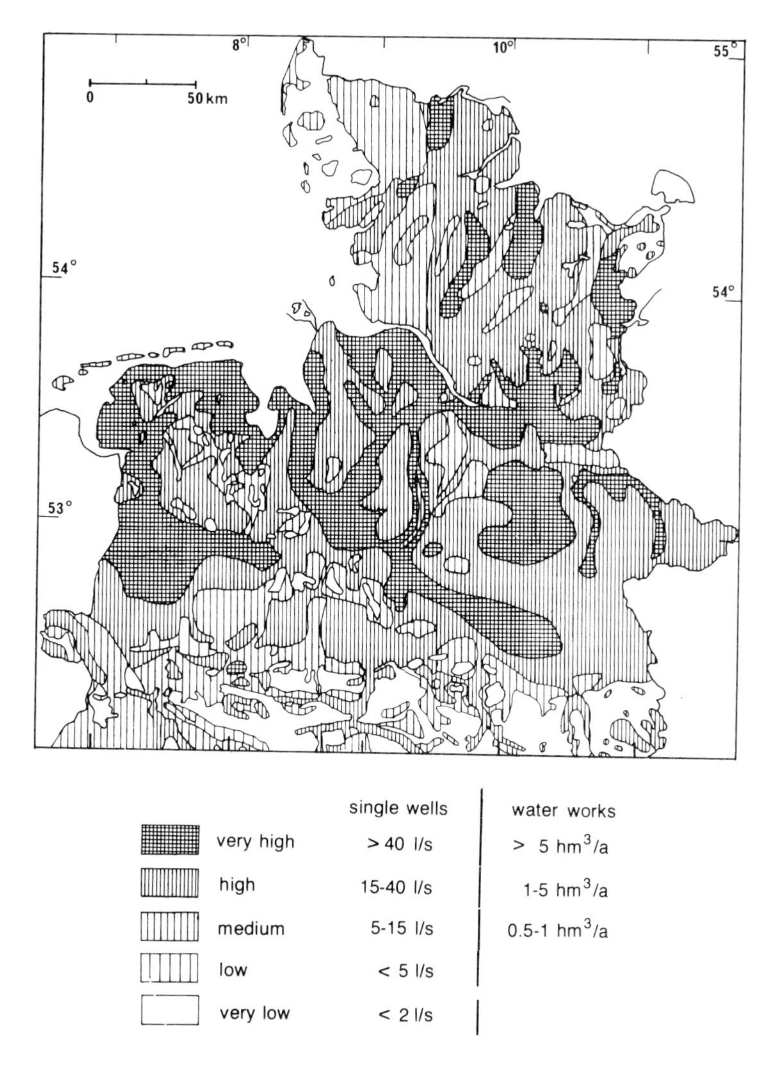

Fig. 2: *Average yields of groundwater resources. Source: Bundesminister for Raumordnung, Bauwesen und Städtebau 1980, modified.*

The bulk of the Pleistocene deposits are outwash sediments, the ratio between meltwater deposits and deposits of glacial origin has been estimated at 3:1 (EHLERS 1981). The difference in grain size of the Saalian and the Weichselian outwash sediments is important hydrogeologically. The former tend to be comparatively fine-grained and well-sorted while the average grain size of the Weichselian meltwater deposits is considerably greater and the variation between individual beds is much higher (EHLERS & GRUBE 1983). Fig.2 summarizes the distribution of aquifers in northern Germany and indicates the yields of exploitable groundwater.

1.2 Soil cover

The soil cover is of great importance in a regionally differentiated assessment of groundwater sensitivity because the physical and chemical properties of the soil, in particular, its filtering, sorption and biodegradation potentials, define the buffering capacity or protective potential.

There is a relatively close relationship between the spatial structure of soil associations and landform development in northern Germany (FRÄNZLE 1981, 1988a, MUTERT 1978, ROESCHMANN 1971, SCHLICHTING 1960). The Weichselian tills are characterized by Luvisol associations comprising Calcaric Regosols or Orthic Luvisols, at the top of moraines. On their flanks, Eutric Cambisols, soil colluvia and shallower Orthic Luvisols are developed which show a tendency to stagno-gleyification near the foot of slopes and in areas of level ground.

The soil cover in intra-Weichselian valleys consists mainly of Humic, Dystric and Histic Gleysols or Histosols whereas the cover sand areas and outwash plains have Dystric Cambisols and, mostly Humic, Orthic or Gleyic Podzols.

Podzols are also characteristic of the sandy Saalian substrates where the groundwater is far enough below the surface. On younger (Holocene) dunes and on slopes liable to strong erosion the Podzols are associated with Dystric Regosols.

In valleys, Humic, Dystric, Histic Gleysols and Histosols predominate. By contrast, areas of poorly-drained Saalian boulder clay have Stagnic Gleysol-Dystric Luvisols-Dystric Cambisols associations with a wide variety of intergrades. In wet depressions, a mosaic of Stagno-gleyic Podzols and Podzolic Gleysols has developed.

Orthic Luvisols are relatively widespread in those parts of Lower Saxony and North Rhine Westphalia where sandy loess was deposited in the transition area between the loess area (Börde) and the central uplands. Ice wedge casts are fairly frequent in the Bt horizons of these Luvisols and may locally be of considerable importance for the enhancement of vertical drainage and the related transport of pollutants. Where thin sandy loesses overlie boulder clay or coarse cover sands on level ground, drainage is impeded and Stagnic Gleysols develop.

The soils of the large meltwater channels (Urstromtäler) separating the Saalian moraines and outwash sediments (Geest) give clear indications of the depth of the water table. Where the water table is close to the surface, Histosols, Histic and Humic Gleysols have formed. In other locations. Dystric and Gleyic Cambisols developed on the normally sandy substrates.

Man has also exerted a considerable influence on soil formation, particularly in the case of Histosols which have been subjected to extensive reclamation. In addition, podzolization has been enhanced considerably by the anthropogenic expansion of heathland and the subsequent afforestation, in general with conifers. Cultosols, especially the "plaggen" soils, may also be important locally because of their unusually thick humic horizons.

2 Assessment of groundwater sensitivity

The sensitivity of a groundwater body to pollution may be defined in terms of the physical-chemical properties of the chemical released and in terms of the soil cover, the local water balance and the texture, minerochemistry and hydraulic properties of the unsaturated and saturated zones at the site under consideration. For practical purposes - planning, for example - an assessment of site characteristics should consist of a comparative review at the regional scale and at the local level. At the regional level a determination of the infiltration, filtering and buffering capacities of the soils and the underlying aeration zone is required and, at the local level, a consideration of the structure and hydraulic properties of the aquifers.

2.1 Regionalization of groundwater hazards

2.1.1 Infiltration

The most important regulators for the predominantly vertical transport of water-soluble chemicals through the unsaturated soil and sediment zones and the sorption processes involved are infiltration and soil moisture. These in turn are controlled by the highly variable spatial pattern of effective precipitation, granulometry, density and pore spectra of soils and sediments.

Infiltration (I) is defined in terms of the generalized water budget approach

$$I = P - R - ET \quad (1)$$

where (P) = precipitation, R = runoff, and ET total evaporation loss (cf. Arbeitskreis Grundwasserneubildung 1977).

In the period from 1891–1938 the values for the Federal Republic of Germany were as follows:

I = 112 mm/a
R = 281 mm/a
P = 803 mm/a
ET = 410 mm/a

These figures are generalized and are based on estimates from many regions of the Federal Republic. The values for individual catchments vary considerably as the following values from neighbouring regions in Schleswig-Holstein show (BAUMANN et al. 1970):

Frestedt (Saalian till, loamy sand to sand, catchment = 3.6 km^2)
I = 313 mm/a
R = 205 mm/a
P = 938 mm/a
ET = 420 mm/a

Hennstedt (Saalian till, sandy loam, catchment = 6.9km^2)
I = 79 mm/a
R = 303 mm/a
P = 812 mm/a
ET = 430 mm/a

For the purposes of this paper, the data for P and ET are derived from a digital evaluation of maps and R is determined on a corresponding grid basis using BARTHEL's et al. (1973) approximation. This combines slope angle with infiltration capacity (as related to granulometry), and land use (FRÄNZLE et al. 1987 b). The quantitative characterization of each grid point in terms of infiltration is summarized on an ordinal scale in which 1 = low, 2 = middle, and 3 = high This indicates that, other things being equal, groundwater sensitivity to

1 low		2 medium		3 high	
Predominant soil	Texture	Predominant soil	Texture	Predominant soil	Texture
Orthic Luvisols of high to medium base saturation, locally associated with Calcaric Regosols and relic Chernozems	silty loam heavy-textured subsoil	Orthic Luvisols of medium to low base saturation	loam, sand	shallow Rendzinas	loam, often stony
Chernozems	silty loam	Eutric Cambisols, locally associated with Stagnic Gleysols	silty loam, locally stony	Dystric Cambisols	loamy sand
Degraded Chernozems	silty loam, frequently with heavy-textured subsoil	Rendzinas	stony loam to clay		loamy sand
		Calcaric Regosols	silty loam	Podzols	sand, partly slightly loamy
		Stagnic Gleysols	loamy sand to silty loam	Gleysols and related intergrades	variable

Tab. 1: *Groundwater hazard as related to the average filtering capacity of north German soils.*

pollution by water-soluble chemicals is directly proportional to the amount of seepage water percolating through the zone of aeration.

2.1.2 Filtering and buffering capacities of soil and aeration zone

Filtering and buffering processes, that is, the entire set of physical and chemical retention processes, are controlled by granulometry and porosity and by the amounts, composition and distribution (fabric) of organic matter, clay minerals and hydrated weakly crystalline iron, aluminium and manganese oxides. In this connection, soil pH is of paramount importance since it defines the non-specific net charge of colloids in relation to their points of zero charge. In the range of low concentrations, soil and sediment properties are the main regulators of migration processes while for higher and, in particular, very high concentrations, the physical-chemical potentials of the actual chemicals transported, together with the filtering properties of the substrate that is percolated, become decisive (FRÄNZLE et al. 1987a, FRÄNZLE 1988b).

2.1.2.1 Filtering capacity of soil

A network of micropores, which are typically 0.2–10 μm in diameter but may be as fine as 10 nm in clay soils, lies between individual particles or between micro-aggregates within the peds. Between major soil aggregrates, macropores occur whose diameter exceeds 10 μm, in addition to larger voids and cracks. The nature, size and distribution of soil aggregates and soil voids and pores together determine the fabric of a soil and consequently its hydraulic conductivity and filtering capacity. Filtering capac-

Metal	Adsorption° by humus	clay	sesquioxides	Enhanced sorption above pH
Cd	4*	2	3	6
Mn	2	3	3	5.5
Ni	3–4	2	3	5.5
Co	3	2	3	5.5
Zn	2	3	3	5.5
Cu	5	3	4	4.5
Cr(III)	5	4	5	4.5
Pb	5	4	5	4
Hg	5	4	5	4
Fe(III)	5	5	—	3.5
(Al)	5	4	4	4.5

° relevant for pH values below the threshold given in column 5; above this value strong bonding due to oxide formation (Al, Fe and Mn) or hydroxo-complexes (others)
* scores: 1 very low, 2 low, 3 medium, 4 high, 5 very high

Tab. 2: *Average intensity of physio- and chemosorption of heavy metal to soil constituents (after BLUME & BRÜMMER 1987, modified).*

pH ($CaCl_2$) Metal	2.5	3	3.5	4	4.5	5	5.5	6	6.5	7
Cd	0*	0–1	1	1–2	2	3	3–4	4	4–5	5
Mn	0	1	1–2	2	3	3–4	4	4–5	5	5
Ni	0	1	1–2	2	3	3–4	4	4–5	5	5
Co	0	1	1–2	2	3	3–4	4	4–5	5	5
Zn	0	1	1–2	2	3	3–4	4	4–5	5	5
Cu	1	1–2	2	3	4	4–5	5	5	5	5
Cr(III)	1	1–2	2	3	4	4–5	5	5	5	5
Pb	1	2	3	4	5	5	5	5	5	5
Hg	1	2	3	4	5	5	5	5	5	5
Fe(III)	1–2	2–3	3–4	5	5	5	5	5	5	5
Al	1	1–2	2	3	4	4–5	5	5	5	5

scores: 0 none, 1 very low, 2 low, 3 medium, 4 high, 5 very high

Tab. 3: *Influence of soil acidity on metal sorption in sandy soils of low (0–2%) humus content (after BLUME & BRÜMMER 1987).*

ity increases with decreasing hydraulic conductivity but the filtering process increases in efficiency with the rate of percolation. Tab.1 shows the average filtering capacity of soils in northern Germany. A value of 1 indicates the highest physical retention or lowest risk potential for the underlying groundwater, 2 indicates an intermediate situation in both respects and 3 characterizes the highest risk potential of groundwater resources in terms of the low filtering capacity of the overlying soil.

2.1.2.2 Buffering capacity of soil

Laboratory experiments made with stan-

Sorptivity to tab.3	2–3	3	3–4	4	5
Humus content (%)					
0–2	0	0	0	0	0
2–8	0	0–1	0–1	0–1	1
8–15	0–1	0–1	1	1	1–2
>15	0–1	1	1	1–2	2
Texture*					
S, u'S	0	0	0	0	0
c'S, l'S, uS, sU, U	0	0	0–1	0–1	0–1
lU, slU, uL, sL, sC	0–1	0–1	0–1	0–1	1
lC, sC, uC, scL, cL	0–1	0–1	1	1	1–2
C	0–1	1	1	1–2	2

* S = sand, U = silt, L = loam, C = clay
s = sandy, u = silty, l = loamy, c = clayey
u' = slightly silty, l' = slightly loamy, c' = slightly clayey

Tab. 4: *Adjustment to sorption values of tab.3 in consideration of humus content and texture (after BLUME & BRÜMMER 1987, modified).*

dardized boundary conditions allow the intensity of heavy metal sorption to soils to be defined in relation to the varying content of organic matter, clay, sesquioxides and pH (FRÄNZLE 1982, GROVE & ELLIS 198O, HERMS & BRÜMMER 1984, KÖNIG et al. 1986, BRÜMMER & HERMS 1986, BLUME & BRÜMMER 1987). Tab.2 summarizes the results. In sandy soils with a low humus content, which occur often in north Germany, the pH value is decisive, as indicated in tab.3.

In order to allow for higher humus content and finer texture, the adjustments shown in tab.4 are required.

Estimates based on tab.3 are valid only for soils with low primary heavy metal concentrations. The effective sorption capacities may be much lower where accumulations due to sewage water or sludges, industrial and traffic emissions exist (SAUERBECK 1987). The detection of such influences requires analyses, as does the detection of antagonistic or synergetic effects of heavy metals. For example, high lead or copper concentration in the soil solution enhance cadmium mobility, while high zinc concentration reduce cadmium toxicity.

A generalized assessment of the (cationic) buffering capacity of soils averages the values for individual heavy metals (FRÄNZLE 1988) and groups sorption classes in relation to the definition of infiltration and filtering capacities. The following groundwater sensitivity scale results: 1 (sorption classes 5 and 4), 2 (sorption class 3) and 3 (sorption classes 1 and 2). A shallow water table reduces the retention capacity of a soil. The class values are lowered in correspondence with the depth of the water table: <0.8 m = -2; 0.8–2 m = -1 (MÜLLER 1975).

2.1.2.3 Filtering and buffering capacities of the aeration zone

A basically similar approach is used to define the filtering and buffering potentials of the aeration zone that underlies the soil mantle. In general, however, the buffering capacity of unweathered rocks is considerably lower than that of

Low permeability (Class 1):	boulder clays, clays, marls
Fine- and medium-grained sediments (Class 2):	loesses, solifluction mantles colluvia, sands, gravels, calcareous tills
Solid rocks and coarse gravels (Class 3):	limestones, dolomites, sandstones, greywacke, phyllites, magmatites

Tab. 5: *Average permeability of rocks in the zone of aeration.*

soils since the specific surface of minerals is smaller and the amount of sorptive constituents, in particular, organic matter and hydrated sesquioxides, is much lower. Tab.5 shows the generalized petrographic characteristics of the various rocks with regard to their permeability or hydraulic conductivity and the related physical-chemical buffering potential.

At the local level, adjustments on the basis of the thickness of the aeration zone are possible.

2.1.3 Map of groundwater sensitivity

Based on the spatial variables infiltration, filtering and buffering capacities of the soil and aeration zones, which can provide five values to characterize any locality, a synthetic map can be constructed. The primary data are derived from the following maps using digitization procedures:

Soil Map of the Federal Republic of Germany, 1:1 million (HOLLSTEIN 1963), including substrate and subsurface geology.

Mean annual precipitation (1931 - 1960), 1:2 million (Hydrologischer Atlas der Bundesrepublik Deutschland, 1979)

Mean annual evaporation (1931 -1960), 1:2 million (Hydrologischer Atlas der Bundesrepublik Deutschland, 1979)

Erodibility of soils 1:1 million (RICHTER 1965)

Land use, 1:2 million (Hydrologischer Atlas der Bundesrepublik Deutschland, 1979)

The interpretation of the soil map in terms of pH values and organic matter content is based on a comparative evaluation of geopedological literature (FRÄNZLE 1988). It provides the basis for the determination of substance-specific sorption rates as defined by BLUME & BRÜMMER (1987) in the form of table functions.

The aquifers in the north German lowlands consist primarily of loose sediments the average hydraulic conductivity of which varies between 1×10^{-4} and 1×10^{-2} m/sec. The pattern of the sensitivity of the groundwater resources shown on fig.3 reflects the complicated regional pattern of climatic, pedological and geological boundary conditions at the scale 1:1 million.

The medium sensitivity of the limited groundwater resources of the coastal marsh (fig.1) is due to the excellent filtering and high buffering capacities of the soil cover which is developed predominantly on marine sediments (FRÄNZLE 1988a). This advantage is, to a considerable extent, offset by the shallow depth (<2 meters) of the water table. Inland from the marshlands wide outwash plains of Weichselian age in Schleswig-

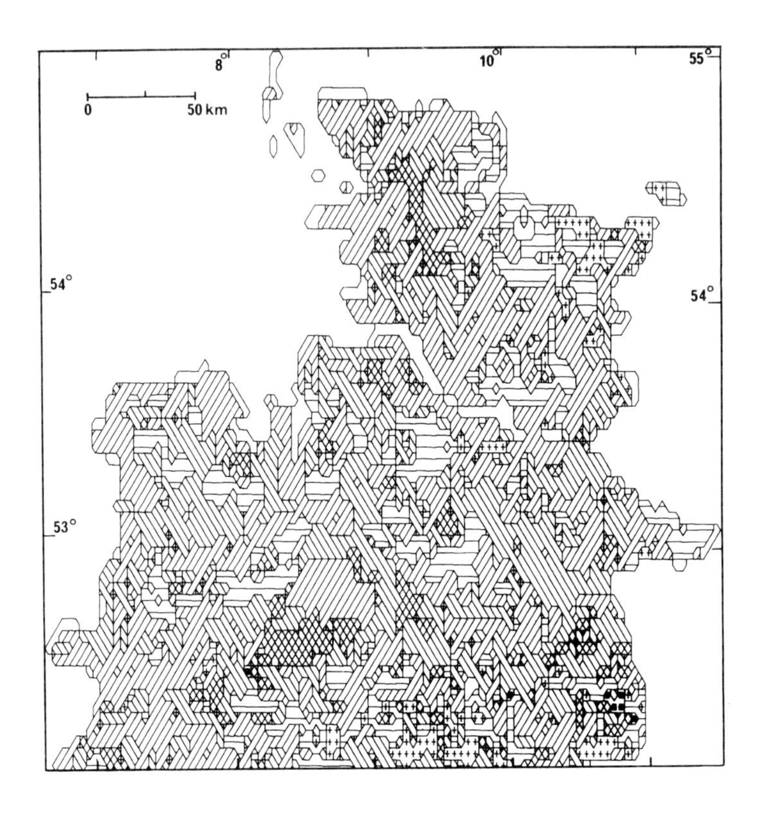

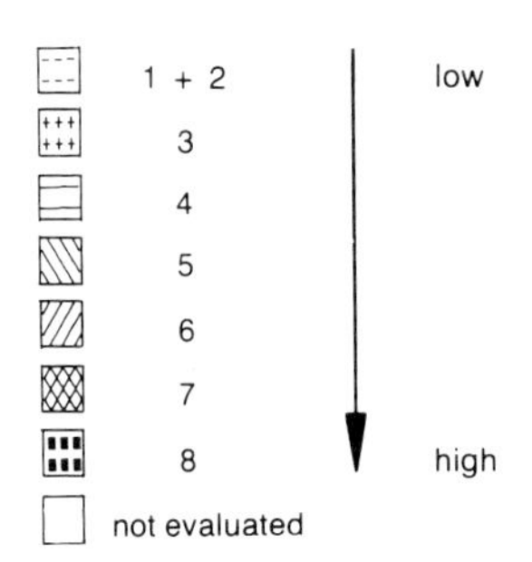

Fig. 3: *Sensitivity of groundwater resources to pollution.*

Holstein and of Saalian age in Lower Saxony occur, on which there are remnants of low moraines. This is a predominantly sandy region (Geest), with patches of moorland, swamp or peaty fens. The water table is often high and this, combined with higher infiltration rates, causes the area to have a high sensitivity to pollution. In contrast to the Geest, the pollution hazard is medium to low on the young moraines to the east and on clayey areas that increasingly replace them towards the Baltic Sea, for example, on the Isle of Fehmarn. In these areas, lower infiltration rates combine favourably with the good filtering and buffering potentials of the soils and a fairly deep aeration zone. The pollution hazard along the edge of the central uplands (Börde) is even lower because the combined thickness of the Saalian and Weichselian loess deposits can exceed 10 meters and they are covered with Orthic Luvisols and locally with Chernozems, which buffer and filter well and lie above deep groundwater.

At the local scale estimates of groundwater sensitivity require, in addition to the criteria considered at the regional scale, information about throughflow and groundwater hydraulics. Because of their very high spatial, and partly also temporal, variability, the generalized statements made in this paper are not applicable to the local scale.

Acknowledgements

Financial support from the Federal Environmental Agency (Umweltbundesamt)/ Bundesministerium für Umwelt, Naturschutz und Reaktorsicherheit is gratefully acknowledged.

References

ARBEITSKREIS GRUNDWASSERNEUBILDUNG (1977): Methoden zur Bestimmung der Grundwasserneubildungsrate. Geol. Jahrbuch, C **19**, 98 pp.

BARTHEL, H., MANNSFELD, K. & SANDNER, E. (1973): Flächen gleicher Abflußbereitschaft bei sommerlichen Starkregen. Petermanns Geogr. Mitt. **117**, 107–116.

BAUMANN, H., SCHEKORR, E. & SCHENDEL, U. (1970): Gebietswasserhaushaltsbilanzen in kleinen Einzugsgebieten in Schleswig-Holstein. Z. deutsch. geol. Ges., Sonderh. Hydrogeol. Hydrogeochemie, 97–107.

BLUME, H.-P. & BRÜMMER, G. (1987): Prognose des Verhaltens von Schwermetallen in Böden mit einfachen Feldmethoden. Mitt. Dtsch. Bodenkundl. Gesellsch. **53**, 111–117.

BRÜMMER, G., GERTH, J. & HERMS, U. (1986): Heavy metal species, mobility and availibility in soils. Z. Pflanzenernähr. u. Bodenkunde **149**, 382–398.

EHLERS, J. (1981): Some aspects of glacial erosion and deposition in North Germany. Ann. Glaciol. **2**, 143–146.

EHLERS, J. & GRUBE, F. (1983): Meltwater deposits in northwest Germany. In: J.EHLERS (ed.), Glacial Deposits in Northwest Europe, Balkema, Rotterdam, 249–256.

EHLERS, J., MEYER, K.-D. & STEPHAN, H.-J. (1984): Pre-Weichselian glaciations of northwest Europe. Quaternary Sci. Rev. **3**, 1–40.

FRÄNZLE, O. (1969): Zertalung und Hangbildung im Bereich der Süd-Ville. Erdkunde **23**, 1–9.

FRÄNZLE, O. (1981): Erläuterungen zur Geomorphologischen Karte 1:25,000 der Bundesrepublik Deutschland, GMK 25 Blatt 8 1826 Bordesholm. Berlin.

FRÄNZLE, O. (1982): Erfassung von Ökosystemparametern zur Vorhersage der Verteilung von neuen Chemikalien in der Umwelt. Schriftenreihe "Texte" des Umweltbundesamtes Berlin.

FRÄNZLE, O. (1988a): Erläuterungen zur Geomorphologischen Karte 1:100,000 der Bundesrepublik Deutschland, GMK 100 Blatt 7 C 1518 Husum. Berlin.

FRÄNZLE, O. (1988b): Sensitivity of soils in relation to pollution. Proc. World Conference on Hazardous Waste, Budapest 25–31 Oct. 1987. Elsevier, Amsterdam, 703–711.

FRÄNZLE, O., KUHNT, D. & KUHNT, G. (1985): Die ökosystemare Erfassung von Bodenparametern zur Vorhersage der potentiellen Schadwirkung von Umweltchemikalien. Verh. Ges. für Ökologie (Bremen 1983) XIII, 323–340.

FRÄNZLE, O., BRUHM. I, GRÜNBERG, K.-U., JENSEN-HUSS, K., KUHNT, D., KUHNT, G., MICH, K., MÜLLER, F. & REICHE, E.-W. (1987a): Darstellung der Vorhersagemöglichkeiten der Bodenbelastung durch Umweltchemikalien. Forschungsbericht 10605026 im Umweltforschungsplan des Bundesministers für Umwelt, Naturschutz und Reaktorsicherheit, Kiel.

FRÄNZLE, O., ELHAUS, D. & FRÖHLING, J. (1987b): Naturwissenschaftliche Anforderungen an die Sanierung kontaminierter Standorte - Standortspezifische Klassifikation im Hinblick auf die Migration von Stoffen im Untergrund. Forschungsbericht 102 03 405/2 im Umweltforschungsplan des Bundesministers für Umwelt, Naturschutz und Reaktorsicherheit, Kiel.

GALBAS, P.U., KLECKER, P.M. & LIEDTKE, H. (198O): Erläuterungen zur Geomorphologischen Karte 1:25,000 der Bundesrepublik Deutschland, GMK 25 Blatt 5 3415 Damme. Berlin

GOLWER, A., KNOLL, K.-H., MATTHESS, G., SCHNEIDER, W. & WALLHÄUSER, K.H. (1976): Belastung und Verunreinigung des Grundwassers durch feste Abfallstoffe. Abh. hess. L.-Amt Bodenforsch. 73.

GROVE, J. H. & ELLIS, B.G. (1980): Extractable chromium as related to soil pH and applied chromium. Soil Sci. Soc. Am. J. **44**, 238–242.

GRUBE, F. (1979): Übertiefte Rinnen im Hamburger Raum. Eiszeitalter und Gegenwart **29**, 157–172.

HERMS, U. & BRÜMMER, G. (1984): Einflußgrößen der Schwermetallöslichkeit und -bindung in Böden. Z. Pflanzenernähr. u. Bodenkunde **147**, 400-424.

HINSCH, W. (1979): Rinnen an der Basis des glaziären Pleistozäns in Schleswig-Holstein. Eiszeitalter u. Gegenwart **29**, 173–178.

HOLLSTEIN, W. (1963): Bodenkarte der Bundesrepublik Deutschland 1:1,000,000. Bundesanstalt für Bodenforschung, Hannover.

KÖNIG, N., BACCINI, P. & ULRICH, B. (1986): Der Einfluß der natürlichen organischen Substanzen auf die Metallverteilung zwischen Böden und Bodenlösung in einem sauren Waldboden. Z. Pflanzenernähr. u. Bodenkunde **149**, 68–82.

KUSTER, H. & MEYER, K.-D. (1979): Glaziäre Rinner im mittleren und nördlichen Niedersachsen. Eiszeitalter u. Gegenwart **29**, 135–156.

MEYER, K.-J. (1974): Pollenanalytische Untersuchungen und Jahresschichtenzählungen an der holstein-zeitlichen Kieselgur von Hetendorf. Geol. Jahrb. **A21**, 87–105.

MÜLLER, H. (1974): Pollenanalytische Untersuchungen und Jahresschichtenzählungen an der eem-zeitlichen Kieselgur von Bispingen/Luhe. Geol. Jahrb. **A21**, 149–169.

MUTERT, E. (1978): Untersuchungen zur regionalen Gruppierung von Böden - durchgeführt an einer Kleinlandschaft im schleswig-holsteinischen Jungmoränengebiet. Doctoral Dissertation, Kiel.

RICHTER, G. (1965): Bodenerosion. Schäden und gefährdete Gebiete in der Bundesrepublik Deutschland. Forsch. Deutsch. Landeskunde **152**. Bundesanstalt f. Landeskunde u. Raumforschung, Bonn-Bad Godesberg.

ROESCHMANN, G: (1971): Die Böden der nordwestdeutschen Geest-Landschaft. Mitt. Dt. Bodenkundl. Ges. **13**, 151–231.

ROESCHMANN, G. (1986): Soil map of the Federal Republic of Germany 1:1,000,000. Federal Institute for Geosciences and Natural Resources, Hannover.

SAUERBECK, D. (1987): Effects of agricultural practices on the physical, chemical and biological properties of soils: part II, Use of sewage sludge and agricultural wastes. In: H. BARTH & P. L'HERMITE (eds), Scientific Basis for Soil Protection in the European Community. Elsevier, London, New York, 181–210.

SCHLICHTING, E. (1960): Typische Böden Schleswig-Holsteins. Parey, Hamburg, Berlin.

Address of author:
Prof. Dr. Otto Fränzle
Geographisches Institut der Universität Kiel
Olshausenstraße 40
D-2300 Kiel 1
Federal Republic of Germany

CATENA SUPPLEMENT 15 p. 107–119 Cremlingen 1989

GEOECOLOGICAL ASPECTS OF THE LATE PLEISTOCENE AND HOLOCENE EVOLUTION OF THE BERLIN LAKES

Hans-Joachim **Pachur**, Berlin

1 Introduction

Berlin depends on groundwater for its water supply. The importance of the quality of this supply is shown by the very high ratio of 274 inhabitants per litre per second of mean minimum flow. The corresponding value for the Rhine is 120. Berlin has a mean annual precipitation of only 590 mm. This unfavourable hydrographic situation is improved only because the Havel (fig.1) and the Spree are lowland rivers with low runoff, less than 150 mm per year and have, therefore, a significant groundwater recharge that is intensified by the artificial impounding of the Havel. There are, however, problems related to waste water control. The flow velocity of the Lower Havel is about 0.013 m/s so that, hydrologically, the Lower Havel is a lake.

The Brandenburg stage of the Weichsel (Würm) glaciation (20,000 BP) extended further south than the Berlin lake basins and the Frankfurt stage of the same glaciation (17,000 BP) stopped north of them. If it is assumed that the lake basins were eroded by the advancing ice, they must have evolved in the period between these two dates. The origin of the Havel lakes is of particular importance because the wells of Berlin's waterworks are located on the eastern and part of the western banks of the Havel and Lake Tegel. The bank-filtrated groundwater extracted from the wells comprises at least 38% of the total discharge and, in some cases up to 66% (KÜNITZER 1956, KLOOS 1986). Owing to the drawdown at the wells the lake sediments are subject to a high hydraulic gradient.

ISSN 0722-0723
ISBN 3-923381-18-2

3-923381-18-4/89/5011851/US$ 2.00 + 0.25

2 The development of the lake basins since the late Pleistocene

2.1 The lake sediments and their ages

The lake sediments along the Havel were sampled down to the Pleistocene material using a Livingstone corer and heavy drilling equipment. Thicknesses of 30 m were found (fig.2) so that the Pleistocene lake floors lie below sea level. The Havel depression is characterized by thick lake sediments which attain their greatest thicknesses on either side of the Spree delta. By contrast, on the Spree delta, fluvial and limnic-telematic formations alternate over short vertical and

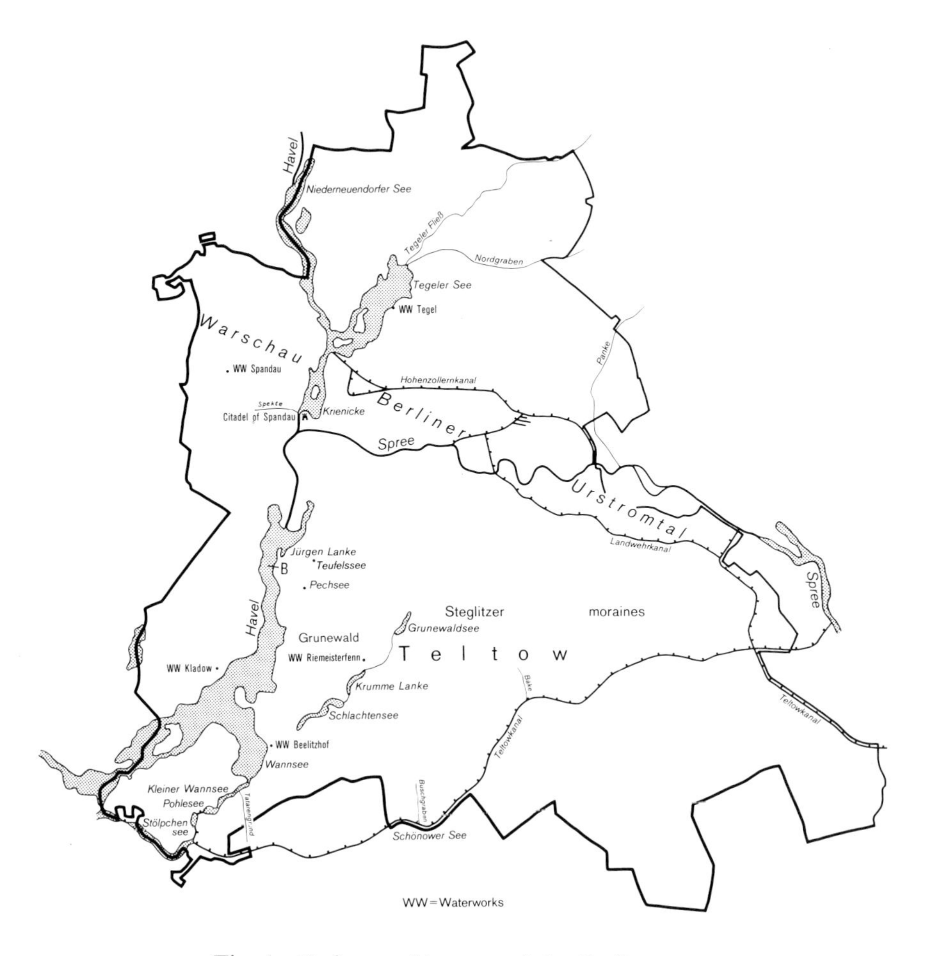

Fig. 1: *Hydrographic map of the Berlin area.*

horizontal distances, indicating that the present difference between the sediment load of the Havel and the Spree that results from different types of flow, has existed since pollen zone IIa (fig. 2.1, No. 26) about 13,000 years BP. The lake sediments were also dated by radiocarbon analysis, pollen analysis and tephrochronology.

The tephra layer shown in fig.2 consists of gas-rich, xenolith-poor pumic tuff from the Eifel. The ash fall buried organic matter dated from 11,300 BP to 10,800 BP (FRECHEN 1959). The tuff is up to 25 mm thick in the Berlin lakes and is a stratigraphical marker, having a sharply delimited boundary to the underlying muds. The tephra layer seems to be homogeneous throughout, indicating that only one eruption occurred. It is contemporaneous with the upper part of pollen zone II, that is, the relatively warm phase of the Alleröd.

The maximum age of the Havel riverlakes can be inferred from pollen analysis of the mud at the bottom of the lacus-

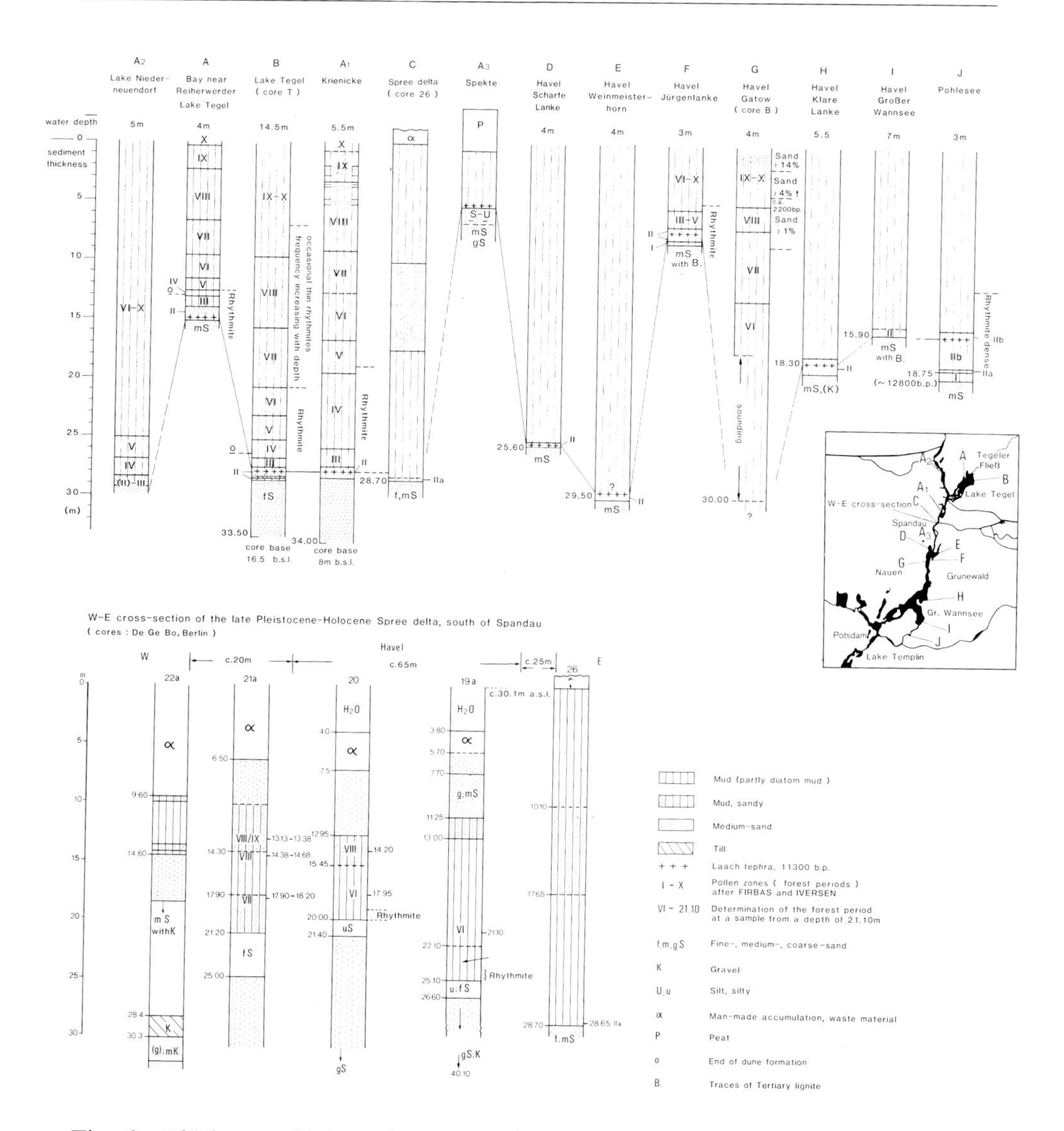

Fig. 2: *Thickness of lake sediments in the Havel. The lake sand at the base of the mud is at least 3 m thick (Spree delta course drilled by K. MEYER, Berlin).*

trine sequence. The mud was deposited in pollen zone I, which has been dated at 13,000 years BP by MANGERUD et al. (1974). The boundary between zone I and the next older zone is not defined palynologically in the Berlin area. Radiocarbon analysis of the mud from Lake Krienicke also produced an age of 13,000±125 BP (UZ 2160) which, given discontinuous sedimentation, could be correct. The underlying lake sands are, therefore, older.

2.2 Dead ice theory and lake basin formation

It has been suggested that the lake depressions were preserved by dead ice masses. In the case of the East Prussian lakes, for example, it has been assumed that lake sediments were accumulated on top of the ice (WIECKOWSKI 1969). No evidence has, however, been found in the Berlin area of small-scale disruptions in the bedding of the fine-layered sediments deposited in the Alleröd that in some areas continue into the Pre-Boreal (Rhythmites). Such disruptions might be expected to occur as a result of slumping on the top of melting dead ice and it is doubtful that processes associated with long-term melting caused the formation of the lake basin. In addition, limnic sedimentation began around 13,000 BP so that the basin was filled, at the latest, by this date. Any ice left in the basin was covered with sand and insulated from low winter temperatures. The ice was also close to water that maintained a temperature of at least 4°C, except perhaps in some winters. In the Havel valley, this water, which included the interstitial water in the sediment layer above the ice, was in motion. Consequently the dead ice in the valley melted more rapidly than in the areas of till-covered stagnant ice outside the valley (PACHUR & RÖPER 1988). This corresponds to observations in permafrost regions where the permafrost table has been found to be lower beneath rivers and lakes (HOPKINS & KARLSSTROM 1965). BOULTON (1972) describes conditions in Spitzbergen "where the active layer of permafrost rarely exceeds 3 m, melting beneath lakes and streams often produces collapse" melting, that is, of the underlying body of ice. He also states that "the high thermal capacities of surface streams have the ability to induce melting beneath at least 20 m of gravels."

The problem of melt under water has been examined by CARSLAW & JAEGER (1959). It is also known as the STEPHAN problem (1890) or "moving boundary problem". Based on CARSLAW & JAEGER, the following sequence has been suggested by G. BRAUN for the formation of the lake basin. It is assumed that there is a semi-finite body of water overlying an ice block, the surface of which is covered by a layer of quartz sand. It is assumed that the other sides of the ice block are fully insulated. Because the thermal conductivity of quartz sand is higher than that of water, the ice under the quartz sand melts more rapidly than an ice block lying directly under water. Since water has its maximum density at 4°C, it can be expected that a temperature of 4°C is reached at the lake bed, at least in summer. As a result the ice melts at an estimated rate of 0.6 m in 100 days, given a water temperature of 4°C and an ice temperature of -15°C. In winter, the water probably does not refreeze because of the heat capacity of the lake water above the ice and, as the water deepens, freezing down to the lake bottom becomes increasingly unlikely. If a maximum dead ice thickness of 30 m is assumed for the Berlin area, at a rate of 0.6 m per 100 days, total thaw would have occurred within a century.

After evaluating algal gyttja at the bottom of lakes in dead ice depressions in Minnesota (Cedarbog Lake) FLORIN & WRIGHT (1969) concluded that "perhaps the entire lake formation after a very slow initial period took only a few decades." The analytical solution of the

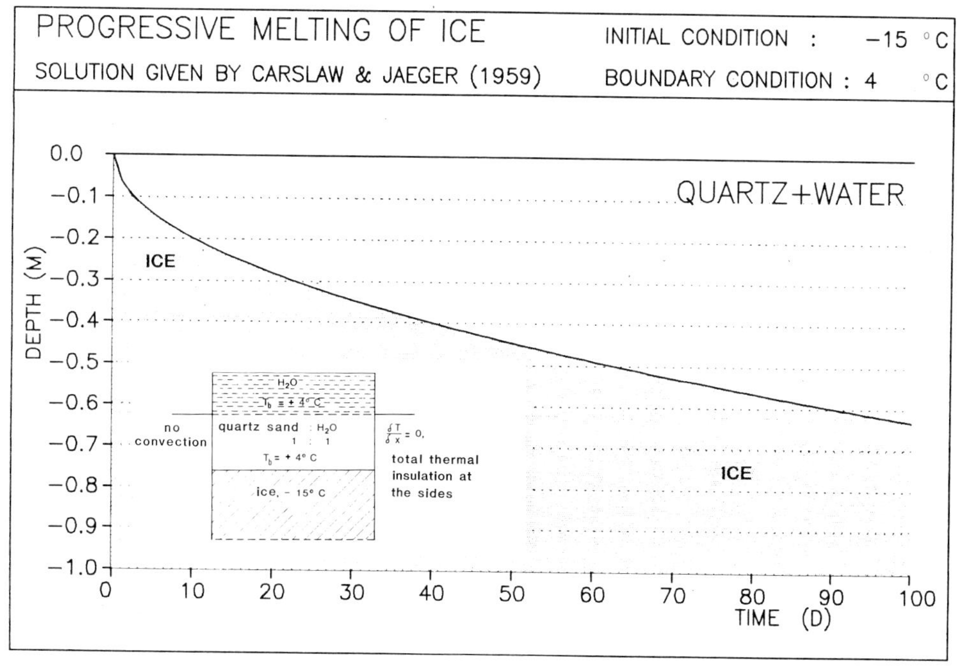

Fig. 3: *Progressive melting of ice.*

moving boundary problem shown in fig.3 does not take into account that water density is a function of temperature and that meltwater rises at the interface. As a result of this convection, temperature gradients would be higher than those shown in the model and the estimate of the rate of melting is probably conservative.

3 Structure of the sediments: sedimentological and geochemical parameters

The groundwater wells are located near the lakes so that the structure, sedimentology and geochemistry of the thick late Pleistocene and early to mid-Holocene lacustrine fills are of hydrological importance. The sediments act as a filter and their physical properties determine how much groundwater percolates per unit time. In addition, their exchange properties influence the substances contained in the groundwater.

3.1 Rhythmites

A varve-like sequence of dark and light beds is present in the bottom few metres of the sediment columns of various Berlin lakes. The light-coloured rhythmites are thought to be the result of calcilutite precipitation in the late spring and summer because the lamina are subdivided into diatom-rich spring layers and calcite-rich summer layers. Fossil rhythmites show a similar differentiation. The dark winter layers contain mainly organic substances, iron hydroxide and other detritus. This sequence cannot have been formed in shallow waters because it would have been destroyed by wave impact and other turbulences. Also the structure should have been destroyed by intensive bioturbation. The preservation of the structure indicates anoxic conditions. The presence of iron sulphide in the dark layers also indicates an O_2 deficit in the sediment. The Berlin lakes contain rhythmites only at the sediment base. They occur in a similar position in other lakes. In southwest Germany, in the Schleinsee (GEYH et al. 1971), for example, the rhythmites were deposited before the end of the Atlantic period. An exception is found in the deepest parts of Lake Tegel where subboreal rhythmites occur within a 50 mm thick layer. In general, however, intensive rhythmite formation in the Berlin lakes seems to have ended in the Atlantic period.

The lamination of the rhythmites was caused perhaps by euxinic phases on the floor of the lake, possibly accompanied by an anaerobic hypolimnion. This hypothesis is supported by the occurrence of calcium/manganese carbonates (calcium rhodochrosite) over a period of about 3000 years in the early phase of mud sedimentation in Lake Tegel. These were formed directly in the hypolimnion, probably during marked euxinic phases. Most of the calcium rhodochrosite was probably formed in the uppermost sediment zone, where sufficient organic material was available to reduce deposited manganese oxide. The presence of rhodochrosite in the coarse silt and fine sand fractions of the sediments provides evidence of intrasedimentary formation of manganese carbonates. It also coats some silicate grains. It would seem that the distribution of oxygen, temperature and light in the lake varied greatly with the seasons, with intermittent oxygen transport to the lake bottom during phases of full circulation and with permanent anoxic conditions in winter. Meromixis may also have prevailed. The sediment structure indicates, therefore, that there was an increase in biological production and organic detritus during the mid-Holocene. The quantity of reducing material rose and the manganese content decreased.

3.2 Granulometry

Fig.4 shows an idealized sequence of limnic deposits in Lake Tegel. In general, the lacustrine deposits in the Berlin lakes consist of lake sand or till at the base, calcareous mud above on which there is a layer of sapropel. The sediments are very fine-grained (tab.1). Permeability is in the range of impermeable beds, that is 10^{-6} – 10^{-7} cm/s at a hydraulic gradient of $>1.5^0/_{00}$. The grain size range of the sandy base of the muds shows a dominant fine sand fraction. The sorting coefficient of 1.2–1.3 at a mean grain size value of 0.2 to 0.15 mm indicates an aeolian origin. This also accounts for the intercalations of calcilutite in mm thick horizons which contain charophytes that occur in the basal sands. These calcilutites cannot have an intrasedimentary origin. The increased proportion of fine gravel and coarse sand shows that during the early period of lake formation, several phases of fluvial deposition of sand, grading upwards into finer aeolian-fluvial deposits must have occurred.

SEM analysis of all grain fractions reveals that the mud particles consist mainly of diatom frustules, carbonate phosphate and iron hydroxide crusts of diatom fragments, as well as quartz, feldspar and calcite grains. Grain size analysis is difficult because of the large number of diatoms present. Entire strata are composed of pure diatom mud. SEM

Sample no.	Sediment depth (m)	W.A.	>1 mm (%)	630–1000 μ (%)	315–630 μ (%)	200–315 μ (%)	125–200 μ (%)	63–125 μ (%)	20–63 μ (%)	6.3–20 μ (%)	2–6.3 μ (%)	$< 2\ \mu$ (%)	$> 63\ \mu$ (%)	Silt (%)	Clay (%)
T 1.1	1.60 – 1.90			0.03	0.06	0.09	0.28	3.82	6	10	25	54.7	4.3	41	54.7
T 1.3	2.10 – 2.30		0.01	0.02	0.18	0.38	0.59	3.04		$< 63\ \mu = 95.8\%$			4.2	95.8%	
T 2-1/2	2.92 – 3.20				$> 63\ \mu = 1.0\ \%$				6	7	17	69	1.0	30	69
T 3.1	3.34 – 3.50				$> 63\ \mu = 0.27\%$				0.4	7.3	43	49	0.3	50.7	49
T 3.5	4.11 – 4.30			0.002	0.01	0.03	0.03	0.18	0.2	5.8	29.7	64	0.3	35.7	64
T 4.1	4.50 – 4.67				$> 63\ \mu = 0.28\%$				0.3	9.8	31.7	57.9	0.3	41.7	58
T 4.3	4.86 – 5.03				$> 63\ \mu = 0.32\%$				0.3	7.6	28.6	63.2	0.3	36.7	63
T 5.2	5.80 – 5.90	IX–X		0.003	0.03	0.04	0.06	0.28	0.2	5.8	30.0	63.6	0.4	35.6	64
T 5.3	5.90 – 6.10				$> 63\ \mu = 0.39\%$				0.3	6.3	30.2	62.8	0.4	36.6	63
T 5.5	6.30 – 6.50			0.006	0.03	0.04	0.04	0.19	0.2	6.7	36.3	56.5	0.3	43.2	56.5
T 6.2	6.70 – 6.88				$> 63\ \mu = 0.50\%$				0.3	14.4	20.7	64.1	0.5	35.5	64
T 7.4	8.25 – 8.45		0.01	0.03	0.11	0.07	0.08	0.27	0.2	15.4	28.4	54.4	0.6	44	54.4
T 8.2	8.90 – 9.08			0.02	0.07	0.06	0.07	0.69	0.3	6.3	25.2	67.3	0.9	31.8	67.5
T 8.4	9.30 – 9.48	—			$> 63\ \mu = 1.00\%$				0.2	7.2	29.2	62.4	1.0	36.6	62.4
T 9.2	9.90 –10.08				$> 63\ \mu = 0.30\%$				0.8	8.3	30.3	60.3	0.3	39.4	60.3
T 10.2	11.02 –11.22				$> 63\ \mu = 0.64\%$				0.3	8.6	26.3	64.2	0.6	35.2	64.2
T 11.3	12.33 –12.54	VIII		0.007	0.03	0.03	0.03	0.14		$< 63\ \mu = 99.7\%$			0.2	99.76%	
T 12.3	13.435–13.635				$> 63\ \mu = 0.23\%$				0.2	6.6	25.0	68	0.2	31.8	68
T 13.4a	14.76 –14.79	—		0.005	0.03	0.04	0.04	0.10	0.2	6.1	29.8	63.7	0.2	36.1	63.7
T 14.4	15.86 –16.08				$> 63\ \mu = 0.31\%$				0.2	5.4	31.4	62.7	0.3	37	62.7
T 15.3	16.70 –16.88				$> 63\ \mu = 0.26\%$				0.3	7.1	32.4	59.9	0.3	39.7	60
T 17.3	18.32 –18.49	VII		0.004	0.05	0.04	0.03	0.08	0.2	6.8	30.1	62.7	0.2	37.1	62.7
T 18.3	19.34 –19.54				$> 63\ \mu = 0.08\%$				0.2	6.6	22.8	70.3	0.1	29.6	70.3
T 19.3	20.44 –20.64	—		0.02	0.07	0.05	0.04	0.05	0.3	8.2	23.8	67.5	0.2	32.3	67.5
T 20.3	21.50 –21.68	VI			$> 63\ \mu = 0.29\%$				0.3	7.1	21.9	70.5	0.3	29.2	70.5
T 21.3	22.54 –22.74	—		0.005	0.02	0.03	0.04	0.05	0.2	6.4	25.6	67.6	0.2	32.2	67.6
T 22.3	23.64 –23.84	V			$> 63\ \mu = 0.14\%$				0.2	6.7	32.5	60.4	0.1	39.5	60.4
T 24.3	25.70 –25.88	IV		0.003	0.009	0.02	0.02	0.19	0.3	7.0	25.8	66.7	0.2	33.1	66.7
T 25.1	26.33 –26.40	—		0.01	0.09	0.07	0.22	0.22	1	9.2	35.2	54.1	0.5	45.4	54.1
T 25.4	26.90 –27.10			0.04	0.17	0.23	0.45	3.56	10.8	16.9	26.1	41.8	4.4	53.8	41.8
T 26.1	27.30 –27.525	III	0.06	0.01	0.32	0.20	0.60	8.73	16.1	18.5	16.6	38.9	9.9	51.2	38.9
T 26.2	27.525–27.75		0.02	0.03	0.095	0.07	0.15	2.72	13.1	18.4	23.4	42.0	3.1	54.9	42.0

Tab. 1: *Granulometric analysis of the carbonate mud, Lake Tegel (Pollen zones after FIRBAS, analysis by BRANDE).*

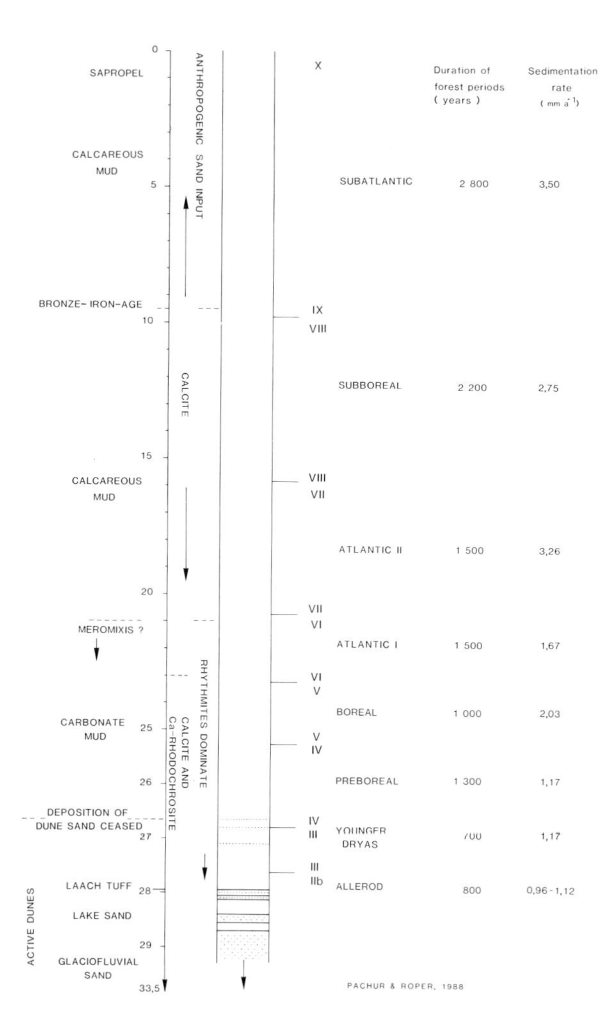

Fig. 4: *Idealized sequence of limnic deposits, water depth 16 m, Lake Tegel. Duration of pollen zones determined by BRANDE (1980).*

photomirographs show that the diatoms have split into filaments, similar in shape and size to asbestos fibres, and are small enough to be inhaled. Open tipping of these lake sediments could be a health hazard although the physiological effects of the deposition of frustules in human organs as a result of long periods of exposure to muds in open dumps is not yet clear.

Organic compounds, an important component of the lake sediments, are referred to as organic carbon (Corg). The Corg contents increase with increasing depth. The maximum percentage in the Havel is 22.83%; 45% was measured in the Grunewald lakes. Corg contents can, however, vary within the same chronological layer in a lake by up to 10%.

3.3 Carbonates

The dominant carbonate mineral in the Berlin lake deposits is calcite, with the exception of a few sediment sections in which siderite and rhodochrosite pre-

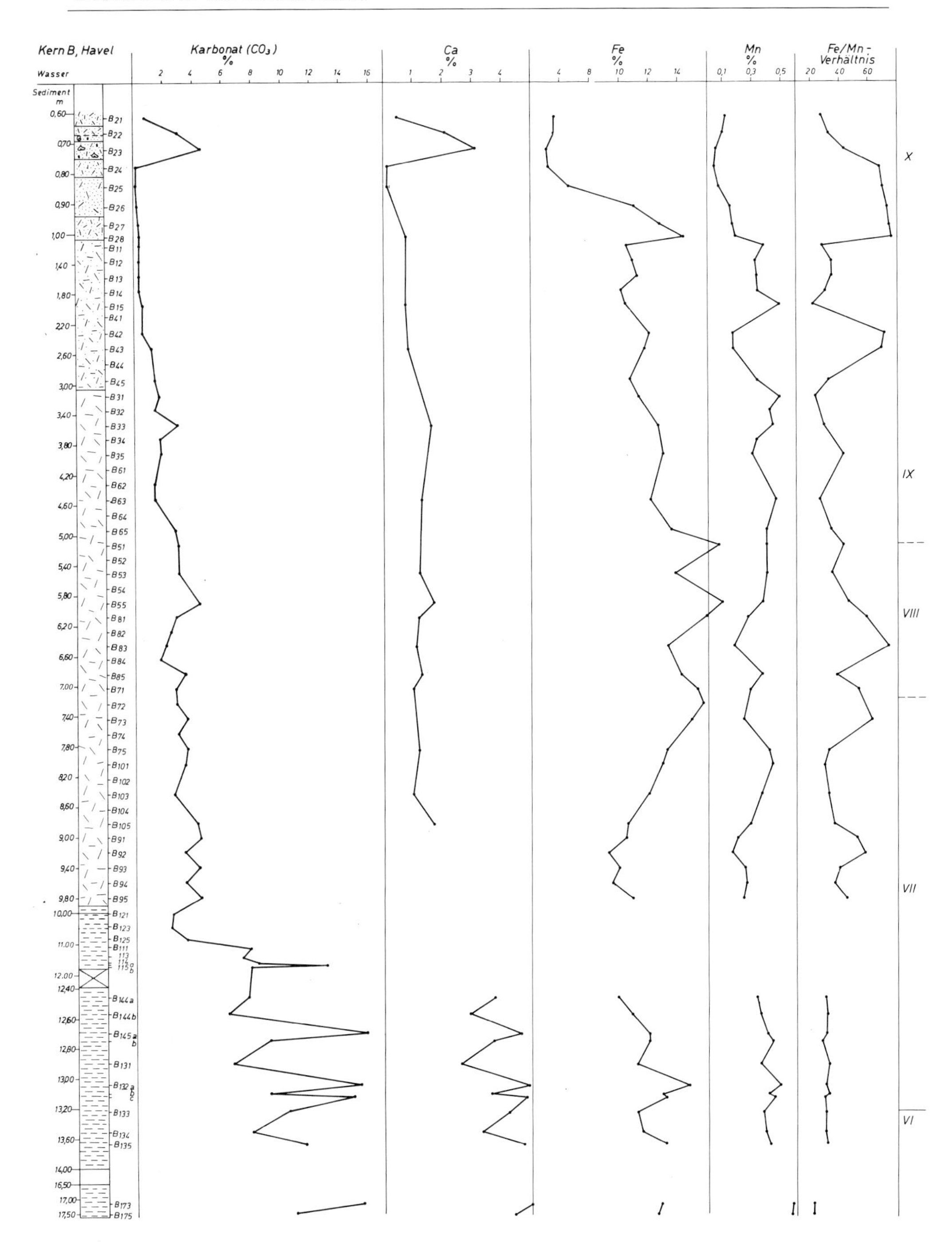

Fig. 5: *Changing carbonate content in muds since the Atlantic period (B on fig.1).*

dominate. The amount of calcite formed in the lake metabolism is regulated mainly by biogenic decalcification and direct precipitation. KOSCHEL et al. (1983) have demonstrated a link between the $CaCO_3$ and the degree of eutrophication in recent sediments in the Mecklenburg lakes where the $CaCO_3$ content of the littoral sediments increased as the lakes changed from the oligotrophic to the eutrophic stage. In the deep sediments, the lowest values were also found in the oligotrophic lake. Lake Stechlin, for example, has 1–2% of $CaCO_3$ and 2.5 g/m^3 PO_4-P in contrast to the mesotrophic lake Lake Breiter Lucin which had 45–63% $CaCO_3$ and 40.0 g/m^3 of PO_4-P. Both can be compared with the Tegel and Havel lakes during the mid-Holocene. The decrease in calcite content (fig.5) that began in the upper Atlantic was due to a rise in the level of the North Sea (GEYH 1966, SINDOWSKI 1973) and consequent backflow of the Elbe. In the Havel and its arms, floods were more frequent and the groundwater table rose. Lakes that previously were connected only intermittently, became river-lakes within the Havel depression.

4 Pollution by xenobiotic substances

The Berlin lakes contain polynuclear aromatic hydrocarbons (PAH), polychlorinated hydrocarbons, chlorinated benzenes, polychlorinated terphenyls (PCT), persistent insecticides of the DDT group and their transformation products and a wide range of aliphatic hydrocarbons of the n-alcanes between C10 and C40. While the heavy metal concentration reaches the geochemical background level below a depth of about 3 m, the xenobiotics migrate further. A relatively high mobility of xenobiotics in sediments of both marine and fluvial systems has also been observed by ZOETEMANN et al. (1980).

In a study of Lake Krienicke, KOFELD (1982) sampled the contents of 19 polynuclear aromatic hydrocarbons using gas chromatography. The surface sample contained fluoranthene, benz-(a)anthracene, chrysene, benzo(e)pyrene, benzo(a)pyrene and benzo(phi)perylene and numerous other compounds. Samples from a depth of 4.78–4.96 m had a smaller number of PAH, with markedly lower peak values. At a depth of 10.39–10.62 m only perylene is still identifiable. Since a uniform mixture of PAH (which would result from biogenic synthesis) has not yet been found in the sediments, most of the chemicals detected are almost certainly anthropogenic. Perylene, on the other hand, increases with depth and is, therefore, either biogenic or was synthesized during diagenesis. Since polychlorinated hydrocarbons were detected at a depth of 6 m in Lake Tegel, there is increasing evidence of a deep-reaching migration of organic chemicals that were deposited initially within the infiltration zone; this migration may impair groundwater quality. In addition, pollution by xenobiotics and heavy metals occurs which is almost independent of the structural and textural parameters of the upper few meters of lake sediment. A significant positive correlation of 0.9199 exists only between lead and gamma HCB. In the lower sections of the sediment, DDT-metabolics from the 1940's and 1950's generally predominate with maximum values of, for example, 1.529 ng/g of DDT. A value of 643 ng/g measured in Schlachtensee is

		Cu		Zn		Cd		Pb	
Langsee		1.2	(0.5)	3.5	(2.3)	20	(7.3)	3.8	(3.4)
Schulensee		7.0	(3.0)	7.0	(4.5)	20	(7.3)	3.4	(3.0)
Gildehaus Venn		1	(0.1)	5.1	(0.2)	2	(0.7)	1.9	(0.1)
Seeburger See		1.1	(0.5)	1.2	(1.1)	3	(0.8)	1.6	(1.6)
Schloß Berge*		—	(2.3)	—	(9.5)	—	(23)	—	(18)
Kletterpoth*		—	(3.4)	—	(32)	—	(67)	—	(50)
Jungferweiher		0.9	(0.9)	2.2	(2.5)	4.2	(3.9)	2.5	(4.3)
Schleinsee		2.9	(1.6)	1.9	(2.3)	6.1	(4.5)	17	(6.4)
Großer Arbersee		0.8	(0.4)	0.6	(0.5)	1.1	(1.6)	8.6	(3.2)
Schlachtensee		36	(4.8)	14	(6.5)	32	(18)	25	(15)
Krumme Lanke:	sample KL2G1/18A	9.5	(1.9)	11	(6.1)	>7	(7.0)	36	(14)
	sample KL2G3/18C	11	(2.2)	17	(9.6)	>9	(9.3)	43	(16)
	sample KL1G1	17	(2.9)	12	(8.2)	13	(12)	59	(16)
	sample KL1G3	24	(4.0)	22	(15)	22	(20)	84	(22)
Pechsee, bank		36	(7.5)	9.3	(13)	17	(22)	180	(21)
Pechsee, mid-lake		66	(14)	8.6	(13)	14	(17)	160	(19)
Lower Havel		139	(10.3)	46	(61)	150	(0.16)	212	(4.3)
Lake Tegel		259	(0.85)	127	(7.79)	—		—	

(after PACHUR & RÖPER 1988)
* thickness of mud <1.5 m

Tab. 2: *Enrichment factors related to the deep part of the sediment (local background) and clay/shale standard.*

due to increased input during the 1950's. The migratory capacity of xenobiotics at sites with thin sediment layers and influent conditions possibly represents a potential hazard to groundwater.

5 Heavy metal pollution

All the investigated lakes show increasing, man-made concentrations of zinc, lead, mercury and copper from a depth of 2 m to the top of the sediment. Beneath this zone, the curve of heavy metal concentration is uniformly low and represents the lake specific background level. Another increase in copper and zinc content in the deepest sediment zones has, as yet, only been detected in Lake Tegel, in association with a manganese accumulation of up to 17%.

Since over 90% of the sediment of the investigated muds is in the <63 μm fraction and the silt fraction also contains diatoms, carbonates, organic matter and iron compounds, the heavy metal contents were analysed with reference to the total solid substance, that is the <63 μm fraction. Only in the uppermost decimetres does the sand content dilute the heavy metal concentration; this effect is eliminated by computation.

Tab.2 shows enrichment factors with reference to the lake- specific background and the shale standard for several Berlin lakes compared to lakes in the Federal Republic of Germany. A comparison of data from FÖRSTNER & MÜLLER (1974) and MÜLLER (1985) with tab.2 shows that the Havel is one of the most heavily polluted rivers.

The permeability coefficients of the muds in the upper half of the sequence are approximately 2.1 · 10^{-7} m s^{-1} compared with 2.8 · 10^{-9} ms^{-1} in the lower half. For the migration of substances or xenobiotics, a variation in conductivity of 2 decimal powers with increasing depth means that the substances reach the sandy bottom in a spatially differentiated pattern which results in their spatially variable migration into the aquifers that are used for groundwater exploitation. Near the shore their residence time in the sediment column is shorter because the water outflow increases owing to the greater permeability and higher hydraulic gradients in the cones of depression round the wells, where the gradient can be about $6^0/_{00}$. It is assumed, therefore, that the effects of accumulation, and hence of filtration, of the muds are less effective here than in zones of thicker sediment. Filtration is restricted to a relatively small area near the shore of the lake. The kF values in thicker muds are insignificant for the volume of filtered water. In the case of pollution in the milligram zone, however, the apparent flow velocity is decisive.

Infiltration into the deeper mud sections is likely since there is qualitative and quantitative evidence of organic chemicals at a depth of 6 m in the centre of Lake Tegel. BALLSCHMITER et al. (1972) detected 1,2,3- and 1,2,4-trichlorobenzene, pentachlorbenzene, polychlorinated terphenyl (0.8 ug/kg, 4.4-DDE (0.12 ug/kg) and 4.4-DDT (0.02 ug/kg). Since the high nutrient content in various lakes leads to a reduction in the redox potential at the sediment surface, a mobilization of the heavy metals adsorbed on the sediments must be expected.

Lake basin morphology and the properties of subhydric sediments (limnites) are clearly important factors in maintaining groundwater quality.

References

BALLSCHMITER, K. & BUCHERT, H. (1985): Die Belastung limnischer Sedimente durch persistente Umweltchemikalien. In: Umwelt-Forschungsplan des Bundesministers des Innern, im Auftrag des Umweltbundesamtes, Forschungsbericht 10605027/1+2, Berlin, 240 pp.

BOULTON, G. S. (1972): Modern Arctic glaciers as depositional models for former ice sheets. Quart. J. Geol. Soc. London **128**, 361–393.

BRANDE, A. (1980): Pollenanalytische Untersuchungen im Spätglazial und frühen Postglazial Berlins. Verh. Bot. Ver. Prov. Brandenburg **115**, 21–72.

BUCHERT, H., BIHLER, S. & BALLSCHMITER, K. (1982): Untersuchungen zur globalen Grundbelastung mit Umweltchemikalien, VII. Hochauflösende Gas-Chromatographie persistenter Chlorkohlenwasserstoffe (CKW) und Polyaromaten (AKW) in limnischen Sedimenten unterschiedlicher Belastung. Fresenius Z. Anal. Chem., Springer, Berlin, Heidelberg and New York, 1–20.

CARSLAW, H.J. & JAEGER, J. C. (1959): Conduction of heat in solids. 2. Aufl., Oxford, 510 pp.

CHROBOK, S., MARKUSE, G. & NITZ, B. (1982): Abschmelz- und Sedimentationsprozesse im Rückland weichselhoch- bis spätglazialer Marginalzonen des Barnim und der Uckermark (mittlere DDR). Petermanns Geogr. Mitt. **126 (2)**, 95–102.

FIRBAS, F. (1953): Das absolute Alter der jüngsten vulkanischen Eruptionen im Bereich des Laacher Sees. Die Naturwissenschaften **40 (2)**, 54–55.

FLORIN, M.-B. & WRIGHT, H.E. (1969): Diatom evidence for the persistence of stagnant glacial ice in Minnesota. Publ. Inst. Quatern. Geol. Univ. Uppsala, Octave Ser. **33**, 695–703.

FRECHEN, J. (1959): Die Tuffe des Laacher Vulkangebietes als quartärgeologische Leitgesteine und Zeitmarken. Fortschritte d. Geologie v. Rheinland u. Westfalen, **4**, 363–370.

GEYH, M. A. (1966): Versuch einer chronologischen Gliederung des marinen Holozäns an der Nordseeküste mit Hilfe der statistischen Auswertung von 14C-Daten. Z. dtsch. Geol. Ges., Jg. 66, **118**, 351–360.

GEYH, M.A., MERKT, J. & MÜLLER, H. (1971): Sediment-, Pollen- und Isotopenanalysen an jahreszeitlich geschichteten Ablagerungen im zentralen Teil des Schleinsees. Arch. Hydrobiol. **69** **(3)**, 366–399.

HOPKINS, D. M. & KARLSTROM, T.N.V. (1965): Permafrost and ground water in Alaska. US Geol. Surv. Prof. Paper, 264-F, 2nd ed., 1982, 113–146.

KLOOS, R. (ed.) (1986): Das Grundwasser in Berlin. Bedeutung, Probleme, Sanierungskonzeptionen. In: Gewässerkundl. Jahresber. d. Landes Berlin, 5–165.

KOFELD, E.-G. (1982): Pleistozän-holozäne Seesedimente und marine Ablagerungen als Senken im Geosystem und ihre Rolle im Kreislauf der Umweltchemikalien. Wissenschaftl. Hausarbeit am Institut für Physische Geographie, FU Berlin, (unpublished), 189 pp.

KOSCHEL, R. BENNDORF, J. PROFT, G. & RECKNAGEL, F. (1983): Calcite precipitation as a natural control mechanism of eutrophication. Arch. Hydrobiol. **98, 3**, 380–408.

KÜNITZER, W. (1956): Uferfiltration — chemisch gesehen. Das Gas- u. Wasserfach **97**, 422–425.

MANGERUD, J., ANDERSEN, S. T., BERGLUND, B. E. & DONNER, J.J. (1974): Quaternary stratigraphy of Norden, a proposal for terminology and classification. Boreas **3**, 109–126.

PACHUR, H.-J. & SCHMIDT, J. (1985): Die Belastung limnischer Sedimente durch persistente Umweltchemikalien. In: Umweltbundesamtes, Forschungsbericht 10605027/1+2, 240 pp.

PACHUR, H.-J. & RÖPER, H.-P. (1987): Zur Paläolimnologie Berliner Seen. Berl. Geogr. Abh. H.44 (in print).

RÖPER, H.-P. (1985): Zur Hydrochemie des Tegeler Sees (West-Berlin). Geoökodynamik **6** **(3)**, 293–301.

SCHMIDT, J. (1987): Belastung limnischer Sedimente durch Schwermetalle. Versuch eines Seenkatasters unter besonderer Berücksichtigung einiger Berliner Grunewaldseen. Dissertation, Institut für Physische Geographie, FU Berlin, (unpublished), 303 pp.

WIECKOWSKI, K. (1969): Investigations on bottom deposits in lakes of NE-Poland. Mitt Intern. Ver. Limnol. **17**, 332–342.

Address of author:
Dr. H.-J. Pachur
Institut für Physische Geographie
Freie Universität Brlin
Altensteinstraße 19
1000 Berlin 33

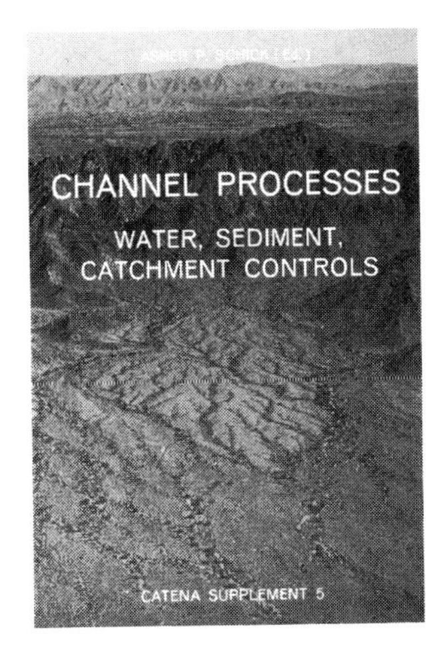

Asher P. Schick (Ed.):

CHANNEL PROCESSES
WATER, SEDIMENT,
CATCHMENT CONTROLS

CATENA SUPPLEMENT 5, 1984

Price DM 110,—

ISSN 0722–0723 / ISBN 3–923381–04–2

PREFACE

Two decades ago, the publication of 'Fluvial Processes in Geomorphology' brought to maturity a new field in the earth sciences. This field – deeply rooted in geography and geology and incorporating many aspects of hydrology, climatology, and pedology – is well served by the forum provided by CATENA. Much progress has been accomplished in fluvial geomorphology during those twenty years, but the highly complex and delicate relationships between channel processes and catchment controls still raise intriguing problems. Concepts dealing with thresholds and systems, and modern tools such as remote sensing and sophisticated tracing, have not decisively resolved the simple but elusive dual problem: how does the catchment shape the stream channel and valley to its form, and why? And: how does the channel transmit its influence upstream in order to make the catchment what it is?

Partial solutions, in a regional or thematic sense, are common and important. In addition to contributing a building block to the study of fluvial geomorphology, they also produce a number of new questions. The consequent proliferation of research topics characterises this collection of papers. The basic tool of geomorphological interpretation – the magnitude, frequency, and mechanism of sediment and water conveyance – is a prime focus of interest. Increasingly important in this context in recent years is the role of human interference natural fluviomorphic process systems. Effects of drainage ditching, transport of pesticides absorbed in fluvial sediment, and the flushing of nutrients are some of the Manconditioned aspects mentioned in this volume. Other contributions deal with the intricate balance, especially in extreme climatic zones, between physical process generalities and macroregional morphoclimatic influences.

The contributions of PICKUP and of PICKUP & WARNER represent two of the very few detailed quantitative geomorphological analyses of very humid tropical catchments. The 8 to 10 m mean annual rainfall in the equatorial mountain areas studied combines with effective landsliding to produce extremely high denudation rates. However, many aspects of channel behaviour are similar to those of temperate rivers. Particularly interesting are the relationships derived between channel characteristics, perimeter sediment and bedload transport.

Several small ephemeral and intermittent streams in Ohio studied by THARP, although variable in catchment area and in peak discharge, have a similar competence; while sorting increases downstream, the coarsest sizes tend to remain constant. Sorting of fluvial sediment, though on a much longer time scale and in an arid climate, also plays an important role in the contribution MAYER, GERSON & BULL. They find that modern channel sediment size ex bits the most rapid downstream decrease in mean particle size, while Pleistoce deposits show the least rapid decrease and are consistently finer than youn deposits. The difference is attributed to climatic change and a predictive mo thereto is presented.

HASSAN, SCHICK & LARONNE describe a new method for the magn tracing of large bedload particles capable of detecting tagged particles redeposi by floods up to several decimetres below the channel bed surface. Their met may considerably enhance the value of numerous experiments with pain pebbles, previously reported or currently in progress. Suspended sediment is subject of the paper by CARLING. He experiments with sampling gravel-bed flashy streams by two methods, and concludes that pump-sampling and 'buc sampling show significant differences only for very high discharges. Suspen sediment concentration is also dealt with by GURNELL & FENN, but in a pro cial environment – a climatic zone about which our knowledge is largely defici They find some correspondence between 'englacial' and 'subglacial' flow com nents and the total suspended sediment concentration.

The effects of human interference by ditching in a forest catchment on se ment concentration and sediment yield is discussed by BURT, DONOHO VANN. A local reservoir afforded an opportunity to monitor in detail the in ence of these drainage operations on the sediment concentration which inc sed dramatically and, after several months, gradually recovered due to reveg tion. TERNAN & MURGATROYD analyse sediment concentrations and sp fic conductance in a humid, forest and marsh environment. Permanent vegeta dams are found to influence sediment concentration directly through filtra and indirectly through changes in water depth and velocity . Changes in sp fic conductance are influenced by marsh inputs as well as by the addition of a of coniferous forest. The relationship between quality of water and fluvial s ment characteristics is dealt with by HERRMANN, THOMAS & HÜBNER, analyse the regional pattern of estuarine transport processes. They conclude high pesticide concentrations are correlated with high concentrations of sus ded sediment. Hydrodynamic rather than physicochemical factors influence regional distribution in the estuary, and the effect of brooklets draining intensi cultivated land is quite evident.

Asher P. Schick

CONTENTS

CATENA SUPPLEMENT 15 p. 121–131 Cremlingen 1989

SOIL EROSION DURING THE PAST MILLENNIUM IN CENTRAL EUROPE AND ITS SIGNIFICANCE WITHIN THE GEOMORPHODYNAMICS OF THE HOLOCENE

Hans-Rudolf **Bork**, Braunschweig

1 Introduction

Knowledge of the history of soil erosion during the past thousand years and the quantification of soil erosion processes are necessary tools for an evaluation of the recent soil erosion processes as well as for making long term estimates.

Soil erosion in central Europe has been studied in detail by VOGT (1953, 1958 a,b, 1960) HEMPEL (1954, 1957), SEMMEL (1961), SCHULTZE (1965), RICHTER & SPERLING (1967), HARD (1976) and MACHANN & SEMMEL (1970). Their research revealed that many of the gullies found in the landscape of the present day were formed in historical times. Using contemporary documents and maps, HARD (1976) estimated that gullies in southwest Germany were formed between 1750 and 1850. The causes of gullying were also analysed. VOGT (1953, 1958 a, b, 1960), HARD (1976) stressed the influence of land use on gullying and BORK & BORK (1987) discussed the importance of heavy rainfall and climatic fluctuations.

ISSN 0722-0723
ISBN 3-923381-18-2

3-923381-18-4/89/5011851/US$ 2.00 + 0.25

These studies deal with the distribution and the formation of gullies. The intensity of sheet erosion during the Holocene is, however, comparatively unknown. It is also uncertain whether or not soil erosion processes in the late Holocene can be compared to those of the middle and the early Holocene. Did soil erosion begin with the clearing of woodland or did erosion occur already in the natural woodland ecosystems of the early Holocene? To answer these questions the extent, the causes and effects of sheet and gully erosion that occurred during medieval and modern times were investigated. Observations were made first in southern Lower Saxony and continued later in other parts of central Europe (fig.1).

2 Methods of investigation

The truncation of soil profiles was determined at more than 800 sites in southern Lower Saxony. From these data the total extent of soil erosion in historical times was estimated. The methods and the results were published by BORK (1983). It was shown that the surfaces of slopes used for agriculture were lowered by more than two meters as a result of soil erosion.

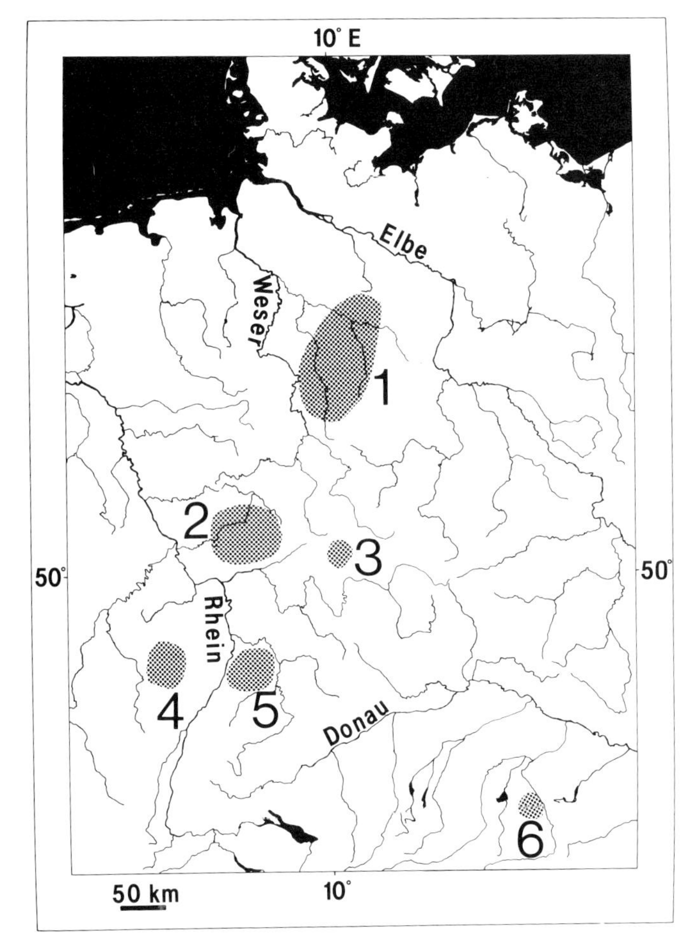

Fig. 1: *Location of the study areas in the Federal Republic of Germany.*
1. Southern and eastern Lower Saxony
2. Central Hessen
3. Western Franconia
4. Western Palatinate
5. Northwestern Baden-Württemberg
6. Southern Bavaria

In order to reconstruct the processes of soil erosion in historical times more precisely, a large number of soil profiles were examined in detail in the research areas shown in fig.1. Investigated were, for example, the distribution and the properties of the sediments and of the soil horizons. Cooperation with archaeologists allowed sediments to be dated with reasonable accuracy.

3 The processes of soil erosion in medieval and modern times

The detailed survey of the soil profiles resulted in the reconstruction of two phases of gully formation. The more recent phase dates from the second half of the 18th century and has been described in a number of publications (for example, VOGT 1960, HARD 1976). The earlier phase could be dated, with the aid of pottery finds, to the first half of the 14th century. The excavations near the deserted village of Drudevenshusen in

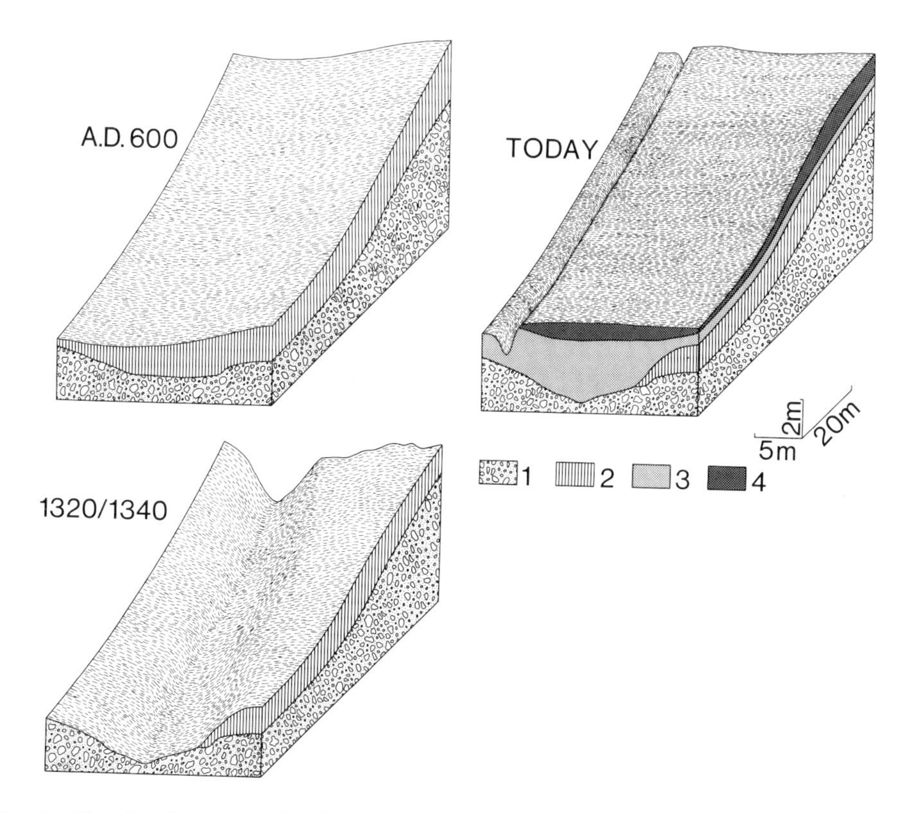

Fig. 2: *The development of sediments and soil surfaces during the younger Holocene near the deserted village of Drudevenshusen (southern Lower Saxony), viewed from the north.*

1. Parent material
2. Holocene soil (typical parabrown earth/orthic luvisol on the slope and chernozem-parabrown earth/mollic luvisol at the valley bottom), developed in loess from early Holocene to the 5th century A.D. under forest.
3. Sediments deposited during the 13th, 14th and 15th centuries A.D.
4. Sediments deposited since the 16th century A.D.

southern Lower Saxony (BORK 1985) showed, for example, that the youngest sediments deposited before the disastrous soil erosion date from the late 13th century, whereas the material at the base of the gully filling dates from the second to the fourth decade of the 14th century.

An extreme rate of linear and sheet erosion was found to have occurred in this earlier phase of gully formation. Along slopes and in valleys deep gullies were cut into arable land. They were often more than 5 meters in depth and sometimes more than 10 meters deep. The forms and the changes in the small valley floors during the recent Holocene are shown in figs.2 and 3. The majority of the streams cut several meters deep into the soil and rock forming narrow gullies. These forms are found unchanged

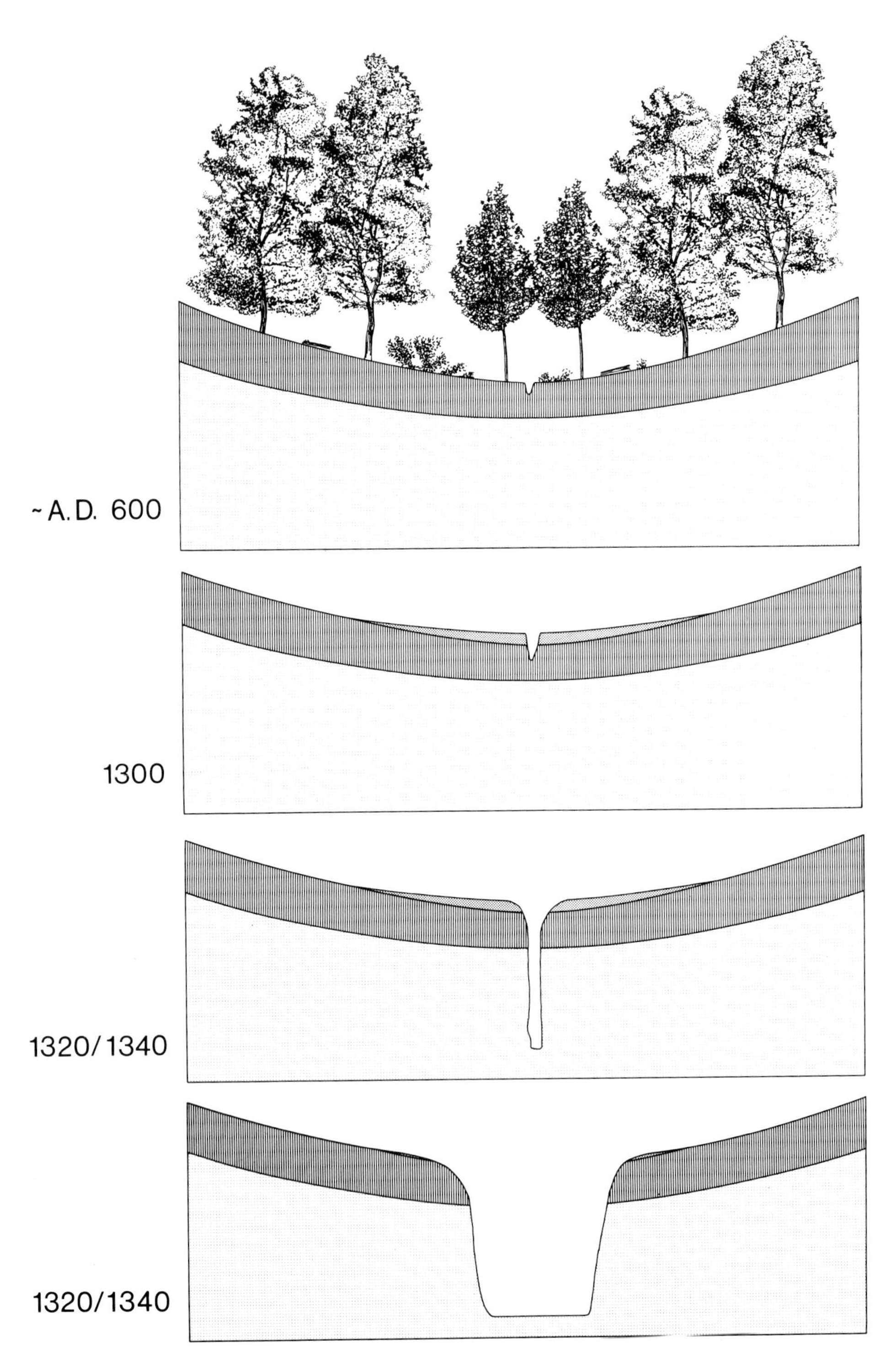

Fig. 3: *The development of small valleys and foot slopes during the last 1,500 years in Central Europe.*

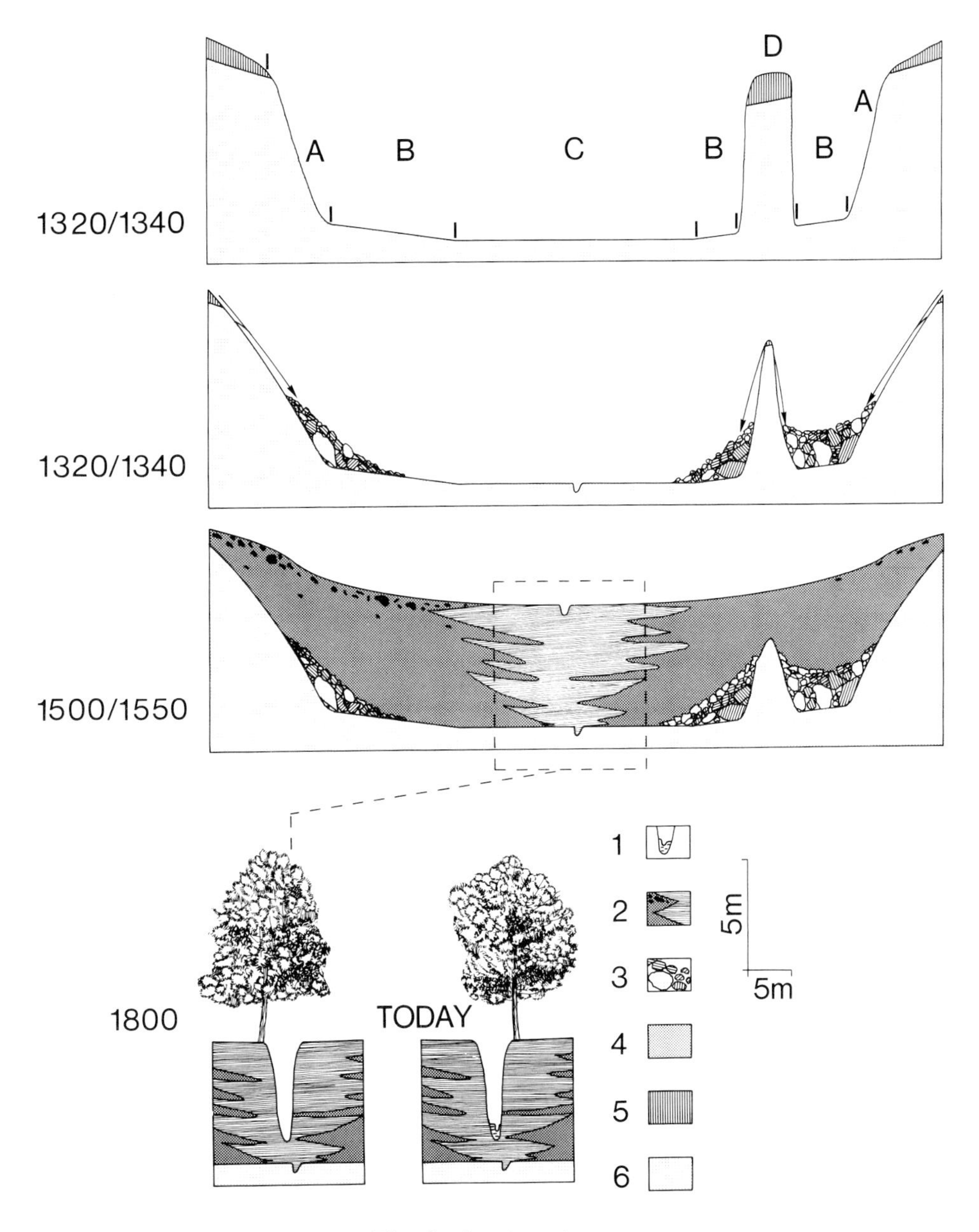

Fig. 3: *Continuation.*

1. Heterogenous material transported and deposited in the 19th and 20th centuries by fluvial processes in the river bed (valley colluvium).
2. Reworked soil material, transported by overland flow and deposited on the foot slope as an unbedded slope colluvium and a bedded valley colluvium, that has been transported in the river bed (alternating silty and loamy beds). Lower part dated to the 14th and 15th centuries, upper part to the 16th and 17th centuries.
3. Slide masses (loess and solum material), deposited by land slides during the first half of the 14th century.
4. Reworked solum material, transported after clearing from forest during early Medieval times until the 13th century by overland flow and deposited on the foot slope as an unbedded slope colluvium.
5. Holocene soil (in loess areas: typical parabrown earth / orthic luvisol and chernozem-parabrown earth / mollic luvisol), developed from early Holocene to 5th century A.D. under forest.
6. Parent material (e.g. calcareous loess).

A. Micro-scarp of the pediment; B. Slope pediment (scarp retreated by overland flow); C. Valley pediment (scarp retreated by lateral planation by the river); D. Outlier.

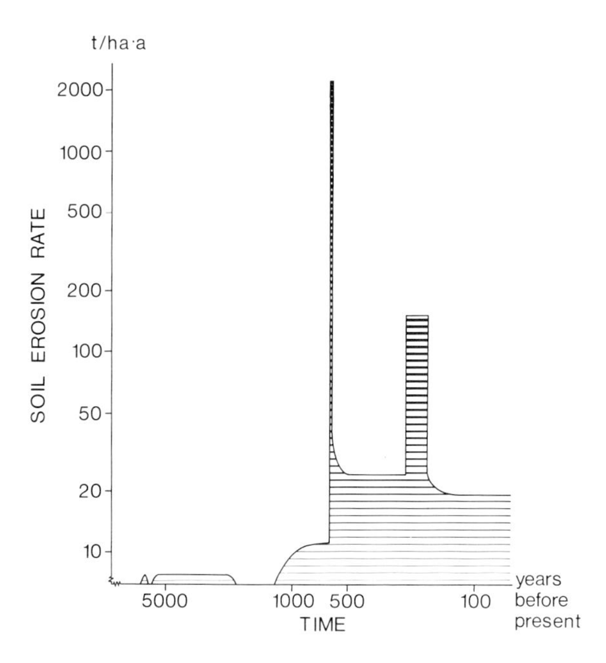

Fig. 4: *The variation of the mean soil erosion rate on agriculturally used slopes in southern Lower Saxony during the Holocene.*

beneath the younger sediments in the Drudevenshusen area (fig.2). Wherever highly erodible sediments outcropped in the stream beds, the erosion processes continued. After the erosion baselevel had been reached, continued excessive runoff led to the widening of the narrow gully floors and to the development of valley pediments, as defined by ROHDENBURG (1969, 1977). Where there were highly erodible sediments with low hydraulic conductivities near the surface of the slopes, large volumes of overland flow eroded the slopes leading to the retreat of the gullies or of the valley pediment scarps. In some cases, slope pediments with a width of 50 meters and a length of several kilometers developed (fig.3). Where the scarps were composed of sediments with low aggregate stability, the steepest parts of the slopes slipped immediately after their formation and fell into the gullies and on to the pediments (fig.3). The length of time required for slide masses to fill the gullies and to cover the pediments can be used as an indicator of the processes and their rates.

High rates of overland flow on arable land created not only gullies and pediments, they also caused extensive sheet erosion so that slopes used for agriculture that were covered with highly erodible sediments were lowered by amounts that ranged from a few centimeters to several tens of centimeters.

In the area investigated in southern Lower Saxony, a ten year phase of gully and pediment formation was assumed to have taken place in the early 14th century. During this period an average rate of soil erosion of 2,250 tons per hectare per year was calculated (fig.4). The rates of sheet erosion were probably much higher in the first year. It seems possible that most of the erosion during this period was due to one single disastrous rainfall event.

The more recent phase of gully formation on slopes and in valleys took place in the second half of the 18th century and lasted about 50 years. Gullies with depths of several meters were formed during this phase (fig.3). The volumes of material eroded by gullying amounted to only 10 to 30 percent of those obtained for the early 14th century. In the research area in southern Lower Saxony, an average of about 160 tons of soil per hectare per year was eroded during the second phase of gullying (fig.4). For both dissection phases, the highest rate of soil erosion was determined for steep slopes with highly erodible rocks (various Mesozoic marls and clays) or soils. Considerable erosion also occurred on less steep slopes with silty and sandy rocks (loess, Tertiary siltstone and sandstone, glacio-fluvial sands) or soils. Even graywacke, basalt and limestone were deeply dissected.

Apart from these two dissection phases, sheet erosion has been moderate and gullying absent in central Europe during the past thousand years. The average annual rate of soil erosion in southern Lower Saxony was about 10 tons per hectare until about 1400, about 25 tons per hectare from the mid-14th century to the mid-18th century and 20 tons per hectare in the 19th and 20th centuries (fig.4).

The gullies and pediments developed early in the 14th century were filled almost completely during the centuries that followed with soil material removed from the slopes by minor sheet erosion (figs 2 and 3). Only a few gullies formed in the early 14th century were afforested and consequently not filled.

The investigation of soil profiles showed that wherever fertile sediments or soils had been thin, they were completely eroded by intensive sheet erosion in the first half of the 14th century. Many such sites could no longer be used for agriculture and have remained wooded until the present day. The gullies that can be found on present-day large-scale maps were developed mainly during the second half of the 18th century. Many of them were also afforested at a later period and remained so until the 20th century (fig.3).

4 Causes of soil erosion during the past millennium

The nature and intensity of soil erosion processes caused by water depend on precipitation, relief, soil and rock type and also the land use, for example, field size, crop rotation, and tillage. Variations in relief, rock type and soil can be eliminated as potential causes of short term intensive sheet and gully erosion and long term minor sheet erosion because they are largely constant. However, in both medieval and modern periods, relief, rock and soil composition changed as a result of soil erosion (BORK 1985). Intensive rainfall and changes in land use, either separately or together, can influence erosion processes decisively at a specific site. Contemporary documents refer to great variations in rainfall in the course of the past 1,000 years (FLOHN 1949/1950, 1958, 1967, WEIKINN 1958, PFISTER 1985 a, b, ALEXANDRE 1987). Land use also underwent substantial changes within this period (HARD 1976).

As indicated above, "normal" soil erosion, that is minor sheet erosion, has been observed for the period from 1350 to 1750 and since about 1800. Extreme rainfall events did not occur during these periods and erosion was, and is at the present time, primarily a consequence of land use and tillage. For example, soil erosion developed mainly on fields that lay completely fallow. In fields with summer and winter crops, soil erosion rarely occurred or was of minor importance.

These "normal" erosion processes were interrupted by periods in the early 14th century and in the second half of the 18th century during which rates of sheet and gully erosion were very high. The reconstructed soil surfaces and the deposits in the gullies indicate the duration of the period of gully filling and, also, therefore, the duration of the processes and the causes of the dissection that took place in the late medieval period. The steep soil scarps of the gullies and of the pediments were often on wet sites and composed of sediments with low aggregate stability. Under these conditions steep slopes can be expected to have collapsed after a short time. Investigations of similar recently-formed steep soil scarps confirm this hypothesis. The slide masses found at the bottom of the steep soil scarps can be assumed to have slid down very soon after formation of the gullies. Consequently, the dissection processes must also have occurred within a short period, possibly during the course of one or of a few catastrophic rainfall events. The absence of any sign of pedogenesis on the former surfaces of the gullies indicates that dissection and filling took place at a rapid rate and confirms the theory that extremely heavy rainfall events were the cause of the gully and pediment formation.

During the first half of the 14th century very exceptional weather conditions prevailed, making the occurrence of such catastrophic rainfall events probable. For example, in the second decade of the 14th century, extremely wet summers led to serious crop shortfalls and famines (LAMB 1977). In addition, the most devastating flood in central Europe of the past 1500 years was recorded during the summer of the year 1342 (FLOHN 1967, BORK & BORK 1987).

Changes in the land use cannot have been a cause of gully formation since no innovations were made which might have increased the erosion. Contemporary documents indicate that in the second half of the 18th century erosion damage was minor in fields with winter crops, more severe in fields with summer crops and very severe on fallow land with no vegetation. In the early 14th century severe erosion damage was also recorded in fields with winter crops and summer crops. No soil erosion processes were reported for forests or meadows.

Recent observations confirm that in forests surface runoff can develop locally only on paths or on sites that have an impermeable, or almost impermeable, layer near the surface. The runoff infiltrates into more permeable sediments after flowing only a short distance. Surface runoff that develops on fields may flow downslope into wooded areas and cause local soil erosion. Clearly an increased density of vegetation results in a decreased rate of soil erosion. On bare soils, however, soil erosion may be very intensive. Tillage changes the microrelief and has, thereby, some influence on the position of gullies.

In order to verify these observations, a large number of measurements were made of soil erosion processes, including data from rainulator experiments in the field and in the laboratory as well as from simulations based on various soil erosion models (BORK 1988). These investigations lead to the conclusion that climatic changes must have been the sole cause of the high rates of soil erosion that apparently occurred during the early 14th century. Changes in land use, such as the clearing of forests and the subsequent use of a three year crop rotation system that included one year of bare soil, may have increased the mean annual soil erosion rate to a maximum of about 50 tons per hectare on moderately inclined slopes with loess soils. But, soil erosion rates of several thousand tons per hectare per year could only be caused by extraordinarily heavy rainfall events and only on soils with little or no vegetation. The intensity of such excessive rainfall events must have been a multiple of the one thousand year rainfall event calculated by various methods. The gullying processes in the 14th and 18th centuries could, therefore, only have been caused by temporary climatic changes.

5 A comparison of soil erosion during the early, middle and late Holocene

Data are available to describe and quantify the processes and degree of soil erosion in southern Lower Saxony for the period prior to the early Middle Ages (BORK 1983). Colluvium, that is soil material which has been eroded, transported and deposited by overland flow, provides evidence of soil erosion processes. Colluvium forms only in periods of intensive land use. In southern Lower Saxony, colluvium has been dated from the early Neolithic age, the period from the early Bronze age to the Roman age and the period since the early Middle Ages. The areas covered with colluvium until the Roman period are usually very small since the amount of arable land was limited. It was concluded from the thickness of the colluvial sediments and from their stratigraphy that only minor sheet erosion occurred before the Middle Ages and that it was restricted to areas used for agriculture. Mass balances indicate that the average soil erosion rates were less than 4 tons per hectare per year (fig.4) For this period prior to the early Middle Ages, there are no indications of rill and gully erosion. It is most probable, therefore, that no excessive rainfall events occurred during the early and middle Holocene. Further investigations are necessary to support this hypothesis.

No colluvial sediments were found dating from the early Holocene or from the period between the 3rd and 6th centuries A.D.. Consequently, there is unlikely to have been any soil erosion during these periods. Pollen analyses confirm that in the early Holocene and between the 3rd and 6th centuries A.D. the surroundings of lakes and areas of peat in southern Lower Saxony were completely covered with woodland (STEINBERG 1944, CHEN 1982). STEINBERG (1944) and CHEN (1982) were also able to show that during the periods in question practically no mineral components were deposited in the lakes of southern Lower Saxony, evidence that soil erosion did not occur in natural woodland. Most probably, central Europe was completely covered in woodland in the early Holocene (FIRBAS 1949) so that soil erosion would have been unlikely during this period. Soil erosion occurred only after the clearing of woodland in the Neolithic, Bronze, Iron and Roman periods and only as long as these areas were used for agriculture.

It is very likely that no excessive rainfall events occurred during these periods of early agricultural activity. Since the cleared areas covered only a few hectares, not much soil was eroded. Only after large areas had been cleared and used for agriculture between the 7th and 14th centuries did the rates of soil erosion increase considerably. However, no evidence of significant gully erosion was found for this period. Gullying occurred exclusively in the 14th and 18th centuries and in almost all parts of central Europe with the exception of the very flat areas of the upper Rhine lowland and the northwest German plains.

References

ALEXANDRE. P. (1987): Le Climat en Europe au Moyen Age. Contribution à l'histoire des variations climatique de 1000 à 1425, d'après les sources narratives de l'Europe occidentale. Recherches d'histoire et de science sociales **24**, Paris. (Ecole des Hautes Etudes en Sciences Sociales), 828 pp.

BORK, H.-R:; (1983): Die holozäne Relief- und Bodenentwicklung in Lößgebieten - Beispiele aus dem südöstlichen Niedersachsen. In: H.-R. BORK & W. RICKEN, Bodenerosion, holozäne und pleistozäne Bodenentwicklung, CATENA SUPPLEMENT **3**, Braunschweig, 93 pp.

BORK, H.-R. (1985): Untersuchungen zur nacheiszeitlichen Relief- und Bodenentwicklung im Bereich der Wüstung Drudevenshusen bei Landolfshausen. Ldkr. Göttingen. Nachr. Nieders. Urgeschichte **54**, Hildesheim, 59–75.

BORK, H.-R: (1988): Bodenerosion und Umwelt. Verlauf und Folgen der mittelalterlichen und neuzeitlichen Bodenerosion, Bodenerosionsprozesse, Modelle und Simulationen. Landschaftsgenese und Landschaftsökologie **13**, Braunschweig, 251 pp.

BORK H.-R. & BORK, H. (1987): Extreme jungholozäne hygrische Klimaschwankungen in Mitteleuropa und ihre Folgen. Eiszeitalter und Gegenwart **37**, 109–118.

CHEN, S. (1982): Neue Untersuchungen über die Spät- und postglaziale Vegetationsgeschichte im Gebiet zwischen Harz und Leine. Dissertation, Universität Göttingen. 102 pp.

FIRBAS, F. (1949): Spät-und nacheiszeitliche Waldgeschichte Mitteleuropas nördlich der Alpen. 1. Allgemeine Waldgeschichte, Jena, 480 pp.

FLOHN, H. (1949/50): Klimaschwankungen im Mittelalter und ihre historisch-geographische Bedeutung. Ber. z. dtsch. Landeskunde **7**, Stuttgart, 347–358.

FLOHN, H. (1958): Klimaschwankungen der letzten 1000 Jahre und ihre geophysikalischen Ursachen. Deutscher Geographentag Würzburg, Tagungsber. und wiss. Abh., Wiesbaden, 201–214.

FLOHN, H. (1967): Klimaschwankungen in historischer Zeit. In: H.v. Rudloff, Die Schwankungen und Pendelungen des Klimas seit Beginn der regelmäßigen Instrumenten-Beobachtung. Braunschweig, 81–90.

HARD, G: (1976): Exzessive Bodenerosion um und nach 1800. In: G. RICHTER. Bodenerosion in Mitteleuropa, Wege der Forschung **430**, Darmstadt, 195–239.

HEMPEL, L. (1954): Tilken und Sieke - ein Vergleich. Erdkunde **8**, 198–202.

HEMPEL, L. (1957): Das morphologische Landschaftsbild des Unter-Eichsfeldes unter besonderer Berücksichtigung der Bodenerosion und ihrer Kleinformen. Forsch. z. dtsch. Landeskunde **98**, Remagen, 55 pp.

LAMB, H. H. (1977): Climate - present, past and future. Climatic history and the future. London, 835 pp.

MACHANN, R. & SEMMEL, A. (1970): Historische Bodenerosion auf Wüstungsfluren deutscher Mittelgebirge. Geogr. Zeitschr. **58**, Wiesbaden, 250–266.

PFISTER, C. (1985a): Bevölkerung, Klima und Agrarmodernisierung 1525–1860. Das Klima der Schweiz von 1525-1860 und seine Bedeutung in der Geschichte von Bevölkerung und Landwirtschaft. Academica helvetica **6/I**, Bern, Stuttgart, 184 pp.

PFISTER, C. (1985b): Bevölkerung, Klima und Agrarmodernisierung 1525–1860. Das Klima der Schweiz von 1525-1860 und seine Bedeuting in der Geschichte von Bevölkerung und Landwirtschaft. Academica helvetica **6/II**, Bern, Stuttgart, 163 pp.

RICHTER, G. & SPERLING, W. (1967): Untersuchungen im nördlichen Odenwald. Mz. Naturw. Arch. **5/6**, Mainz, 136–176.

ROHDENBURG, H. (1969): Hangpedimentation und Klimawechsal als wichtigste Faktoren der Flächen- und Stufenbildung in den wechselfeuchten Tropen an Beispielen aus Westafrika, besonders aus dem Schichtstufenland Südost-Nigerias. Giessener Geogr. Schr. **20**, 57–152.

ROHDENBURG, H. (1977): Beispiele für holozäne Flächenbildung in Nord- und Westafrika. CATENA **4**, 65–109.

SCHULTZE, J.H. (1965): Bodenerosion im 18. und 19. Jahrhundert. Grundsätzliche Möglichkeiten für die Feststellung der Rolle der Bodenerosion im 18. und 19. Jahrhundert. Forschungs- und Sitzungsber. d. Akademie f. Raumforschung u. Landesplanung **XXX**, Historische Raumordnung 5/1, Hannover, 1–16.

SEMMEL, A: (1961): Beobachtungen zur Genese von Dellen und Kerbtälchen im Löß. Rhein-Mainische Forschungen **50**, Frankfurt/M, 135–140.

STEINBERG, K. (1944): Zur spät- und nacheiszeitlichen Vegetationsgeschichte des Untereichsfeldes. Hercynia **3**, Leipzig, 529–587.

VOGT, J. (1953): Erosion des sols et techniques de culture en climat tempéré maritime de transition (France et Allemagne). Revue de Géomorphologie Dynamique **4**, 157–183.

VOGT, J.(1958a): Zur Bodenerosion in Lippe. Ein historischer Beitrag zur Erforschung der Bodenerosion. Erdkunde **12/2**, 132–134.

VOGT, J. (1958b): Zur historischen Bodenerosion in Mitteldeutschland. Peterm. Mitt. **102**, 199–203.

VOGT, J. (1960): Hardt et nord des Vosges au XVIIIe siècle. Le déclin d'une moyenne montagne. Bull. de la Section de Géographie, Comité des Travaux Histoiques et Scientifique, Actes du Congrés National des Sociétés Savantes, Dijon, 181–207.

WEIKINN, C. (1958): Quellentexte zur Witterungsgeschichte Mitteleuropas von der Zeitwende bis zum Jahre 1850. I/I, Hydrographie, Zeitwende bis 1500, Berlin, 531 pp.

Address of author:
Prof. Dr. Hans-Rudolf Bork
Abt. Physische Geographie und
Landschaftsökologie
Technische Universität Braunschweig
Langer Kamp 19c
D-3300 Braunschweig
Federal Republic of Germany

H.-R. BORK u. W. RICKEN

BODENEROSION, HOLOZAENE UND PLEISTOZAENE BODENENTWICKLUNG

SOIL EROSION, HOLOCENE AND PLEISTOCENE SOIL DEVELOPMENT

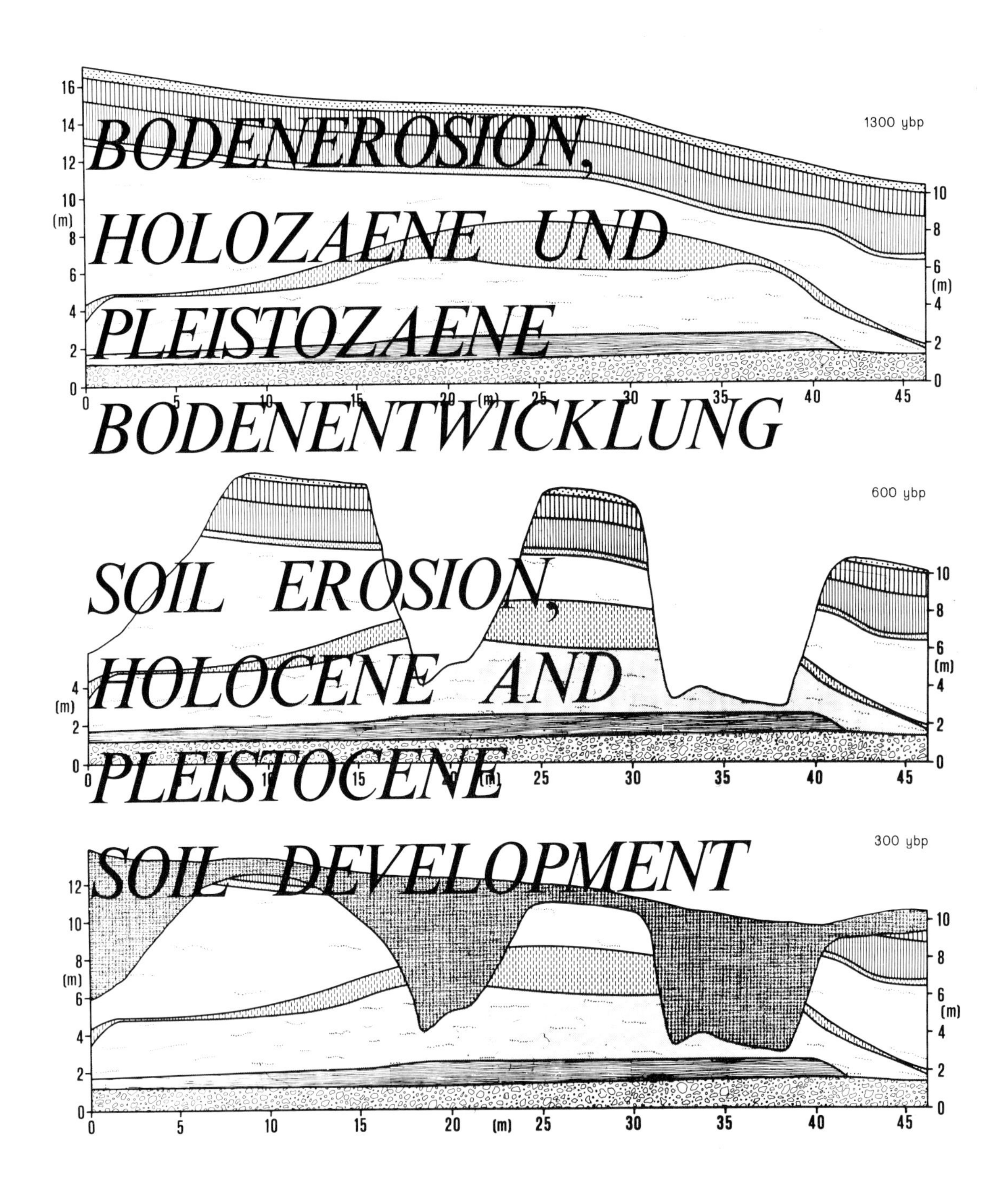

CATENA SUPPLEMENT 3

CATENA SUPPLEMENT 15 p. 133–149 Cremlingen 1989

HOMOCLINAL RIDGES IN LOWER SAXONY

Jürgen **Spönemann**, Göttingen

1 Introduction

Structural landforms made up of Mesozoic rocks occupy most of the area between the North German Lowland and the Central Uplands. In the south of the area the dip angles of the Mesozoic rocks are low and only cuestas occur. In the north, the strata dip steeply and homoclinal ridges predominate. The transition from cuesta to homoclinal ridges is gradual but both forms have characteristic features. Because the resistant rocks dip steeply and outcrops are narrow, the homoclinal ridges are elongated and contact springs are absent. The cuestas, by contrast, are embayed and contact springs occur. The retreat of the cuesta scarp can result in the development of residual outliers. A critical angle of 10° to 12° is used to distinguish between homoclinal ridges and cuestas (fig.3: 1–3).

The Weser-Leine hills north of the Solling between the Harz and the Lippe hills, is a region of homoclinal ridges (fig.1). The alternation of anticlines and synclines results in the juxtaposition of anticlinal and synclinal ridges separated by lowland strips of varying width. In addition, block faulting has complicated the pattern.

The main research problems relating to homoclinal ridges are the cause of valley deepening and widening and the significance of tectonic and climatic conditions for downwearing of the area between the ridges. First, however, the relation between the structure and the form of homoclinal ridges is discussed and, secondly, the structural and non-structural characteristics are differentiated so that the development of the homoclinal ridges can be examined.

2 Form

The form and distribution of homoclinal ridges is controlled mainly by rock structure. Structure includes the dip and the thickness of the strata and rock type. During the development of a homoclinal ridge, the most resistant strata are increasingly exposed. The initial form is a truncation surface, perhaps part of a peneplain, which becomes a swell, then a round-crested ridge and develops finally into a hogback, if the strata are highly resistant. In Lower Saxony all these stages of development are present and the related profiles can be described in terms of their structural geomorphology.

2.1 Structure-controlled features

The shape of homoclinal ridges and cuestas is a function of the dip, thickness and relative resistance of the strata. Schematic profiles of the major form types are shown in fig.2. Tectonic and

ISSN 0722-0723
ISBN 3-923381-18-2

3-923381-18-4/89/5011851/US$ 2.00 + 0.25

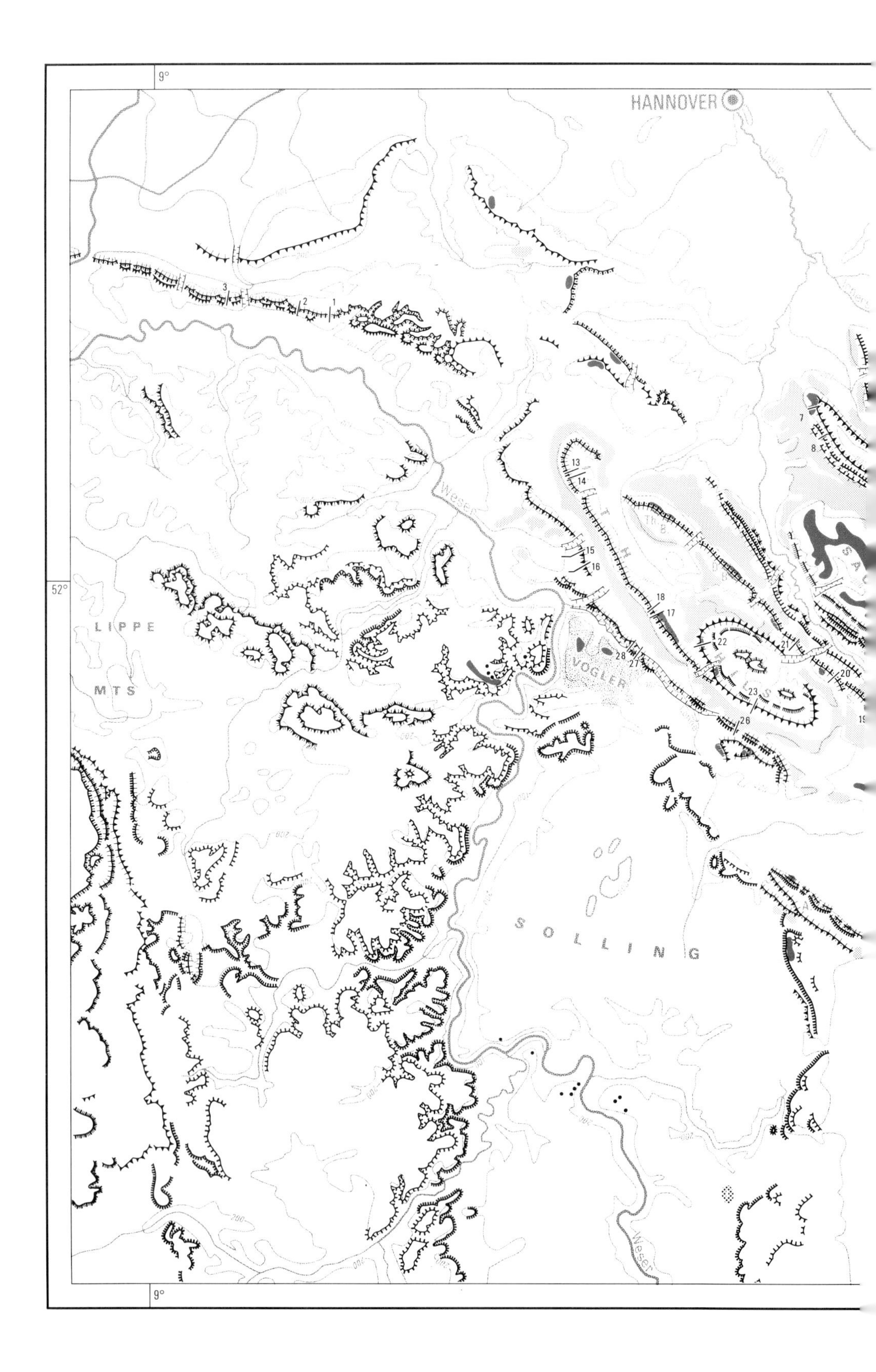

9°
HANNOVER
52°
LIPPE
MTS
Weser
VOGLER
SOLLING
9°

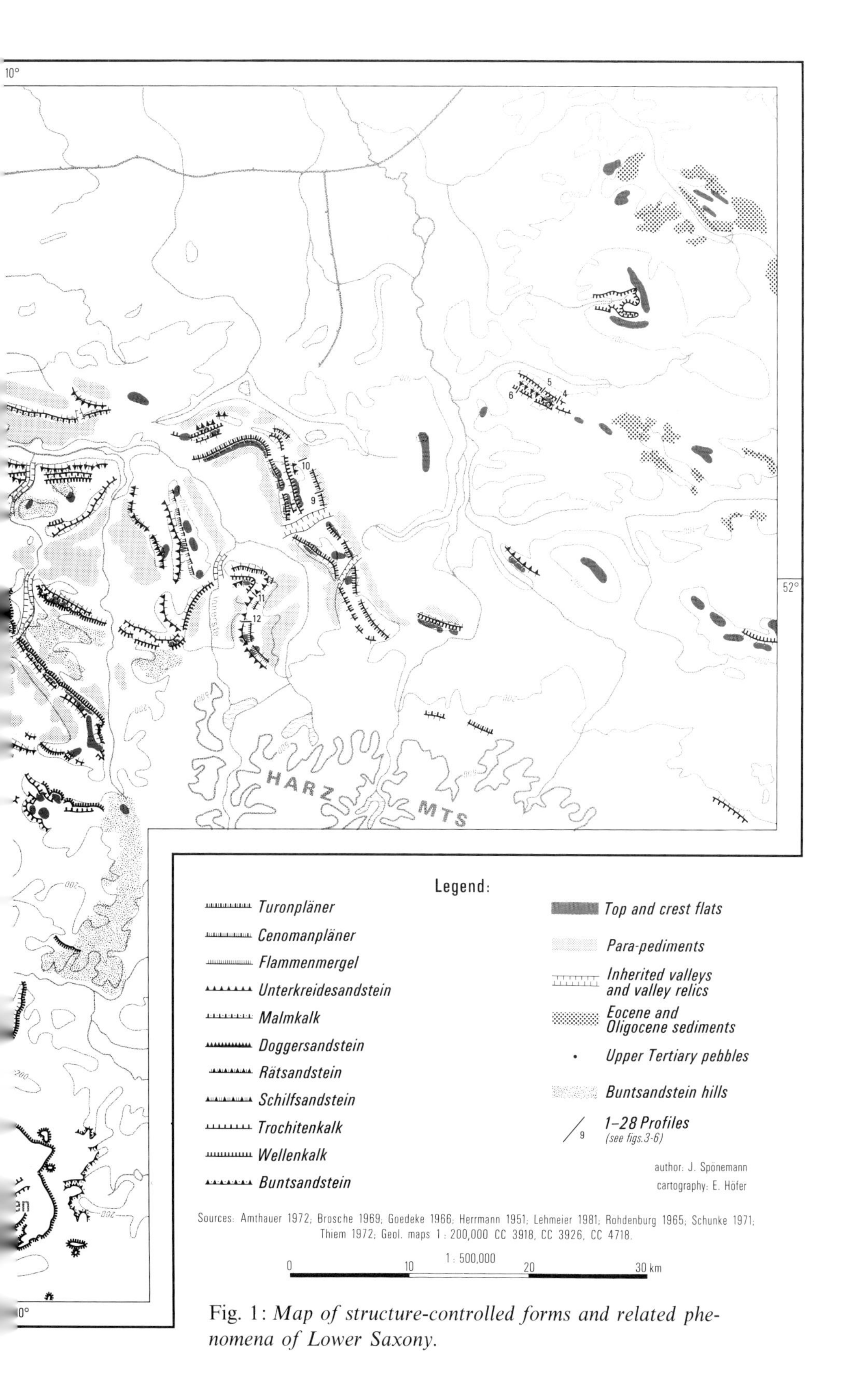

Fig. 1: *Map of structure-controlled forms and related phenomena of Lower Saxony.*

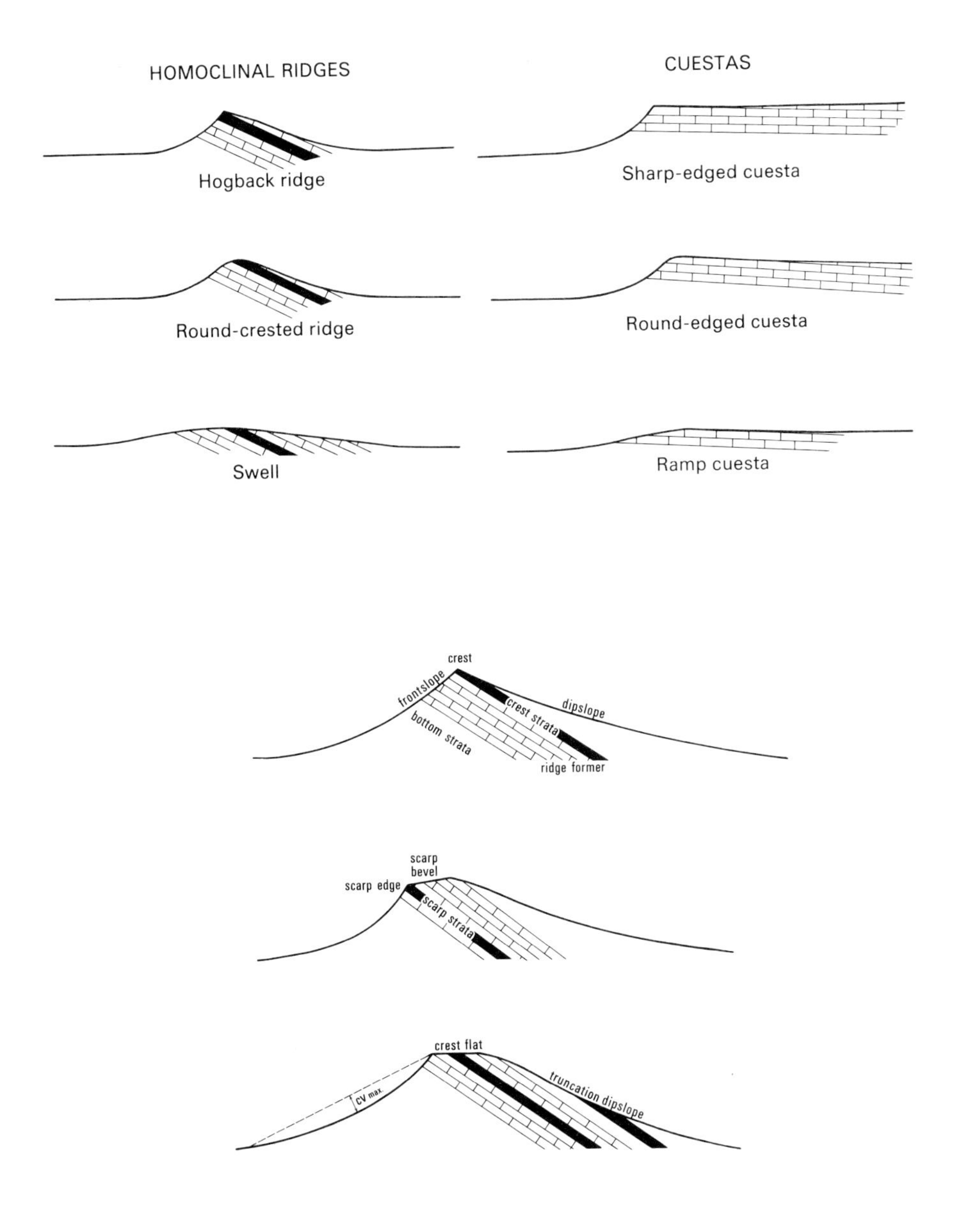

Fig. 2: *Structure-controlled forms and their nomenclature.*

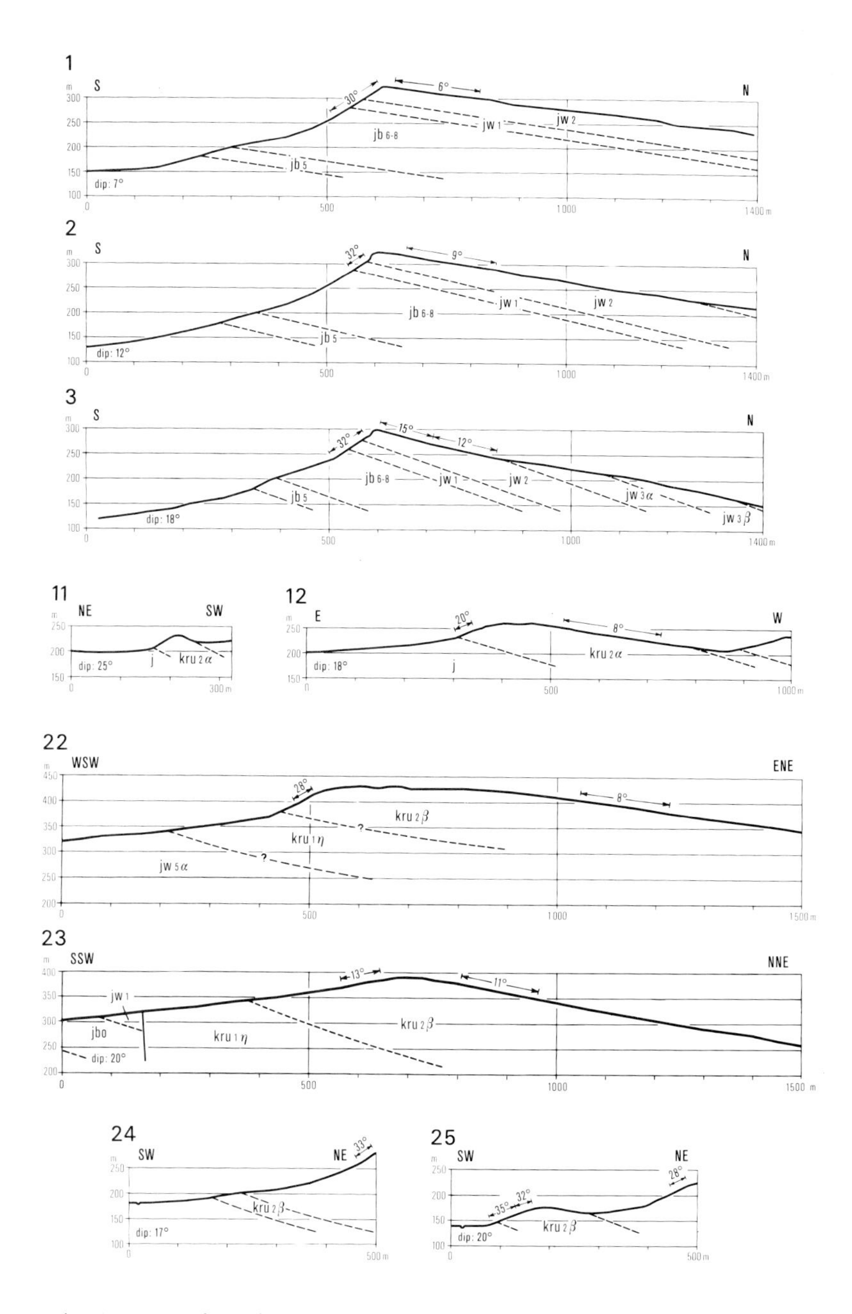

Fig. 3: *Examples of structure-controlled features of homoclinal ridges.*

climatic factors also influence the form, their relative importance has, however, not been determined in detail.

2.1.1 The relationship between the dip of the strata and the cross-profile

Generally structural forms are asymmetric with steep scarp slopes and flat dipslopes at a low angle. On cuestas, the contrast between the slopes decreases with increasing dip of the strata. In the case of homoclinal ridges the contrast is still present unless a vertical or near vertical dip occurs. The front slope (scarp slope) with its sequence of weak beds at the base and resistant scarp-forming beds at the top, is formed by undercutting. The size of the angle of dip has no effect on the shape of the front slope. The denudation on the dipslope is determined by the bedding plane. The greater the dip, the steeper is the dip slope. Normally the dip slope angle is slightly lower than the angle of dip (fig.3: 1–3).

2.1.2 Relation between rock resistance and the cross-profile

The maximum steepness of front slopes depends on the resistance of the scarp rock and their concavity on the difference in the resistance of the basal beds and the scarp-forming beds. A morphometric index, CVmax, describes the concavity (fig.2; tab.1; SUCHEL 1954 and SCHUNKE 1968: imax). The most resistant layers are composed of platy to massive limestones of the upper and lower Muschelkalk (mu, mo1, mo2), massive oolitic limestones and dolomites of the Malm (jw2) and thick-layered or massive sandstones, particularly those with siliceous cement. Cretaceous (Hils sandstone, kru2) and Keuper (Rät sandstone, ko) quartzitic sandstones are highly resistant to weathering. The nature of the cement can influence strongly the resistance of the scarp rock. For example, where the Hils sandstone has a siliceous cement, the maximum slope is about 30° but where argillaceous cement is present, it is about 15° (fig.3: 22, 23). The influence of the scarp rock on the concavity of the front slopes is also demonstrated by this example. If the resistance of the scarp-forming rock is low, or the lithological difference between scarp and basal strata is small, the slope profile is nearly straight. In this case, the cross-profile of a homoclinal ridge can be quasi-symmetrical (fig.3: 23). The concavity of the front slope of the Malm ridges of the Ith-Hils syncline demonstrates the predominance of the effect of rock resistance compared to the effect of the dip of the strata (tab.1).

The shape of the crest is also determined, in general, by rock resistance and can be expressed as a rate of curvature that is measured by its radius (photo 1). The Malm ridges usually have sharp crests. The ridges of the Rät and Hils Sandstones are more rounded (tab.2). If a thick sequence of strata is resistant as a whole but lacks a highly resistant layer, broad round-crested ridges (fig.3: 23) or hills of irregular shape with no distinct crest result. Many of the Bunter Sandstone hills have these types of crests.

2.1.3 The relationship between the thickness of the scarp rock and the cross-profile

The thickness of the resistant strata determines the height of the ridge and their width at the base (fig.3: 11, 12). The thickness can also determine whether an independent homoclinal ridge or a struc-

No.	ridge former	area	n	cv_{max} (m)	dip angle (°)
1	Hilssandstein (Cretaceous, kru)	Hils, SW section	8	21	20
2		Hils, N section	8	43	?
3	Malmkalk (Jurassic, jw2)	Thüster Berg (Th.B.)	15	64	10
4		Duinger Berg (D.B.)	7	46	12
5		Selter (S.)	7	58	22
6		Ith, middle section	8	49	28

sources: 1, 2, 4, 5, 6: calculated from data of SUCHEL (1954, Fig.11);
3: SCHUNKE (1968, Tab.1)

Tab. 1: *Concavity of the front slopes of various ridge formers.*

Photo 1: *Demonstration of the curvature of the Ith crest. The model of the segment has a radius of 10 m.*

No.	ridge former	area	n	cv_{max} (m)	mean dip angle (°)
1	Hilssandstein (Cretaceous, kru)	Hils, SW section	7	166	20
2	Malmkalk (Jurassic, jw2)	Ith, middle section	15	3	28
3	Rätsandstein (Keuper, ko)	west of northern Ith	13	24	18
4	Wellenkalk (Muschelkalk, mu)	Ahlsburg (A.) SE section	10	28	17
5	Buntsandstein (sm)	Ahlsburg (A.) SE section	7	69	27

sources: 2, 3: unpublished figures of LEHMEIER (cf. LEHMEIER 1981, Abb. 23); 1, 4, 5: unpublished measurements of C. ETZLER and A. SIEBERT

Tab. 2: *Radius of crest curvature of various ridge formers.*

tural bench is formed (fig.3: 24, 25).

2.2 Structure-independent forms

A structure-independent form does not reflect the characteristics of the rock type. The surface of the form truncates structural differencs and, in general, structure-independent forms are truncation surfaces. The most noticeable are crest flats, small surfaces at the crest that truncate the resistant strata. Scarp bevels occur in a similar position but truncate the scarp-forming strata at a low slope angle. If resistant strata are exposed on the back slopes of homoclinal ridges, there are structural benches, in some cases in the form of flatirons. However, where the resistant strata are truncated, they did not form benches. In addition, there are truncation surfaces which form part of the footslopes of the homoclinal ridges. These are discussed in section 2.2.3.

2.2.1 Crest flats and the evolution of homoclinal ridges

Crest flats that are remnants of planation surfaces can be used to reconstruct former surfaces. The homoclinal ridges of the Ith-Hils area between the Weser and the Leine valleys have elongated crest flats (fig.4: 17, 19). They can be correlated with former valleys, with larger truncation surfaces to the east and with the highest bevels in the Bunter Sandstone hills to the west (fig.1). There are a large number of crest flats in the Innerste hills north of the Harz. These crest flats have been interpreted as the remnants of an old planation surface. Not only are they truncated but their heights are independent of the younger valley systems. The evidence indicates that these homoclinal ridges were developed as a result of the dissection of an earlier planation surface. The age of this planation surface is discussed in section 3.

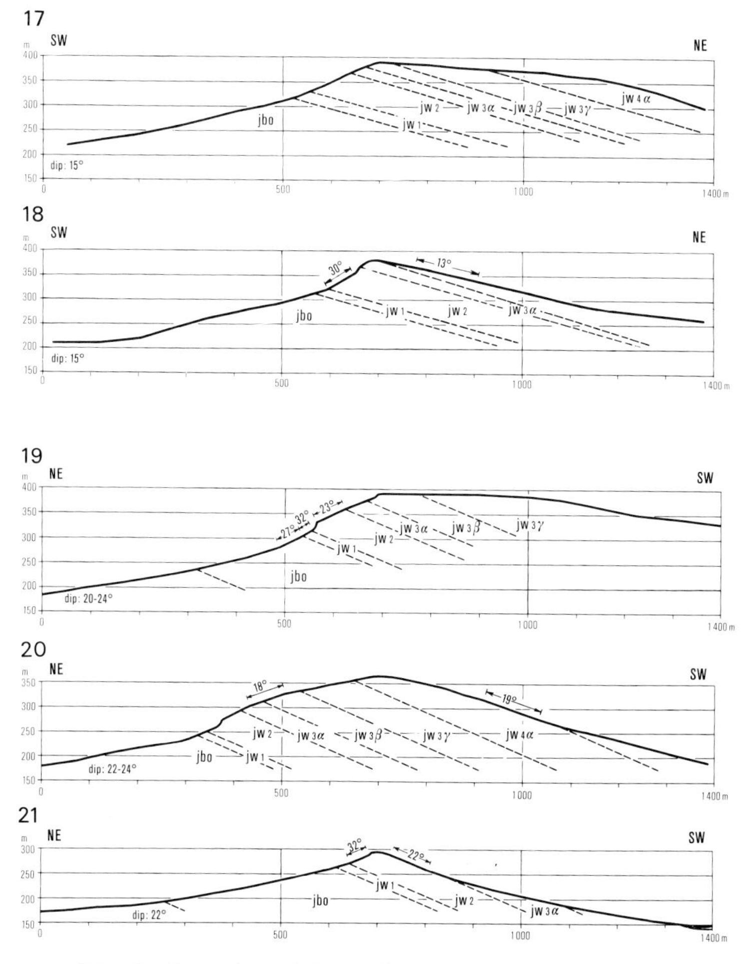

Fig. 4: *Examples of the evolution of homoclinal ridges.*

2.2.2 Scarp bevels and truncation slopes and their relationship to slope evolution

In many cases scarp bevels are similar to crest flats, especially if the scarp-forming rocks are resistant and the scarp bevel is next to a sharp-crested section of the ridge (fig.5: 5, 6). If scarp bevels are also associated with crest flats, they may be relics of swell ridges that had developed from, or been incorporated into a planation surface (fig.5: 4).

Truncation slopes are the equivalent of scarp bevels on dip slopes and have a similarly low inclination. In some areas, however, steep truncation slopes do occur (fig.5: 13). LEHMEIER (1981), BRUNOTTE (1978) suggested that truncation slopes represent a stage of development because they are found next to slope benches and flatirons (fig.5: 14) which can be interpreted as resulting from intensified erosion. The same explanation can be applied to the development of front slopes in the area where

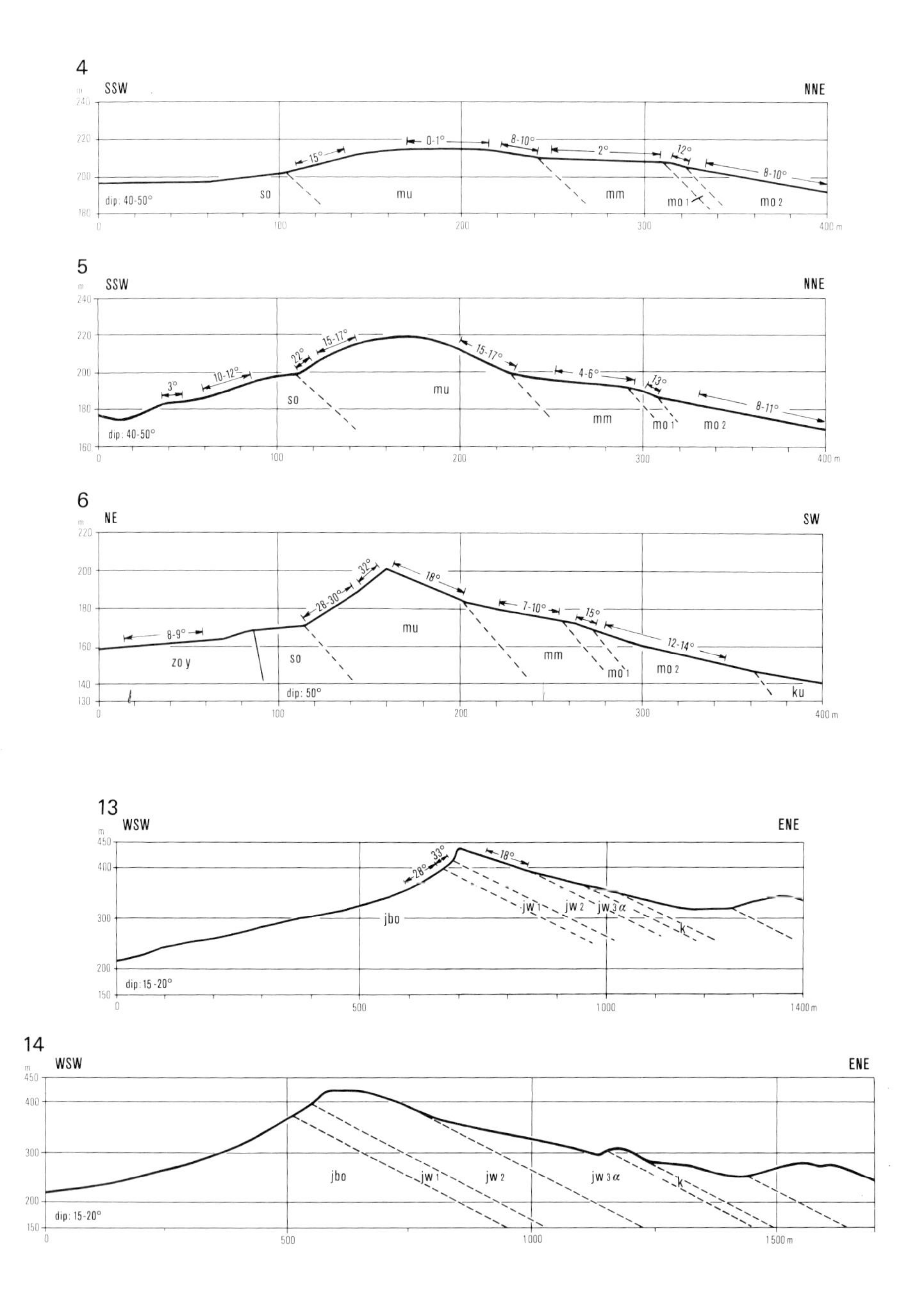

Fig. 5: *Examples of slope evolution of homoclinal ridges (4–5 from BROSCHE 1968, modified).*

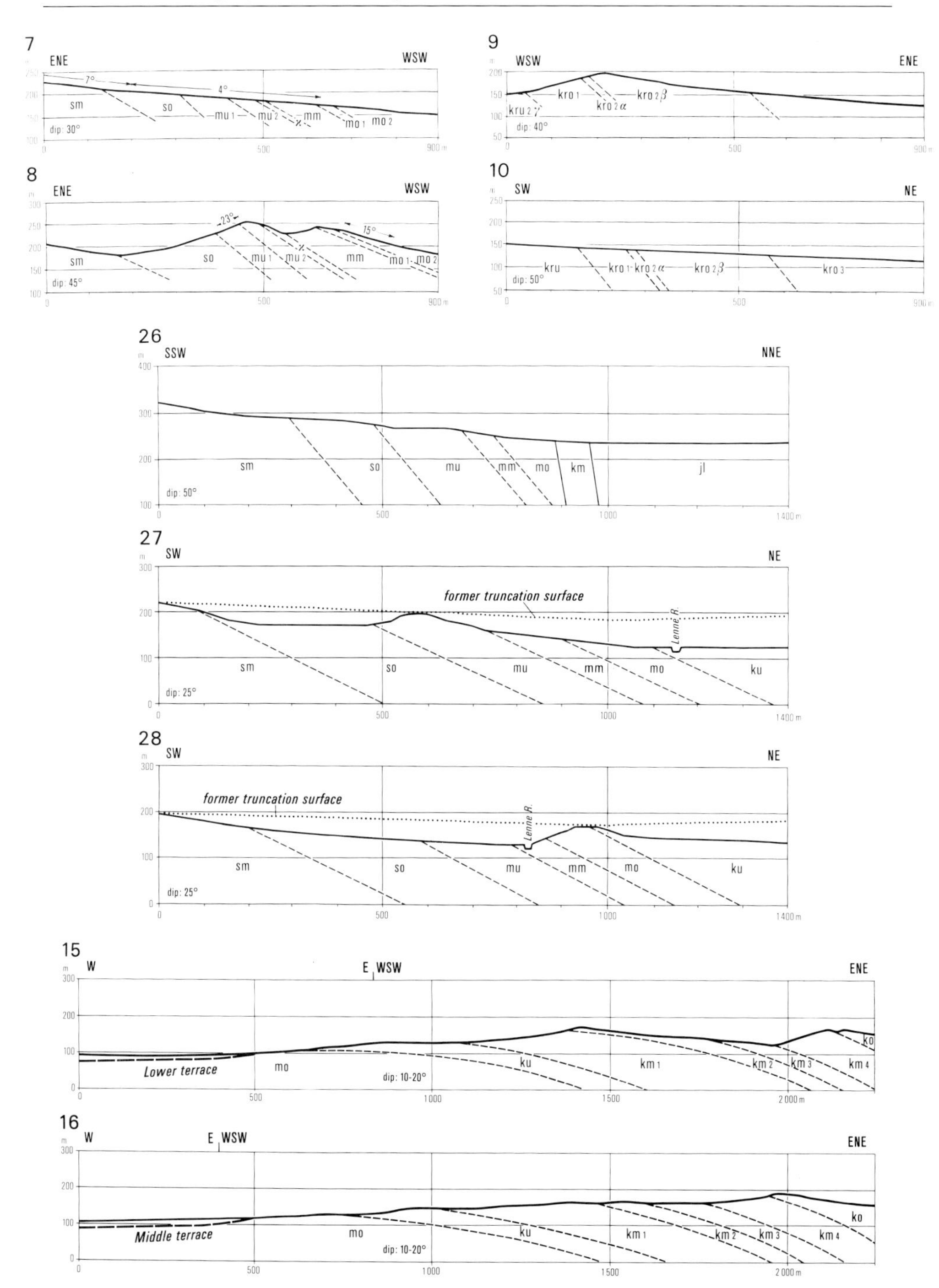

Fig. 6: *Examples of para-pediments and related ridges (15,16 from LEHMEIER 1981, modified).*

the scarp bevels become scarps.

2.2.3 Para-pediments and downwearing

The truncation surfaces in the area investigated are at two levels. In some areas they appear to be stages of relief development (fig.4: 17 with fig.6: 27). The lower surfaces can be related to either declining or evolving ridges (fig.6: 7, 8 and 9, 10 resp., 26–28). They are components of piedmont plains which are pediment-like when they are truncation surfaces. However, since they are developed mainly in weak rocks, such as shales and marls, and do not form a sharp pediment angle, BRUNOTTE (1986) has classified them as para-pediments. The truncation of potentially ridge-forming rocks, the relative lowering of areas of weak rocks and the development of ridges formed of resistant rock, all indicated, that planation tendencies and tendencies towards differential denudation must have varied in time as well as in space during the evolution of the present landforms.

3 Morphogenesis

An explanation of the morphogenesis of homoclinal ridges must be based on morphochronological analysis. There are, however, very few datable sediments on the crests of homoclinal ridges. Apart from Quaternary deposits, sediments are rarely present on para-pediments either, except in areas that have subsided as a result of faulting or sagging. Any attempt to establish a morphochronology of the ridges in Lower Saxony must, therefore, be based on their correlation with Tertiary sediments that occur to the north and to the south of the area of homoclinal ridges.

In the foreland north of the uplands, north of the Harz, relics of pre-Miocene marine sediments are widespread (fig.1). They lie on a planation surface that truncates them and must, therefore, be of Neogene age (HERRMANN 1929). In many cases, the pre-Miocene subsurface of the sediments is also a truncation surface. This is a Paleogene planation surface. In some areas the two surfaces coincide (BROSCHE 1968) and planation was probably continuous in the region. Some low ridges and swells, on which there are also surfaces, rise above the Neogene planation surface. These belong to domal or anticlinal structures. BROSCHE (1969a) has suggested that they are of Paleogene age. Some have been domed up later, others may have been components of the planation surface in the form of swell ridges (fig.5: 4).

South of the ridge region, Tertiary sediments on the Solling Plateau and surrounding areas west of Göttingen have also been used to correlate the truncation surfaces. Extensive relics of a Paleogene planation surface have been preserved that have been dated from Oligocene marine sediments (LOHMANN 1961). A regolith composed of kaolinitic clay and tens of meters thick is widespread and can be related to the pre-upper Oligocene surface because marine sediments of this age, in some areas capped and preserved by Miocene basalt, are deposited on top of it (VON GAERTNER 1968). The Paleogene surface in this area also includes structurally controlled components such as ramp cuestas (MORTENSEN 1949, ROHDENBURG 1965, SCHUNKE 1968). Recent investigations have also proved the existence of more well-defined cuestas (see BRUNOTTE & GARLEFF in this volume). The post-Oligocene development was characterized by tectonic disturbances and downwearing which resulted in piedmont plains, mainly of Pliocene age (HÖVERMANN 1953).

The morphogenesis of the homoclinal ridges between these two areas north and south of the uplands began on an old Tertiary planation surface. The Ith-Hils Syncline between the Weser and the Leine rivers is of major importance to the reconstruction of the history of landforms in the uplands for the following reasons (HERRMANN 1969).

First, about 500 m of Cretaceous sediment that lay above the Neocomian beds, outside the area of the central Hils syncline, have been removed. Eocene sediments of both marine and terrestrial origin were deposited unconformably on these beds indicating that the post-Cretaceous to lower Tertiary denudation surface must have been developed near sea level. Since the Eocene beds do not contain pebbles from the surrounding area, the lower Tertiary surface must have been rather flat with no well-defined ridges, that is, a planation surface (photo 2). The truncation surfaces and crest flats of the Ith and Hils ridges can be related to this planation surface. Former valleys are also present and there may have been swells of resistant quartzitic sandstone (SPÖNEMANN 1966).

Secondly, the coastal lowlands persisted until the upper Oligocene, when marine sands were unconformably deposited on Eocene sediments. After this stage, the regression of the Oligocene sea and the upwarping of the central European uplands began and a period of erosion was initiated that continues at the present time. The solution and removal of salt have caused the subsidence of the Tertiary strata above a salt dome northwest of the Hils ridge. West of the Ith ridge downwearing has resulted in the development of a para-pediment with truncation surfaces on which the meanders of the early Lenne river developed (photo 3). This para-pediment is probably correlated with the course of the Weser river during the Tertiary. Miocene or lower Pliocene pebbles from the early Weser valley have been found west of the area (HERRMANN 1951). The height difference between the Ith ridge and its western para-pediment is about 150 m so that the main period of development of the Ith ridge and the neighbouring ridges must have occurred during the Miocene (HERRMANN 1969).

Thirdly, the incision of the meandering Lenne river resulted in the development of an epigenetic valley. The river cut down into resistant Muschelkalk rocks dissecting them into short swells and hogback ridges (fig.6: 26–28). Climatic change cannot have caused such a development since valley deepening has been limited to the Weser river and its lower tributaries, also the formation of para-pediments has continued until the Quaternary (fig.6: 15, 16, LEHMEIER 1981 and BRUNOTTE 1979 and in this volume). Structure-controlled forms, such as swells, round-crested ridges or hogback ridges, have, however, developed at the same time. The variation in these forms can be related to their location within the drainage basin and is the result of variations in the intensity of downcutting by the rivers (LEHMEIER 1981, 1988).

The morphogenesis of the neighbouring regions has been similar. In the northeastern part of the area of ridges, north of the Harz, the para-pediments can be related to the Neogene planation surface (BROSCHE 1968) and can be traced continuously along the broad

Photo 2: *The Malmkalk hogback of the Ith ridge has developed from a Paleogene planation surface, the residuals of which can be found as crest flats at the rear (cf. fig.4:17).*
Location: NE of the Vogler Mts., view to SE. (Copyright: Landesmedienstelle Hannover).

Photo 3: *The Wellenkalk ridges situated in the middle distance have developed from a Neogene truncation surface at the foot of the Ith ridge, which is lying at the rear. The former para-pediment has been dissected by the Lenne river which follows an inherited valley and joins the Weser river at the foreground (cf. fig.6:26–28).*
Location: N of the Vogler Mts, view to E. (Copyright: Landesmedienstelle Hannover; Luftaufnahme freigegeben unter Nr. 9742-4 durch Reg.-Bez. Mainz).

valleys of the Innerste and Leine and their tributaries until they merge with the para-pediments of the Ith-Hils ridges (fig.1). The para-pediments lie, in general, above Pleistocene river terraces and date, therefore, from the Pliocene or early Pleistocene. Some para-pediments have been formed during the Pleistocene because, along the larger rivers, some areas of the para-pediments are related to the younger terraces (SPÖNEMANN 1966). Relics of an older planation surface in the form of crest flats occur frequently above the Pliocene para-pediments. The vertical distance between the height of the crest flats and the para-pediments increases from the northern foreland in a southerly direction (BROSCHE 1969b). Paleocene relics in the northern foreland and in the Ith-Hils region indicate that this old planation surface is pre-Miocene.

The Mindel and Riss glaciations reached the middle Leine valley near Freden and the western and eastern foreland of the Ith. Both deposited morainic sediments that contained a high proportion of local material but there is no evidence that the ice moved over the homoclinal ridges (HEMPEL 1955, GOEDEKE 1966, BARTELS 1967, LEHMEIER 1981).

4 Conclusion

Planation and valley formation were formerly thought to alternate (MORTENSEN 1949). The differential erosion of the ridges was attributed primarily to humid phases of the upper Tertiary or to the periglacial climate of the Pleistocene (BARTELS 1967, BROSCHE 1968). However, the interrelated development of para-pediments and ridges indicates that denudation by planation and the formation of ridges and cuestas must have occurred simultaneously (BRUNOTTE 1978, LEHMEIER 1981, SCHUNKE & SPÖNEMANN 1972). BROSCHE (1969a) and BRUNOTTE & GARLEFF (1979, 1980) have shown that the development of structural forms was related to an intensity of denudation that was largely independent of the prevailing climatic conditions. The main impulse for landform development on the northern fringe of the central European uplands was initiated by the post-Oligocene uplift. The hinge line of the uplift is the northern border of the ridge area, to the north of which are swells dating from the lower Tertiary. South of the line the relative height of the ridges is at its maximum in the Ith-Hils; the same is true of the vertical distance between the upper and lower levels of the truncation surfaces (BROSCHE 1968). Regional tectonics and subrosion modified the Tertiary landforms (BLENK 1960, BROSCHE 1968, BRUNOTTE 1978) and influenced the shape of the ridges. Downcutting in the valleys accentuated the relief. The distance to the main streams which formed the base-levels of erosion has determined the rate of downwearing of the para-pediments and the increase in the height of the ridges (BROSCHE 1968, BRUNOTTE 1978, SPÖNEMANN 1966). In general, the shape of homoclinal ridges and their slope form can be related to the intensity of valley deepening. This also agrees with computer simulations of the development of structural forms. AHNERT (1971) has demonstrated that the more resistant of two steeply-dipping ridge formers can be reduced to a truncation surface, if it lies next to the stream and if the period of erosion is long enough (fig.7). On the basis of these theoreti-

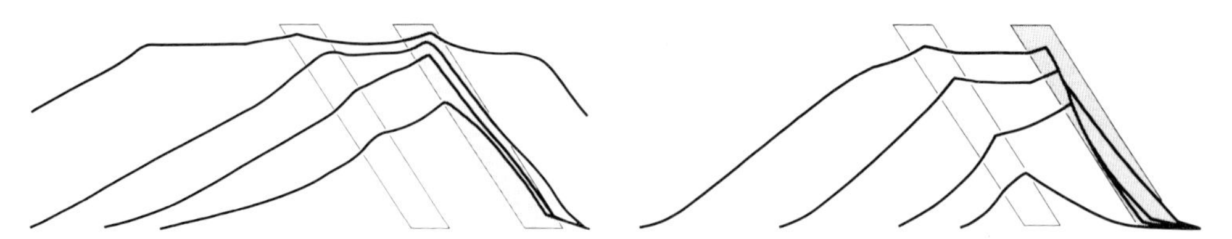

Fig. 7: *Computer simulated development of homoclinal ridges with two resistant beds, the right bed more resistant than the left. The model on the right is closer to the eroding stream (from AHNERT 1971, modified).*

cal but realistic models of relief development, it can be questioned whether processes such as those of planation or valley formation are climatically determined or whether they represent different stages in a long-lasting development. BROSCHE (1968) has considered the apparent contradiction of widespread indications of planation in the Pliocene and the simultaneous existence of a humid climate. Recent investigations of structure-controlled landforms, such as homoclinal ridges and gneissic residual ridges, in various parts of the world (GARLEFF & STINGL 1987, HAGEDORN 1988, SPÖNEMANN 1988) agree that, in general, these landforms are transitional stages within a landform development sequence that includes valley deepening, valley widening, pediment formation, pediment widening and (pedi-)planation. Such a sequence is tectonically induced but develops under stable conditions. Climate may influence the rate of development.

References

AHNERT, F. (1971): A general and comprehensive theoretical model of slope profile development. University of Maryland Occasional Papers Geogr. **1**, College Park, Md.

AMTHAUER, H. (1972): Untersuchungen zur Talgeschichte der Oberweser. Gött. Geogr. Abh. **59**. Göttingen.

BARTELS, G. (1967): Geomorphologie des Hildesheimer Waldes. Göttinger Geogr. Abh. **41**, Göttingen.

BLENK, M. (1960): Morphologie des nordwestlichen Harzes und seines Vorlandes. Göttinger Geogr. Abh. **24**, Göttingen.

BROSCHE, K.U. (1968): Struktur- und Skulpturformen im nördlichen und nordwestlichen Harzvorland. Göttinger Geogr. Abh. **45**, Göttingen.

BROSCHE, K.U. (1969a): Über die Beziehungen von Rumpfflächen zu Schichtkämmen und Schichtstufen sowie Betrachtungen an einigen wichtigen Strukturformtypen erläutert an Beispielen aus dem nördlichen und nordwestlichen Harzvorland. Z. Geomorph. N.F. **13**, 207–216.

BROSCHE, K.U. (1969b): Zum Problem der Auffindung und Deutung von Reliefgenerationen in Schichtkamm- und Schichtstufenlandschaften erläutert am Beispiel des nördlichen Harzvorlandes und des nördlichen Niedersächsischen Berglandes. Z. Geomorph. N.F. **13**, 434–505.

BRUNOTTE, E. (1978): Zur quartären Formung von Schichtkämmen und Fußflächen im Bereich des Markoldendorfer Beckens und seiner Umrahmung (Leine-Weser-Bergland). Göttinger Geogr. Abh. **72**, Göttingen.

BRUNOTTE, E. (1979): Quaternary piedmont plains on weakly resistant rocks in the Lower Saxonian Mountains (W. Germany). CATENA **6**, 349–370.

BRUNOTTE, E. (1986): Zur Landschaftsgenese des Piedmont an Beispielen von Bolsonen der mendociner Kordilleren (Argentinien). Göttinger Geogr. Abh. **82**, Göttingen.

BRUNOTTE. E. & GARLEFF, K. (1979): Geomorphologische Gefügemuster des Niedersächsischen Berglandes in Abhängigkeit von Tektonik und Halokinese, Resistenzverhältnissen und

Abflußsystemen. In: Hagedorn, J., Hövermann, J. & Nitz, H.-J. (eds.), Gefügemuster der Erdoberfläche. Festschr. 42 dt. Geographentag Göttingen 1979, 21–42, Göttingen.

BRUNOTTE, E. & GARLEFF, K. (1980): Tectonic and climatic factors of landform development on the northern fringe of the German Hill Country (Deutsche Mittelgebirge) since the early Tertiary. Z. Geomorph. N.F. Suppl. Bd. **36**, 105–112.

GAERTNER, H.R. VON (1968): Entstehung der heutigen Oberflächenformen. In: Erläuterungen zu Blatt Hardegsen Nr. 4324 Geologische Karte von Niedersachsen 1:25,000, Hannover.

GARLEFF, K. & STINGL, H. (1987): Struktur- und Skulpturformen am argentinischen Andenrand unter randtropischen bis subantarktischen Bedingungen. Z. Geomorph. N.F. Suppl. Bd. **66**, 49–63.

GOEDEKE, R. (1966): Die Oberflächenformen des Elm. Göttinger Geogr. Abh. **35**, Göttingen.

HAGEDORN, J. (1988): Aktuelle und vorzeitliche Morphodynamik in der westlichen Kleinen Karru (Südafrika). In: Hagedorn, J. & Mensching, H. (Hrsg.), Aktuelle Morphodynamik und Morphogenese in den semiariden Randtropen und Subtropen. Abh. Akad. Wiss., Math.-Phys. Kl., 3. Folge, Nr. **41**, Göttingen, 168–179.

HEMPEL, L. (1955): Studien über Verwitterung und Formenbildung im Muschelkalkgestein. Göttinger Geogr. Abh. **18**, Göttingen.

HERRMANN, R. (1929): Erdgeschichtliche Grundfragen der Oberflächenformung in Mitteldeutschland. Festschr. des 23. Dt. Geographentages in Magedeburg, 71–108, Braunschweig, Berlin, Hamburg.

HERRMANN, R. (1951): Das Durchbruchstal der Weser zwischen Holzminden und Bodenwerder. Geol. Jahrb. **65**, 611–619.

HERRMANN, J. (1969): Die Schlüsselstellung des Malm-Schichtkammes der Hilsmulde in der Morphologie des nordwestdeutschen Berglandes. Geol. Rundsch. **58**, 41–51.

HÖVERMANN, J. (1953): Die Oberflächenformen um Göttingen. Göttinger Jb. 1953, 63–74.

LEHMEIER, F. (1981): Regionale Geomorphologie des nördlichen Ith-Hils-Berglandes auf der Basis einer großmaßstäbigen geomorphologischen Kartierung. Gött. Geogr. Abh. **77**.

LEHMEIER, F. (1988): Zum Formenschatz der Schichtkammlandschaft im Niedersächsischen Bergland (insbesondere im Ith-Hils-Bergland). Berlin. Geogr. Abh. **47**, 49–61.

LOHMANN, H.H. (1959): Zum Bau des Oberweserberglandes zwischen Hannoversch-Münden und Karlshafen. Diss. Univ. Hamburg, Hamburg.

MORTENSEN, H. (1949): Rumpffläche — Stufenlandschaft — Alternierende Abtragung. Petermanns Geogr. Mitt. **93**, 1–14.

ROHDENBURG, H. (1965): Die Muschelkalk-Schichtstufe am Ostrand des Sollings und Bramwaldes. Göttinger Geogr. Abh. **33**, Göttingen.

SCHUNKE, E. (1968): Die Schichtstufenhänge im Leine-Weser-Bergland in Abhängigkeit vom geologischen Bau und Klima. Göttinger Geogr. Abh. **43**, Göttingen.

SCHUNKE, E. (1971): Die Massenverlagerungen an den Schichtstufen und Schichtkämmen des Leine-Weser-Berglandes. Nachr. Akad. Wiss. Göttingen, II. Math.-Phys. Kl., Nr. **3**, 43–77.

SCHUNKE, E. & SPÖNEMANN, J. (1972): Schichtstufen und Schichtkämme in Mitteleuropa. Göttinger Geogr. Abh. **60** (Hans-Poser-Festschrift), 65–92. Göttingen.

SPÖNEMANN, J. (1966): Geomorphologische Untersuchungen an Schichtkämmen des Niedersächsischen Berglandes. Göttinger Geogr. Abh. **36**, Göttingen.

SPÖNEMANN, J. (1988): Formungstendenzen im semiariden Kenia: Morphodynamik und Morphogenese im Samburuland. In: Hagedorn, J. & Mensching, H. (Hrsg.), Aktuelle Morphodynamik und Morphogenese in den semiariden Randtropen und Subtropen. Abh. Akad. Wiss., Math.-Phys. Kl., 3. Folge, Nr. **41**, Göttingen, 126–149.

SUCHEL, A. (1954): Studien zur quartären Morphologie des Hilsgebietes. Göttinger Geogr. Abh. **17**, Göttingen.

THIEM, W. (1972): Geomorphologie des westlichen Harzrandes und seiner Fußregion. Jahrb. geogr. Ges. Hannover, Sonderheft **6**.

Address of author:
Prof. Dr. Jürgen Spönemann
Geographisches Institut der Universität Göttingen
Goldschmidtstraße 5
D-3400 Göttingen
Federal Republic of Germany

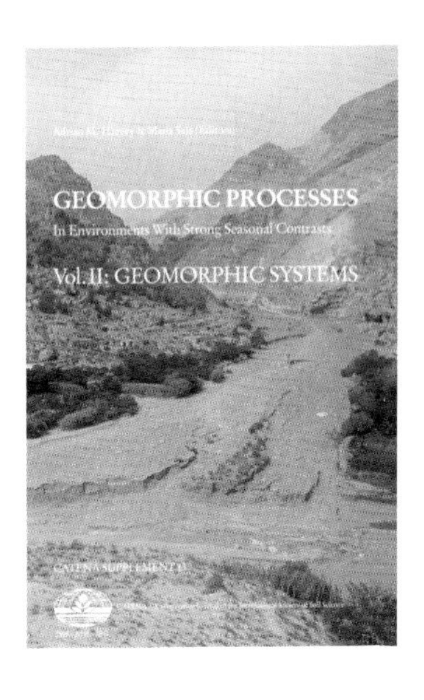

Adrian M. Harvey & Maria Sala:

GEOMORPHIC PROCESSES

In Environments With Strong
Seasonal Contrasts
Vol. II: GEOMORPHIC SYSTEMS

CATENA SUPPLEMENT 13, 1988

Price: DM 126,– / US $74.–

ISSN 0722-0723 / ISBN 3-923381-13-1

CONTENTS

CATENA SUPPLEMENT 15 p. 151–164 Cremlingen 1989

STRUCTURAL LANDFORMS AND PLANATION SURFACES IN SOUTHERN LOWER SAXONY

Ernst **Brunotte**, Köln, and Karsten **Garleff**, Bamberg

1 Morphographic and geologic-tectonic characteristics

The upland of southern Lower Saxony is located on the northern margin of the Central European Uplands. In the north of this area there are long, NW-SE striking ridges, most of which are hogbacks (see SPÖNEMANN in this volume) and in the south extensive plateaus, hilly areas and intervening broad basins and valley zones (fig.1). The plateaus, which are partially dissected, generally lie between 300 and 500 meters above sea level on subhorizontal Mesozoic sedimentary beds and between 500 and 900 meters on the Paleozoic basement rocks of the Harz Mountains in the east. The southwestern part of the area is characterized by volcanic hills which rise 100 to 150 meters above the general level of the plateaus. In the region of Mesozoic rocks, structural landforms and extensive structure-independent planation surfaces occur not far from each other on the same rocks and geological structures. The reasons for such diverging landform developments are examined in this paper.

Pediments and terraces are often present on the floors of the basins and major valleys that for the most part lie at between 100 and 250 meters above sea level. Most of the area is drained by the Leine and the Innerste which flow north to the Weser. Some stretches of these rivers and, more particularly, their minor tributaries, are structurally controlled, predominantly with a Hercynian (NW-SE) orientation but in some cases with Rhenish (N-S to NNE-SSW) and Variscan (SW-NE) orientations.

The extensive Mesozoic sedimentary cover is made up of beds of varying resistance and a germanotype structure, that is, a structure of broad warps, open folds and faults. The northern part, where there are halotectonic anticlines and synclines with predominantly Hercynian strikes, contrasts with the southern part in which there are broad upwarps, subsidence basins and grabens. The upwarps are frequently bordered by monoclines and, in some cases, by faults. The Solling block in the west of this southern area and the Eichsfeld block in the east are separated by the Leine Rift Valley (Leinetalgraben) and the Markoldendorf Basin, both areas of relative subsidence.

In the upwarped areas, the older rocks, such as the middle Bunter Sandstone (sm), and the Muschelkalk beds of the

ISSN 0722-0723
ISBN 3-923381-18-2

3-923381-18-4/89/5011851/US$ 2.00 + 0.25

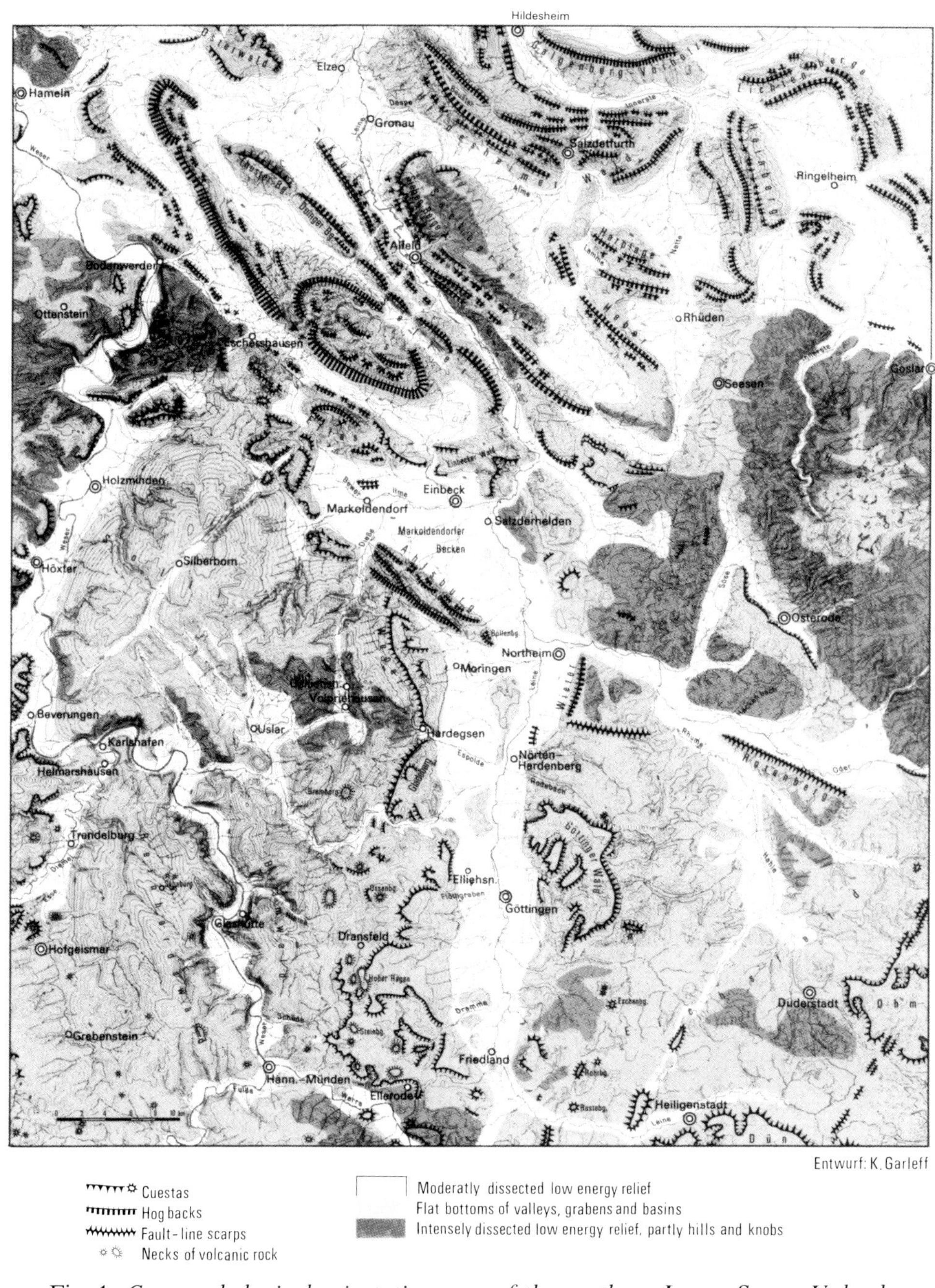

Fig. 1: *Geomorphological orientation map of the southern Lower Saxon Upland.*

Wellenkalk (mu) and the Trochitenkalk (mo1), form most of the land surface. In the basins and grabens, younger, less resistant rocks, including claystones and siltstones of the Röt (so, upper Bunter Sandstone), the Lias (jl) and the Keuper (ku and km) and also Tertiary and Quaternary unconsolidated sediments predominate. The Tertiary and Quaternary sediments have been deposited mainly in areas lowered by subrosion. Since most of the basins and rift valleys are not the result of absolute subsidence but of a relatively smaller uplift, large areas on the basin floors have also been subjected to denudation for a long period. The development of the structures was accompanied by halotectonics, in particular, by movements of Zechstein (upper Permian) salt deposits from the basin areas to the upwarped, upfolded or upfaulted areas.

In a few areas the tectonic and halotectonic movements began during the Jurassic but most areas were affected first during the upper Cretaceous "late Kimmerian" phase of tectonic activity. The most severe deformations occurred during the Tertiary. The Solling block was low enough to be affected by the Eocene and Oligocene transgressions that extended from the Hessian Depression. Only sediments of the last (upper Oligocene) transgression and deposits of the limnic, fluvial and thelmatic phases of the lower Miocene have been preserved over wide areas. Uplifts after the mid-Tertiary amounted to 500 to 700 meters in the centre of the Solling relative to the Leine Rift Valley and the Eichsfeld block to the east.

Eruptions of basaltic volcanics dated at 12 to 19 million years BP accompanied the tectonic movements. Some of the volcanics cover and protect Tertiary sediments; others lie directly on Mesozoic sedimentary rocks, indicating that at the time of the eruption the younger sediments had been removed. During the late Tertiary and the Quaternary, denudation reduced the volcanoes to volcanic necks which, because of their greater resistance, rise above their surroundings.

The plateaus in the southern Lower Saxon Uplands occur mainly in the area of broad upwarps on the subhorizontal resistant Bunter Sandstone and Muschelkalk beds. On their margins where there are monoclines and the dip is greater, the resistant beds, in particular the mu and mo1, form cuestas or hogbacks. However, there are also truncation surfaces over wide areas of the plateaus which extend across strata of differing resistances. These surfaces can be termed peneplains. In the basins, several levels of denudation surfaces are developed as stepped pediments, footslopes or terraces on relatively weak rocks. All such surfaces are attached only very locally to rocks of particular resistance. Generally, they truncate the strata and are, therefore, sculptural, not structural forms.

Frequently structural and sculptural landforms occur in proximity to one another. On the eastern flank of the Solling block, for example, the Dransfeld Plateau (figs. 1 and 2) extends from west to east, across the sandstones of the middle Bunter Sandstone (sm), the weak claystones and siltstones of the upper Bunter Sandstone (so), the resistant limestones of the lower Muschelkalk (mu), the weak marls of the middle Muschelkalk (mm), the resistant limestones of the upper Muschelkalk (mo1) and the shales and Ceratite limestones of the uppermost Muschelkalk (mo2). In some areas on the plateau, cuestas have developed on rocks that nearby are capped by peneplains or gen-

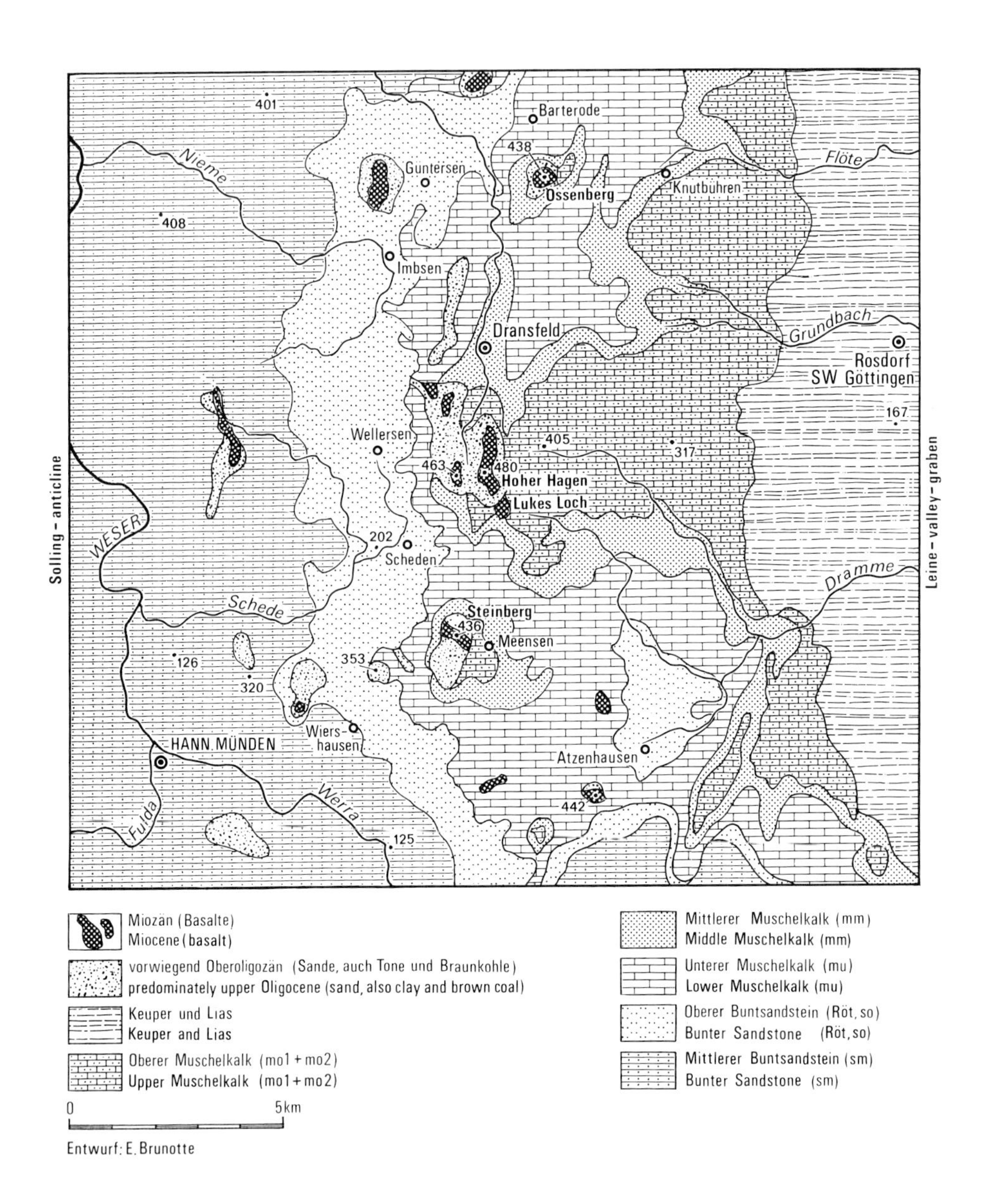

Fig. 2: *Geological map of the Dransfeld Plateau.*

tly sloping (less than 2°) "ramp scarps" (BROSCHE 1968, SPÖNEMANN in this volume). The Dransfeld Plateau has been used by many authors to examine the association of structural and sculptural landforms in this region (BRINKMANN 1932, MORTENSEN 1949, HÖVERMANN 1953, ROHDENBURG 1965a, SCHUNKE 1968, BRUNOTTE & GARLEFF 1980).

The sediments of the Eocene and the Oligocene transgressions and the Miocene sands, clays and lignites covered earlier landforms in the area. Some of these landforms are still buried, others have been exhumed and in part modified by later denudation. The removal of the loose sediments has often left remnants of silicified horizons of the sands in the form of scattered "Tertiary quartzite" blocks. The subsurface of the Tertiary sediments and of the quartzite blocks provides a basis for the reconstruction of the landforms that existed in the early to middle Tertiary. In some areas tectonic movements and subrosion have changed the position of the Tertiary sediments so that a reconstruction of the early to mid-Tertiary relief has to consider all landforms in the area. For this purpose comprehensive geomorphological maps have been made (BRUNOTTE, GARLEFF & WAHLE 1980, GARLEFF, BRUNOTTE & STINGL 1985).

2 Truncation surfaces and cuestas in the southern Lower Saxon Upland

The investigations to reconstruct the early and middle Tertiary landforms attempted to answer two questions:

1. What are the relationships between the old Tertiary relief and the tectonic- structural and climatically-controlled morphodynamic conditions that led to their development?

2. What changes took place in the relief after the removal of the Tertiary sedimentary cover as a result of spatially differentiated tectonics and climatic conditions which varied temporally?

2.1 The physiognomy of the early to middle Tertiary landforms on Dransfeld Plateau

The Dransfeld Plateau has been described variously as a cuesta landscape and as a peneplain. Using the Dransfeld Plateau as an example, MORTENSEN (1949) termed the spatially associated sculptural and structurally controlled landform development an "exchange landscape" (Austauschlandschaft) and its development "alternating denudation". By contrast, HÖVERMANN (1953) described the area as a piedmont benchland with several levels between 200 and 500 meters above sea level that has developed by areal lowering from a late Tertiary peneplain.

From geomorphological maps that show the Tertiary sediments and scattered Tertiary quartzite blocks it is apparent that the major part of what was thought to be a late Tertiary land surface is an exhumed pre-Oligocene to pre-Miocene old relief which, in the area of the Dransfeld Plateau, had considerable elevation differences and which also had both level and dissected areas (GARLEFF, BRUNOTTE & STINGL 1985).

The landforms at the boundary between the upper Bunter Sandstone (so) and the lower Muschelkalk (mu) are closely related to the pattern of valleys.

At some distance from the trunk valleys, truncation surfaces occur, some of which have low ridges or ramp scarps where resistant beds outcrop. Near the main valleys and on the plateau margins, forms are structurally controlled. Using a climatic-geomorphological approach, ROHDENBURG (1965a) suggested that cuesta scarps developed from the gently sloping truncation surface relief as a result of valley deepening during the Pleistocene. SCHUNKE (1968) thought that the deepening could have begun in the late Pliocene. Remnants of the truncation surface exist on the divide between the Nieme and Schede drainage basins, for example, where they extend over several kilometers across the strata from the upper Bunter Sandstone (so) to the upper Muschelkalk (mo2) and have a mean slope of 2° and a total elevation difference of 80 to 100 meters. The increasing differentiation of the scarp profiles is interpreted as a temporal sequence of scarp development that resulted from a change from a Tertiary climate in which planation took place to a periglacial climate in the Quaternary in which valley formation predominated.

The geomorphological maps of the area indicate, however, that there were slopes of from 10° to more than 20° on Muschelkalk in the early to middle Tertiary in, for example, the basins of the rivers that preceded the Werra and the Schede. Gently sloping ramps and steep scarp slopes existed simultaneously, therefore, during the mid-Tertiary. The occurrence of well-defined scarp slopes on the old relief was limited to areas in which the predecessor streams of the present-day drainage systems were incised deeply into the strata near the outcrop boundary of a resistant potentially scarp-forming bed.

Resistance differences resulted, therefore, in scarp formation in areas of fluvial incision, despite the otherwise general dominance of planation. The sequence of different scarp profile forms between the divide and the main trunk valley cannot be interpreted solely in climatic geomorphological terms, although the slopes undoubtedly have been steepened as a result of Pleistocene valley incision. No succession from undifferentiated planation to differential denudation occurred. The development of peneplains in the vicinity of divides took place simultaneously with the development of cuesta scarp slopes near the valleys as described by MENSCHING (1984) and similar to the structural and sculptural landform development discussed by MORTENSEN (1949).

The dating of the development of the scarp profiles and their rates of retreat are closely related. ROHDENBURG (1965a) and SCHUNKE (1968) estimated, from the presence of remnants of upper Muschelkalk (mo) below the Miocene basalts, that the post-basaltic retreat of the upper Muschelkalk varied from 2 to 6 kilometers. From maps it is apparent that this cannot be generally correct. For example, south of the Hoher Hagen, a basaltic sheet that overlies Tertiary sands extends from the crest of the upper Muschelkalk scarp into the middle Muschelkalk in a small valley. The high steep edge of the Dransfeld Plateau east of Scheden was, therefore, deeply embayed before the basaltic eruption and has retreated maximally by only a few tens of meters since the middle Miocene although considerable tectonic uplift has also taken place. At other locations, near Meensen and Atzenhausen, for example, or near Barterode and Knutbühren (fig.2), the foreland of the upper Muschelkalk scarp contains scattered Tertiary quartzite blocks indicating that the scarp existed before the Miocene, so that a scarp retreat of several kilometers cannot have occurred here.

If the retreat of the upper Muschelkalk scarp was so limited, were there significant differences between the scarp profile in the pre-upper Oligocene and that of the present-day? At one location on the south side of the Hengelsberg, the slope of about 15° continues downward under the upper Oligocene Kassel Marine Sands (BRUNOTTE 1987). Moreover, the footslopes of 5° to 10° at the base of the upper Muschelkalk cuesta scarp on the west side of the Hengelsberg are also partially covered by these sands which have quartzitic layers and are glauconitic in the lower beds. At several locations near the Hengelsberg, gray clays have been found which, according to an oral communication by Dr. RITZKOWSKI, are probably Eocene. If this dating is correct, the old relief in this area would date from the late Cretaceous to early Tertiary. The presence of Eocene clays would also indicate that the later transgressions hardly abraded the old relief.

The profiles on the south side of the Hengelsberg shown in fig.3 were based on borings in the slope zone of the upper Muschelkalk outcrop and on geomorphological mapping. Profile 2 and profile 3 show that the upper Muschelkalk scarp had a backslope covered by upper Oligocene glauconitic sands. The upper Muschelkalk outcrop was, therefore, a hogback at the time of the old relief. Should the clays and the dark sands that lie below the upper Oligocene sands be Eocene, the hogback would be older still.

2.2 Plio-Pleistocene pedimentation in the Leine Rift Valley and in the Markoldendorf Basin

The Leine Rift Valley and the Markoldendorf Basin are relatively old structural features that were reactivated by the post-Oligocene tectonic and halotectonic movements. The progressive eustatic lowering of sea level during the late Tertiary and the sea level and climatic variations during the Quaternary together with the crustal movements caused several phases of erosional and denudational lowering. In the areas of weak rock in the Leine Rift Valley and the basin this resulted in the formation of several pediment levels. On the high blocks, especially at their margins and in the vicinity of major valleys, the valleys became more deeply incised and the valley density increased. There are also some pediments on the high blocks but they are more characteristic of the rift valley and basin floors.

The morphological properties of pediments in southern Lower Saxony and northern Hessen have been investigated and mapped in detail (BRUNOTTE 1978, BRUNOTTE, GARLEFF & WAHLE 198O, GARLEFF, BRUNOTTE & STINGL (1985).

The slopes of the pediments are several kilometers long, have inclinations of about 2° and weakly concave longitudinal profiles. Only near the pediment angle do inclinations of 5° or more occur. Frequently there are several pediment levels, separated by elevation differences of a few tens of meters (fig.4). The lower pediment levels have been cut into and are usually larger than the higher levels. Reconstruction of the higher levels has shown, however, that they were much larger in the past and extended formerly

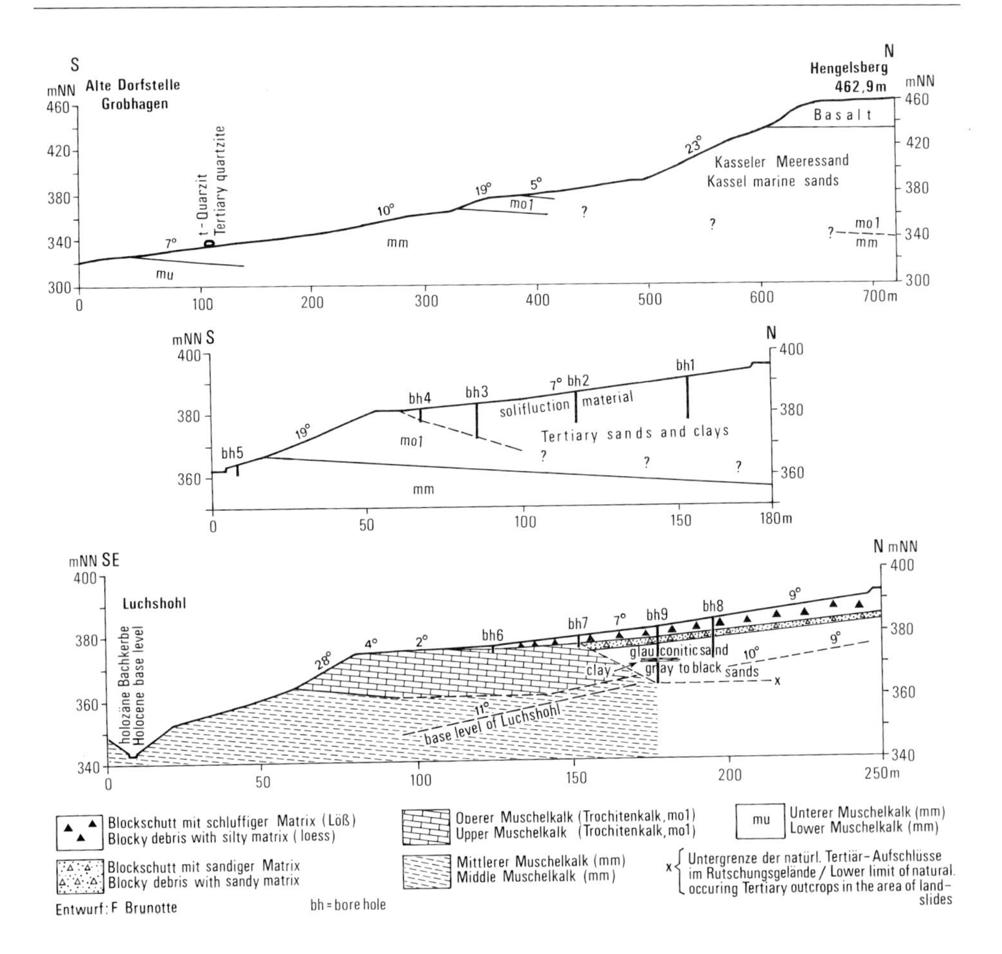

Fig. 3: *Transversal profiles of a pre-upper Oligocene hogback at the southern slope of the Hengelsberg.*

across the minor valleys and also that the major trunk streams were their baselevels. On the basis of their transversal profiles the following sub-types of pediments can be distinguished:

a) smooth inclined planes without transversal convexity

b) pediments with interfluves in which the valleys and interfluves of the upland continue into the pediment

c) cone-shaped pediments, the apexes of which lie at the valley exit from the upland and from where the pediments slope outward in the form of a cone or fan.

These three morphographic pediment types result from differences in morphodynamic development (STINGL, GARLEFF & BRUNOTTE 1983). The smooth inclined plane indicates complete adjustment of the autochthonous pediment development to the dynamics of the local baselevel. On the pediment with

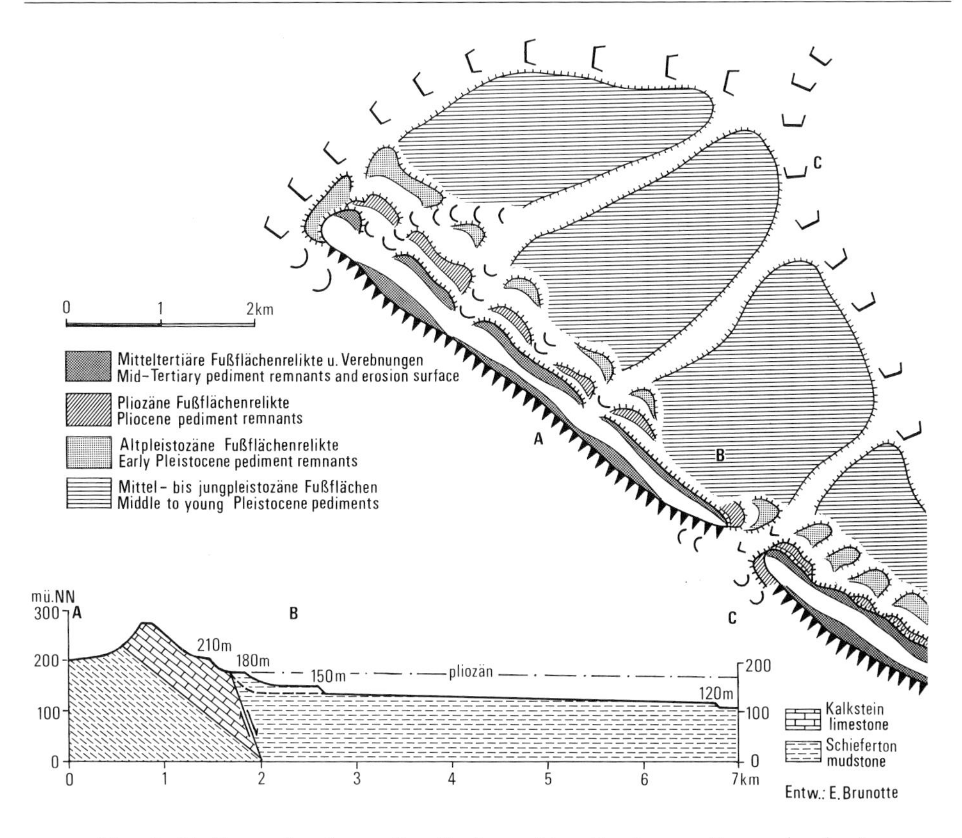

Fig. 4: *Pediment levels on the dipslope side of a Lower Saxon hogback.*

interfluves, by contrast, incision by the streams from the upland dominated over the autochthonous lowering of the pediment surface. Loess accumulations modified this pattern on the pediments on the western side of the Leine Rift Valley. The cone-shaped pediments developed in the basins of southern Lower Saxony where the autochthonous pediment lowering predominated over the lowering of the stream channels because the upper parts of the pediments, near the valley exits from the upland, were covered by resistant layers of coarse gravels (BRUNOTTE 1979).

There are also differences between pediments that lie at different relative elevations. Older, higher pediments, or their remnants, are predominantly denudational landforms that on weak rocks are usually covered by fanglomerates of up to several meters thickness, and on resistant rocks by waste covers with a thickness of only a few tens of centimeters. In the distal parts of the pediments, the fanglomerates are frequently interbedded with fluvial sediments of the trunk streams or grade into gravel terraces, indicating the morphodynamic and morphochronological relationship between the pediment formation and the fluvial dynamics of the major valleys.

The lowest pediments also include large areas of accumulation. At the western side of the Leine Rift Valley and in the Markoldendorf Basin, for example, broad pediments extend across late Quaternary, primarily loess, sediments up to more than 10 meters thick (BRUNOTTE 1978, 1979; figs 5 and 6). The loess contains sandy layers and, near the upper margin of the pediment, thin layers of wash-transported debris composed of weathering products of the adjacent upland rocks. These inclusions indicate that in addition to the aeolian supply of loess, wash processes also played an essential part, particularly in the planation of the surface.

Such pediments, developed by accumulation, also occur in which there is a predominance of aeolian material on the sides of the rift valley or on the leeward of small rises in the terrain. In the latter case, the loess surfaces frequently border on older pediment remnants formed on more resistant bedrock without any break in slope. Locally loess covers the dissected or lowered remnants of these older pediments.

The remnants of weathering zones and of soil horizons in the loess (ROHDENBURG & MEYER 1968) indicate that loess accumulations on the leeward began on the higher parts of the surfaces that were covered, spread laterally to fill small valleys and eventually extended on to the margins of the larger gravel-filled valleys (fig.5).

By contrast, on the eastern flanks of the basins and rift valleys, almost all pediments, including the lowest, are denudational landforms (ROHDENBURG 1965b, BRUNOTTE, GARLEFF & JORDAN 1985). Their surfaces are made up of wash denudation debris and solifluction material, in some areas with a thin loess cover, on a subsurface of claystones, siltstones or gravels of the older terraces.

The successive pediment levels or their remnants in the basins and near the major valleys indicate that a progressive surface lowering took place in distinct phases. Dating of the levels is possible only locally. On the basis of an analysis of landform associations, the lowering phases seem to have been synchronous over wide areas but rates of lowering appear to have differed regionally. The oldest pediment level to be dated is part of the early to middle Tertiary old relief at the base of the old cuesta scarps. The second oldest level dated undercuts the remnants of the oldest by 50 to 60 meters on the southern margin of the Dransfeld Plateau and by 35 to 50 meters in the Leine Rift Valley. In the Kassel basin, south of the area investigated in this paper, the distal parts of this surface are made up of gravels over wide areas that have been dated as late Pliocene to earliest Pleistocene (FINDEISEN 1952). The second oldest pediment in the Leine Rift Valley is probably of a similar age.

In the Markoldendorf basin, the pediment level formed in the Pliocene and early Pleistocene was undercut by a younger level which lies 20 to 30 meters lower. Its distal parts bear gravels and terraces that belong to the early Pleistocene "Upper Terrace Complex" (LÜTTIG 1960). They have, however, not yet been dated in detail. In the Kassel basin, this early Pleistocene pediment level occurs widely but in the Leine Rift Valley only relatively small remnants have been found.

In the Markoldendorf Basin, the next lower middle Pleistocene level lies only 10 to 30 meters lower than the level above and is linked in its distal parts to gravels of the Middle Terrace that date from the Drenthe stadial of the Saale glaciation (see LIEDTKE in this

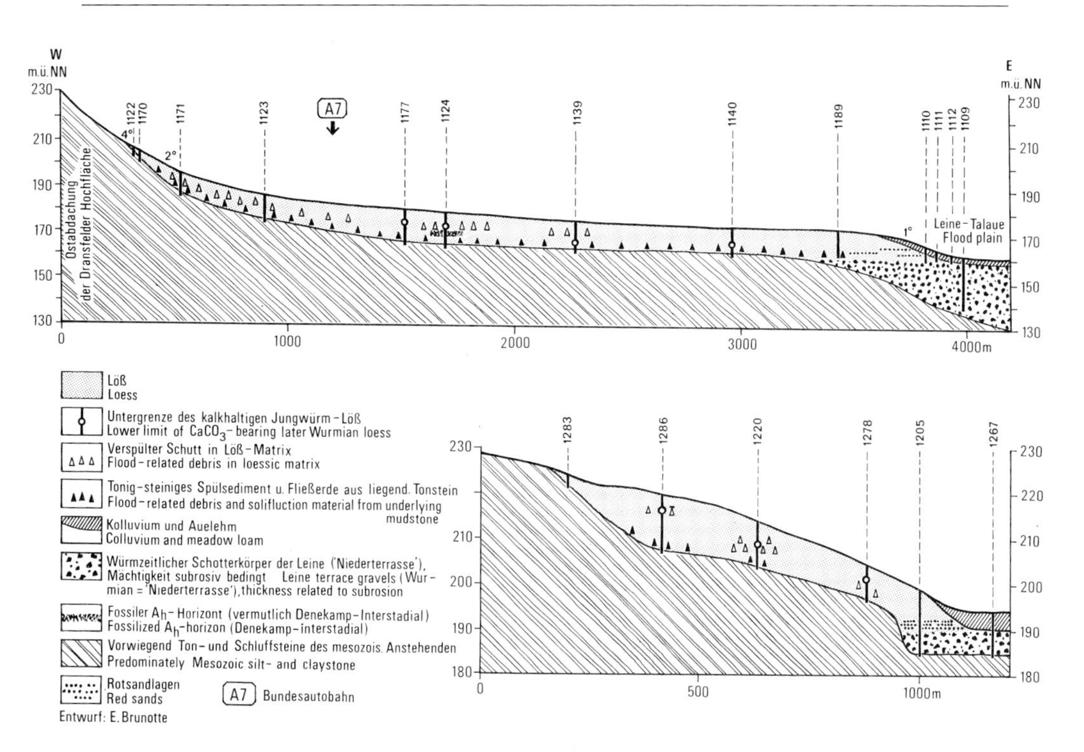

Fig. 5: *Longitudinal profiles of pediments from the southern part of the Leine Rift Valley.*
top: composite pediment surface profile south of Göttingen Autobahn service station
bottom: leeward loess accumulation on the western Leine valley slopes (south of Niedergandern)

volume). It is also present in the Leine Rift Valley but after partial lowering and dissection took place it was covered by sediments. The middle Pleistocene pediments are the result of the last phase of widespread areal lowering of planation surfaces and of the formation of denudation pediments.

During the Weichsel glaciation and the late Pleistocene in general, landform development by areal lowering was of minor importance although it was intensive enough to remove the greater part of old cover sediments, weathering crusts and soils, and to dissect, and partially lower, the Mesozoic rocks that form the substrate of the pediments. The larger landforms were not significantly changed. The processes were clearly of low intensity because on the eastern flank of the Leine Rift Valley, for example, relatively thin debris covers that resulted from solifluction and wash denudation in the Weichsel glacial lie on the upper areas of the middle Pleistocene pediment (BRUNOTTE, GARLEFF & JORDAN 1985).

The pediments in southern Lower Saxony have developed during the Tertiary and Pleistocene at different times and under different geologic, tectonic and climatic conditions. The climatic conditions during the phases of pediment formation are not known precisely. During the middle and late Pleistocene phases of pedimentation, the climates were cold and probably arid to semiarid. The earlier pedimentation phases seem to have occurred under warm temperate to marginal tropical conditions which were probably also semiarid. The areal extension and lowering of pediments required climates in which "areal activity" as defined by ROHDENBURG (1983) could take place. The spatial and morphodynamic linkage between pediments and valleys shows that, under such climates, planation and valley formation could occur simultaneously.

The spatial differentiation of valley formation and pediment lowering was due to other, non-climatic factors such as tectonics and lithology. Slow uplift and low rock resistance in conjunction with near-horizontal bedding tended to result in pediment formation, and steeply dipping rocks of varying resistance were more conducive to an accentuation of differences in resistance, especially in the early phase of pediment lowering and extension.

3 General aspects of structural landforms and truncation surfaces in southern Lower Saxony

The spatial association of structural landforms and truncation surfaces in southern Lower Saxony has existed, at least, since the early Tertiary. During the late Tertiary and Quaternary lowering of truncation surfaces continued, particularly in the basins. The younger surfaces are smaller in extent, more closely spaced vertically and geology and tectonics have tended to be of greater importance in their development. The decrease of the effects of successively later phases of form development was thought by ROHDENBURG (1983) to be the result of the intensity selection principle or effectiveness selection principle, which suggests that relic landforms from past phases of geomorphological development are preserved only until phases of greater effectiveness occur during which these relics are removed.

The effectiveness of a phase is determined by the climatically controlled potential for planation surface formation, that is, by the product of intensity and duration of the development phase, and is modified by the geologic-tectonic boundary conditions.

The early to mid-Tertiary truncation surfaces, which can be classified as pediplains or panplains (ROHDENBURG 1983) originated, therefore, during development phases that were very effective because, probably, of their high intensity and long duration. The pediments formed in the Pliocene and early Pleistocene were developed in a period of lesser but nevertheless still considerable planation effectiveness. The early, middle and late Pleistocene pediments attain a significant extent, however, only where the geologic and tectonic conditions were especially favourable, in those places and at those times at which the incision of the trunk streams was slow due to retarded crustal movement. Examples are the early Pleistocene pediments in the Kassel Basin and the middle Pleistocene pediments in the Markoldendorf basin and in the Leine Rift Valley.

The development of pediment sequences and terrace series in the later Tertiary and Quaternary during successive phases of lowering of erosion and denudation process systems is a worldwide phenomenon caused by the interference of tectonics, sea level changes and climatic cycles. The progressive restriction of planation to areas of weakly resistant rock during the late Tertiary and the Quaternary led to a greater contrast between pediments and cuesta scarps and hogbacks. The younger pediments in southern Lower Saxony should, therefore, be termed parapediments (BRUNOTTE 1983), that is, landforms the areal lowering of which accentuates the contrast between the basins or piedmonts and the adjacent upland, quite different from normal pedimentation which tends towards the development of pediplains by progressive expansion until the adjacent upland has been removed.

During the parapedimentation process, structural landforms are etched out of the next higher pediment level. The structural landforms are morphographic and morphogenetic components of the denudation surface system. In phases of sufficient planation potential and a sufficiently stable baselevel, the structural landforms are again progressively eliminated depending upon the resistance and thickness of the rocks that support them and on hydro-geomorphological situation. This concept of cyclic relief development in which there is first, an initial surface, second, a development of structural landforms as a transitional stage during the lowering of the denudation surface and third, a planation of the lowered surface, is based on the investigation of pediments and structural landforms in Argentina (STINGL 1975) and is supported by research results and geomorphological mapping in South Africa (HAGEDORN & BRUNOTTE 1983) and central Europe (GARLEFF, BRUNOTTE & STINGL 1985).

In the Lower Saxon Uplands the late Tertiary and Quaternary phases of pedimentation were not of sufficient intensity and/or duration to eliminate large structural landforms. Only small hogbacks and cuestas were developed on the moderately resistant rocks as a result of the Quaternary lowering of pediments on the margin of the Markoldendorf Basin. These were subsequently removed, in part by planation of the lower level (BRUNOTTE 1978).

These areas in southern Lower Saxony provide examples of the spatial association and synchronous evolution of structural and sculptural landforms in which, under the influence of short term changes of tectonic and climatic conditions in the recent geological past, the development of extensive truncation surfaces was increasingly restricted and superseded by the development of parapediments and by the accentuation of structural landforms.

References

BRINKMANN, R. (1932): Morphogenie und jüngste Tektonik im Leinetalgrabengebiet. Abh. preu. geol. Anst. NF, **139**, Berlin, 102–135.

BROSCHE, K.-U. (1968): Struktur- und Skulpturformen im nördlichen und nordwestlichen Harzvorland. Gött. geogr. Abh. **45**, 1–236.

BRUNOTTE, E. (1978): Zur quartären Formung von Schichtkämmen und Flächen im Bereich des Markoldendorfer Beckens und seiner Umrahmung (Leine-Weser-Bergland). Gött. geogr. Abh. **72**, 1–138.

BRUNOTTE, E. (1979): Quaternary piedmont plains on weakly resistant rocks in the Lower Saxonian Mountains (W.Germany). CATENA **6**, 349–370.

BRUNOTTE, E.(1983): Parapedimente - zum Formungseffekt planierender Abtragung an Beispielen aus West-Argentinien und Süd Afrika. In: STÄBLEIN, G.: Ferdinand v.Richthofen Symposium Oct. 1983, Berlin, 57–58.

BRUNOTTE, E. (1987): Strukturformen im Altrelief der Dransfelder Hochfläche. Z. Geomorph. NF, Suppl. Bd. **66**, 37–47.

BRUNOTTE E. & GARLEFF, K. (1980): Tectonic and climatic factors of landform development on the northern fringe of the German Hill Country (Deutsche Mittelgebirge) since the early Tertiary. Z. Geomorph. NF, Suppl.-Bd **36**, 104–112.

BRUNOTTE, E., GARLEFF, K. & JORDAN, H. (1985): Die Geomorphologische Übersichtskarte 1:50,000 zu Blatt 4325 Nörten-Hardenberg der Geol. Karte von Niedersachsen 1:25,000. Z. dt. Geol. Ges. **136**, 277–285.

BRUNOTTE, E., GARLEFF, K. & WAHLE, H. (1980): Neue morphographische und geomorphologische Karten aus dem südniedersächsischen Bergland. N. Arch. f. Nds. **29**, **1**, 85–96.

FINDEISEN, H.K. (1952): Pleistozäne und tertiäre Flußablagerungen in der Umgebung von Kassel. Unpublished Dissertation. Heidelberg, 141 S.

GARLEFF, K., BRUNOTTE, E. & STINGL, H. (1985): Erläuterungen zur Geomorphologischen Karte der Bundesrepublik Deutschland. GMK 100 Blatt 5, C 4722 Kassel, Berlin.

GARLEFF, K., BRUNOTTE, E. & STINGL, H. (1989): Fußflächen in zentralen Teil der Hessischen Senke. In: GMK - Beitr. **8**, Berlin (in print).

HAGEDORN, J. & BRUNOTTE, E. (1983): Flächen- und Talentwicklung im südöstlichen Kapland/Südafrika und ihre Faktoren. Z. Geomorph. NF, Suppl-Bd. **48**, 235–246.

HÖVERMANN, J. (1953): Die Oberflächenformen um Göttingen. Gött. Jb. **193**, 63–74.

LÜTTIG, G. (1960): Neue Ergebnisse quartärgeologischer Forschung im Raum Alfeld-Hameln-Elze. Geol. Jb. **77**, 337–390.

MENSCHING, H. (1984): Grundvorstellungen zur Geomorphologie der Trockengebiete. Z. Geomorph. NF, Suppl-Bd **50**, 47–52.

MORTENSEN, H: (1947): Rumpffläche-Stufenlandschaft — alternierende Abtragung. Peterm. Geogr. Mitt. **93**, 1–14.

ROHDENBURG, H. (1965a): Die Muschelkalk-Schichtstufe am Ostrand des Sollings und Bramwaldes. Gött. geogr. Abh. **33**, 1–91.

ROHDENBURG, H. (1965b): Untersuchungen zur pleistozänen Formung am Beispiel der Westabdachung des Göttinger Waldes. Gießener geogr. Schr. **7**, 1–76.

ROHDENBURG, H. (1983): Beiträge zur allgemeinen Geomorphologie der Tropen und Subtropen. CATENA **10**, 393–438.

ROHDENBURG, H. & MEYER, B. (1968): Zur Feinstratigraphie und Paläopedologie des Jungpleistozäns nach Untersuchungen an südniedersächsischen und nordhessischen Lößprofilen. Gött. bodenkdl. Ber. **2**, 1–135.

SCHUNKE, E. (1968): Die Schichtstufenhänge im Leine-Weser-Bergland in Abhängigkeit vom geologischen Bau und Klima. Gött. geogr. Abh. **43**, 1–219.

STINGL, H. (1975): Schichtkämme und Fuflächen als Stadien zyklischer Reliefentwicklung. Z. Geomorph. NF, Suppl-Bd. **23**, 130–136.

STINGL, H., GARLEFF, K. & BRUNOTTE, E. (1983): Pedimenttypen im westlichen Argentinien. Z. Geomorph. NF., Suppl.-Bd **48**, 213–224.

Addresses of authors:
Prof. Dr. Ernst Brunotte
Geographisches Institut der Universität Köln
Albertus-Magnus-Platz
D-5000 Köln 41
Federal Republic of Germany
Prof. Dr. Karsten Garleff
Lehrstuhl II für Geographie
Universität Bamberg
Postfach 1549
D-8600 Bamberg
Federal Republic of Germany

CATENA SUPPLEMENT 15 p. 165–177 Cremlingen 1989

GEOMORPHOLOGY OF LIMESTONE AREAS IN THE NORTHEASTERN RHENISH SLATE MOUNTAINS

Karl-Heinz **Schmidt**, Berlin

1 Introduction

The Rhenish Slate Mountains (Rheinisches Schiefergebirge) have long been of interest to geomorphologists. A large number of papers have been published since PHILIPPSON's work first appeared in 1899. DAVIS (1912) regarded the Rhenish Slate Mountains as an excellent example of the erosion cycle. Later publications have dealt mainly with the formation of Tertiary erosion surfaces (summarized by HÜSER 1973 and SEMMEL 1984) and subsequent valley incision and terrace formation (BIBUS 1980). Recent research has been promoted by a priority programme of the Deutsche Forschungsgemeinschaft (DFG) to study vertical movements of the Rhenish Shield (FUCHS et al. 1983).

This paper concentrates on the limestone areas of the northeastern Rhenish Slate Mountains in which there are large karst depressions near Wuppertal, Iserlohn, Attendorn/Helden, Warstein and Brilon (fig.1). The limestone areas are surrounded by slates and sandstones, and show typical lithologically-controlled relief features (WENZENS 1974a, SCHMIDT 1975, 1976, HOFFSTÄTTER-MÜNCHEBERG & PFEFFER 1982, PFEFFER 1984) that have been formed by both past and present-day processes. Although the limestone areas contrast strongly with the areas surrounding them, they must be examined within the framework of the development of the Rhenish Slate Mountains as a whole.

ISSN 0722-0723
ISBN 3-923381-18-2

3-923381-18-4/89/5011851/US$ 2.00 + 0.25

2 Regional setting

The karst areas investigated are made up of Devonian Massenkalk, a massive limestone which was formed in reefs in the shelf region of the Old Red continent. Development of the reefs began in the late middle Devonian and terminated in the early upper Devonian. Together with the impermeable rocks of the Devonian and Carboniferous, they were tectonically deformed during the Variscan orogeny and folded into anticlines and synclines in which rocks of varying resistance are next to each other. These structures are still generally apparent in the distribution of uplands and basins, although the limestone areas form intermontane basins or plateaus in synclinal (Wuppertal, Iserlohn, Attendorn/Helden) and anticlinal (Brilon, Warstein) positions. Their lower elevation relative to the surrounding areas

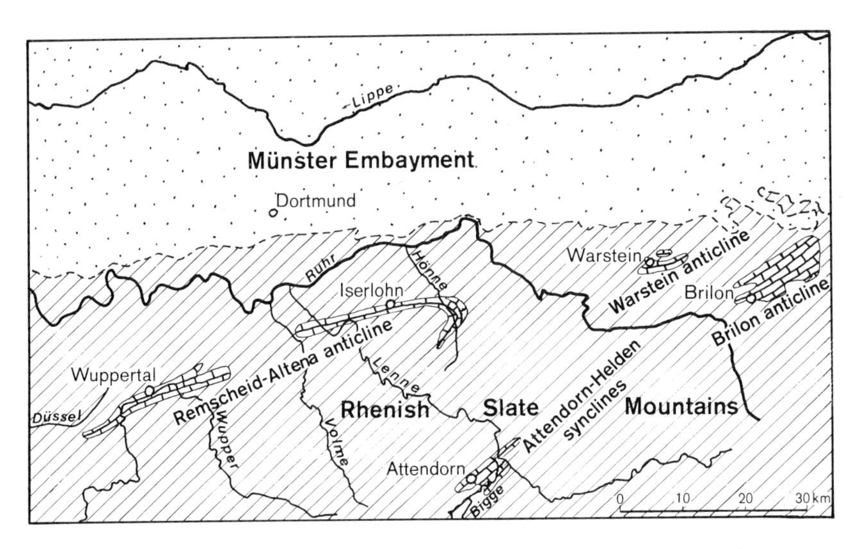

Fig. 1: *Location of the limestone areas in the northeastern Rhenish Slate Mountains. Dotted area in the Münster Embayment is underlain by Cretaceous sediments.*

of impermeable rock is, however, much more pronounced in synclinal positions.

The massive calcareous rocks are generally steeply inclined and attain a thickness of several hundred meters. They are very pure limestones or dolomites with a carbonate content of more than 90% (SCHMIDT 1975). Also present are soft slates and marls and more resistant sandstones of Devonian and Carboniferous age and highly resistant cherts of the lower Carboniferous.

The limestone outcrops in the study area have been worn down to lower elevations than the impermeable rocks (photo 1), a relief pattern that is similar to the other limestone areas in the western and southern Rhenish Slate Mountains. Owing to the general decrease in elevation towards the Cretaceous basin of the Münster embayment in the north, the areas surrounding the limestone depressions are higher on the southern margins of the depressions where there are also remnants of high level planation surfaces (fig.2). The drainage direction of the consequent rivers (Volme, Hönne, Bigge) also follows the regional slope (fig.1). The differences in altitude between the limestone depressions and the uplands in the south totals 250 meters in the Iserlohn and Attendorn basins but only 50 to 100 meters in the plateaus of Warstein and Brilon. The intermontane basins in the limestones form slightly undulating plains with some residual limestone hills in the central area (for example, Brilon, fig.2d) or on the margins (for example, Iserlohn, fig.2a). They are only moderately dissected by shallow, saucer-shaped valleys which have no surface runoff. Their overall valley density is not significantly lower than that in the impermeable rocks because of periglacial valley formation, but their drainage density at the present time is much lower. Allogenous rivers cross the limestone outcrops in deeply incised valleys and canyons, such as that of the Hönne river in the Iserlohn depression.

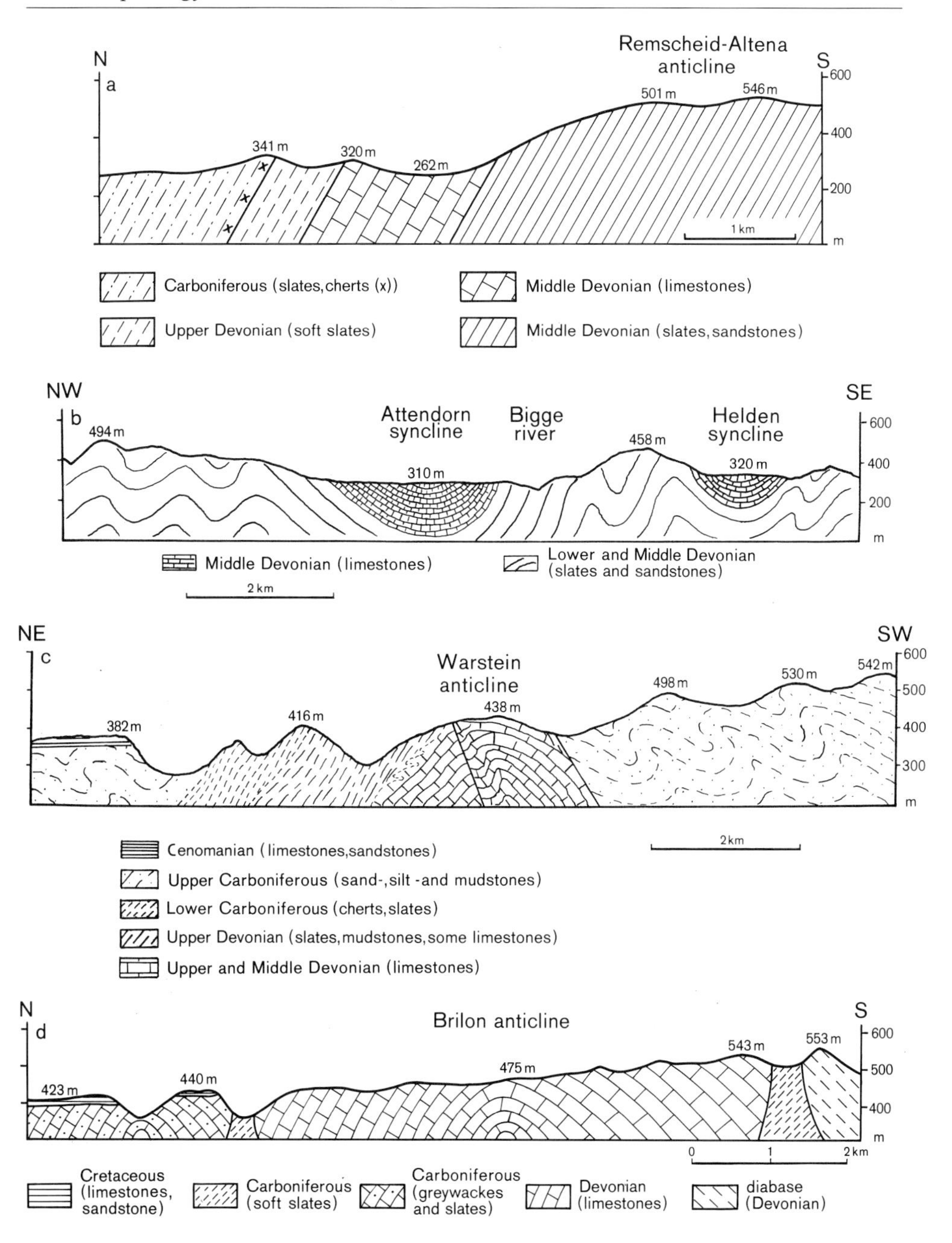

Fig. 2: *Cross-sections of the limestone plateaus in the northeastern Rhenish Slate Mountains. The Wuppertal depression is not shown. It has the same geologic and structural setting as the Iserlohn depression at the northern flank of the Remscheid-Altena anticline.*

a) Iserlohn limestone plateau
b) Attendorn/Helden limestone plateaus (based on PFEFFER 1984)
c) Warstein limestone plateau (modified after WENZENS 1983)
d) Brilon limestone plateau (modified after WENZENS 1983)

Photo 1: *Eastern part of the Iserlohn karst depression.*
The photo shows the approximate location of the cross-profile of fig.2a and looks south to the Remscheid-Altena anticline. At this point, in the Balver Wald area, the anticline has an elevation of more than 500 m in the impermeable middle Devonian slates and sandstones. The karst plateau has an altitude of about 280 m. A flat saucer-shaped valley leads eastward (left) to the Hönne river.

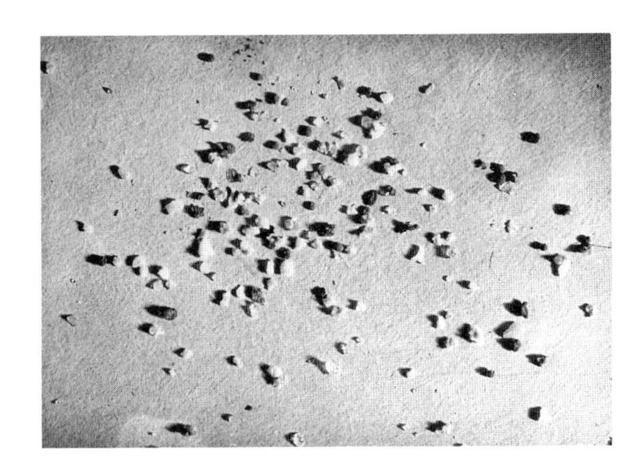

Photo 2: *Medium sand fraction (0.355–0.63 mm) from a doline filling of unconsolidated poorly sorted sands on the Iserlohn karst plateau. The quartz grains are angular, only the glauconite grains (dark) are rounded. The glauconites have been reworked from Cenomanian marine sandstones during Tertiary times.*

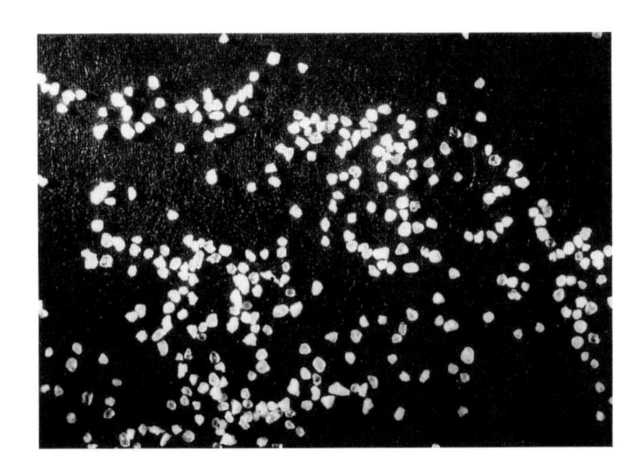

Photo 3: *Medium sand fraction (0.355–0.63 mm) from upper Oligocene marine sands from the Wuppertal depression, which was a bay of the Chattian sea. The quartz grains are well-rounded, the sands are well-sorted and contain no glauconite.*

3 Development and age of the karst plateaus

3.1 Introduction

The basic form of the karst area as intermontane depressions and plateaus is not influenced by different structural settings or by differences in the post-Variscan geologic history of the individual regions in relation to marine transgressions. The close similarity of the karst areas is an expression of the lithologically-controlled processes that were responsible for their formation. What was the nature of these processes and how old are these large karst depressions?

3.2 Earlier investigations

Owing to a lack of datable material the highest planation surfaces in the northeastern Rhenish Slate Mountains have been tentatively assigned to the early and middle Tertiary. They had been thought to be of Oligocene age (KÖRBER 1956) and the lower planation surfaces of the northern margin of the Slate Mountains to be of Miocene age (STEINMANN 1953, HEMPEL 1962). Since the limestone depressions are at lower elevations than the upland surfaces, they had been assumed to be of Pliocene or even early Pleistocene age (for example, STORK 1958). More recent work in other parts of the Rhenish Slate Mountains (Eifel, Westerwald), where datable material has been available, has shown that the older planation surfaces are Cretaceous to early Tertiary and that they were formed in a subdued relief close to baselevel (MEYER et al. 1983, GLATTHAAR & LIEDTKE 1984). Sedimentological evidence has shown that the karst depressions are also much older than previously assumed (SCHMIDT 1975, 1976). Various interpretations of the origin of the depressions have been suggested. They have been described as broad valley floors of fluvial origin (STORK 1958 for the Iserlohn depression; QUITZOW 1978 for the Attendorn depression), as plains shaped by denudation processes on deep weathering profiles (PAECKELMANN 1938) or as poljes and intermontane basins formed by corrosive and erosive processes (WIRTH 1970, WENZENS 1974a, SCHMIDT 1975, 1976). The latter interpretation has found general acceptance (PFEFFER 1984, BURGER 1987).

3.3 Age of the karst deposits

The karst regions in the northeastern Rhenish Slate Mountains are particularly suitable locations for the preservation of marine or terrestrial deposits from the Mesozoic and Tertiary periods because the karst relief offers depositional traps in the form of karst pits and dolines. Only in few cases is there a more widespread depositional cover on the limestones. Unfortunately, the position of the sediments, sometimes deep in the karst hollows, provides no reliable evidence from which the surface elevation of the karst areas at the time of deposition can be determined. This has led to a controversy as to when the large karstic intermontane basins were formed (WIRTH 1964, 1970, 1976, CLAUSEN et al. 1982, CLAUSEN & LEUTERITZ 1984, WENZENS 1983, PFEFFER 1984).

Because of their different spatial and structural settings, the individual karst basins did not experience the same geological and depositional development. When the Rhenish Slate Mountains were exposed after the Variscan orogeny, the limestone areas were subject to solution processes. The oldest palynologically dated karst pit deposits have been found in the karst areas of Iserlohn, Brilon and Warstein. They are lower Cretaceous (Aptian, Albian) and of fluvial origin (WIRTH 1964, CLAUSEN et al. 1982, OEKENTORP 1984).

A major environmental change was caused by the transgression of the Cenomanian (early upper Cretaceous) sea on the northern margin of the Rhenish Slate Mountains. The coastline lay south of the limestone plateaus of Brilon and Warstein, where Cenomanian deposits in karst hollows are common. On the Warstein plateau, marine glauconite sandstones of upper Albian age are also found in karst pits (CLAUSEN & LEUTERITZ 1984).

The ultimate extent of the Cretaceous sea south of the Münster embayment is not known precisely, particularly for the western part of the area investigated. No Cretaceous deposits are known from the Wuppertal or Attendorn/Helden depressions.

It is doubtful whether the upper Cretaceous sediments, dated by their fauna, in a doline on the Iserlohn plateau (WIRTH 1976) are original marine deposits. No detailed description of the fauna and of the location of the doline filled with glauconite sands was given but the material seems to be of the kind investigated by WENZENS (1974b) and SCHMIDT (1975, 1976) in the Beckum doline in the vicinity. The filling of the latter doline has been associated with marine Oligocene glauconite sands from the Cologne embayment (WENZENS 1974a, 1974b). This is improbable for a number of reasons. The quartz grains are not rounded. Only the glauconite grains are rounded (photo 2). Moreover the sands have a high median diameter when compared to the Oligocene marine sands from the Wuppertal depression. Also limitations imposed by topography and, above all, the presence of glauconite, which should be absent in the marginal facies of a marine sediment from the shallow waters of a narrow bay, disprove an Oligocene marine origin (SCHMIDT 1975). It also seems unlikely that the sands are primary Cretaceous marine deposits because the material consists of unconsolidated sands and not of sandstones as they are known from Cretaceous karst deposits in the Warstein and Brilon area (CLAUSEN 1979, CLAUSEN et al. 1982). Clays with a high kaolinite content, which are found in the doline, are not known from primary upper Cretaceous karst deposits in the area investigated.

The sands are very similar to the Tertiary doline sands from the Brilon plateau described by SCHRIEL (1954). The karst deposit of Beckum was interpreted as fluvially reworked Cretaceous material similar to a second, almost identical doline filling in the western part of the Iserlohn depression (SCHMIDT 1975, 1976). This would mean that the Cenomanian transgression reached as far south as the Iserlohn depression but did not cover the karst plateau with sediments at its present level of karst development. The dolines were filled from nearby sources during the Tertiary period.

Recent investigations have shown that the Oligocene marine transgression covered large areas of the western Rhenish Slate Mountains (MEYER et al. 1983, ZÖLLER 1984, SEMMEL 1984, PFEFFER 1984), and also the western margin of the eastern Rhenish Slate Mountains

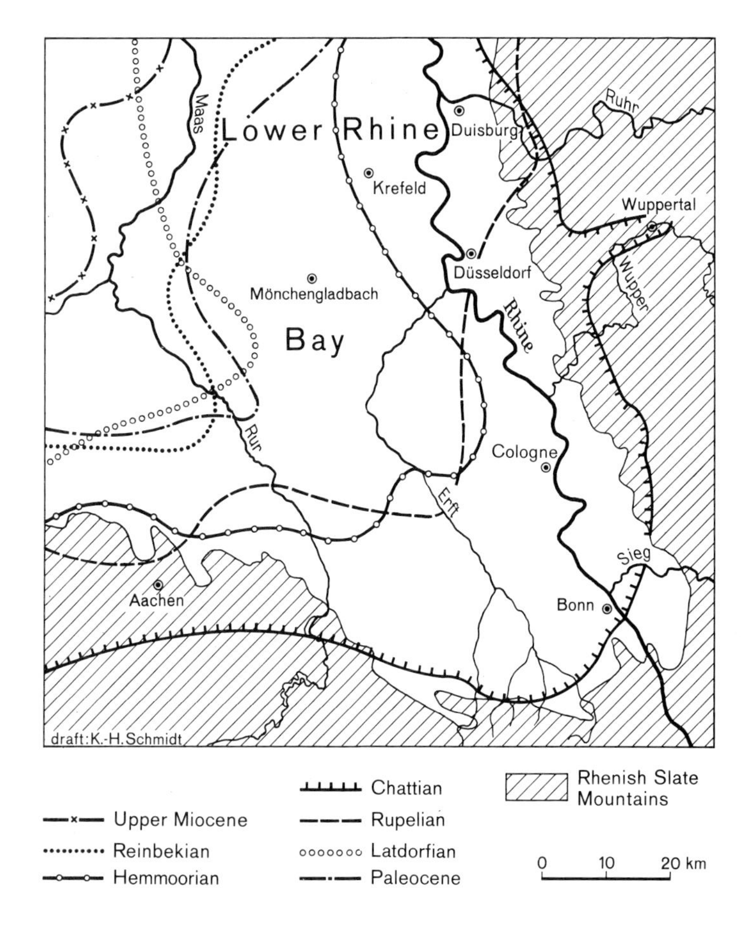

Fig. 3: *Extent of Tertiary transgressions into the Lower Rhine Bay and vicinity (based on QUITZOW 1971 and SCHMIDT 1975).*

(JUX & STRAUCH 1965). The sedimentology of the well-rounded and well-sorted Tertiary sands in the Wuppertal depression (photo 3), which contain well-rounded flint beach gravels from a source in Belgium, indicates that they are equivalent to the Chattian marine sediments of the Lower Rhine Bay (Cologne Embayment) (SCHMIDT 1975, 1976). The Tertiary material of the Wuppertal depression was formerly thought to be an isolated lacustrine depression deposit of Miocene age but it has become clear that the limestone depression of Wuppertal, like the Paffrath depression (HELAL 1958) east of Cologne, was a bay of the Chattian sea (fig.3). The sandy material is not only found in karst hollows, but also covers wider areas of the limestone surface. Further to the east no marine Oligocene sediments are found. After the Oligocene transgression the karst areas in the Rhenish Slate Mountains experienced no further marine periods.

3.4 Implications for the development and age of the karst plateaus

On the basis of their investigations in the Warstein and Brilon area and in the eastern part of the Iserlohn depresssion, CLAUSEN et al. (1982) and WIRTH (1976) draw the general conclusion that, because of the lack of dated early and middle Tertiary deposits in the karst depressions, covered karst conditions prevailed from the Cenomanian transgression until the Pliocene. They infer also that the karst depressions existed already in pre-Cenomanian times.

The lack of early to middle Tertiary deposits in karst cavities may be more apparent than real. Deposits of that age may not be completely missing and all pre-Pliocene deposits may not necessarily belong to the Mesozoic period. The fact that early Tertiary karst deposits are less frequent may be due to the high baselevel elevation of that time which would have prevented the formation of deep karst cavities. Nevertheless, there are examples of Tertiary karst deposits. SCHRIEL (1954) describes a Tertiary doline deposit from the Brilon plateau and PFEFFER (1984) gives additional examples. Doline fillings (SCHMIDT 1975) and nearby palaeosols (PAECKELMANN 1938), which contain bleached cherts, at the northern end of the Iserlohn depression, were assigned to the Tertiary because products of intensive tropical chemical weathering in deep bleached profiles are so far only known from upper Cretaceous and early Tertiary palaeosols. CLAUSEN et al.(1982) state that bleached cherts in lower Cretaceous karst deposits in the Iserlohn depression point to intensive tropical weathering at that time. However, this assumption needs further confirmation since hydromorphic alterations cannot be excluded (WIRTH 1964). Moreover, MNICH (1979) in his detailed study of the pre-Cretaceous palaeo-relief concludes that the Cenomanian sea transgressed a fresh unweathered surface.

The deep karst pits, which were filled with lower Cretaceous material, require a period of deep karstification with a low baselevel. WIRTH (1964) indicates a minimum depth of a lower Cretaceous karst pit of 70-80 meters. The formation of the deep karst pits does not necessarily mean that the limestone areas already lay at lower elevations than the impermeable rocks. The cavities were filled with material from local sources. The cherts in the sediments only indicate that this very resistant lower Carboniferous material stood at higher elevations than the limestone outcrops. The positions of the slates and sandstones in the vicinity are not known.

A difference in elevation of 100–200 meters between the limestone outcrops and the impermeable rocks is assumed for the period of the Cenomanian transgression on the Warstein and Brilon limestone areas (CLAUSEN et al. 1982). The relief difference was then reduced by the marine abrasion processes of the advancing Cenomanian sea. A strong modification of the relief by littoral processes and the pre-Cenomanian existence of greatly lowered karst outcrops seem unlikely for a number of reasons (WENZENS 1983, PFEFFER 1984). The pre-Cenomanian surface in the Brilon area shows a hilly relief with local differences in elevation of 50 meters, making strong littoral action improbable. Moreover, littoral processes are not effective on a subsiding continent. Also, if the limestone surface was 100–200 meters lower than the surrounding area, the drainage from the karst depressions had no outlet from the closed system karst aquifer. If the karst areas were at a lower altitude, they would have become a bay of the Albian sea, the coastline of which was very close to the limestone areas of Brilon and Warstein and touched their northern margins (WENZENS 1983, fig.2). There is no evidence of the actual elevation of the limestone surface at the time of the transgression because deposits are found only in karst cavities. The elevations of the Warstein and Brilon plateaus may well have been above the general surface of the surrounding area (WENZENS 1983).

Because the Cenomanian transgression was not limited to the limestone outcrops and the deposits on the outcrops are only found in isolated cavities below the original surface elevation, there is no evidence of a pre-Cenomanian lowering of the limestone areas. Covered karst conditions (CLAUSEN et al. 1982) must have prevailed for quite a long time after the Cenomanian transgression, longer in the Brilon and Warstein areas than in the vicinity of Iserlohn, because there the present southern limit of the Cretaceous sediments is situated much further to the north. The existence of Tertiary karst deposits (SCHRIEL 1954, SCHMIDT 1975, PFEFFER 1984) indicates, however, that covered karst conditions ceased to exist before the beginning of the Pliocene.

In the Attendorn/Helden karst depressions there is no evidence of either a pre-Cenomanian or a pre-Oligocene corrosive or erosive lowering, because of the general lack of relevant deposits (SCHMIDT 1975). In the northeastern Rhenish Slate Mountains, sediments of a marine transgression are limited to a limestone area only in the Wuppertal karst plateau, which was a bay of the Chattian sea (fig.3). This is the only evidence that a limestone outcrop was relatively lower than the surrounding impermeable rocks. In addition to filling the karst cavities, the Oligocene sedimentary cover overlies parts of the pre-Chattian surface. The Wuppertal karst depression existed as a lithologically-controlled landform in pre-Chattian times and there is reason to suppose that the Iserlohn depression probably developed at the same time (SCHMIDT 1975, 1976). The Stromberg karst depression in the southern Rhenish Slate Mountains (Hunsrück) was also a bay of the Oligocene sea (ZÖLLER 1984) and since there is no evidence of a Mesozoic development of the intermontane basins, it is concluded by analogy that all karst depressions were formed in pre-Chattian times, probably in the early to middle Oligocene periods. This is in agreement with palaeohydrologic and palaeo-climatic evidence.

Karst areas are subject to more rapid degradation than impermeable rocks only in a humid climate with equable runoff and when they do not stand too high above the regional baselevel. In the latter case, the karst hydrography is activated and limestones become resistant to surface lowering, for instance as caprocks of cuesta scarps. In Jurassic and early Cretaceous times the limestone areas probably lay at a higher elevation above baselevel, as demonstrated by the deep karst pits of that period. In the upper Cretaceous and early to middle Tertiary periods, the areas lay below or close to the baselevel elevation. The latter position was suitable for lateral corrosion and erosion and for downwearing of the limestone surfaces. Humid climates prevailed throughout most of the

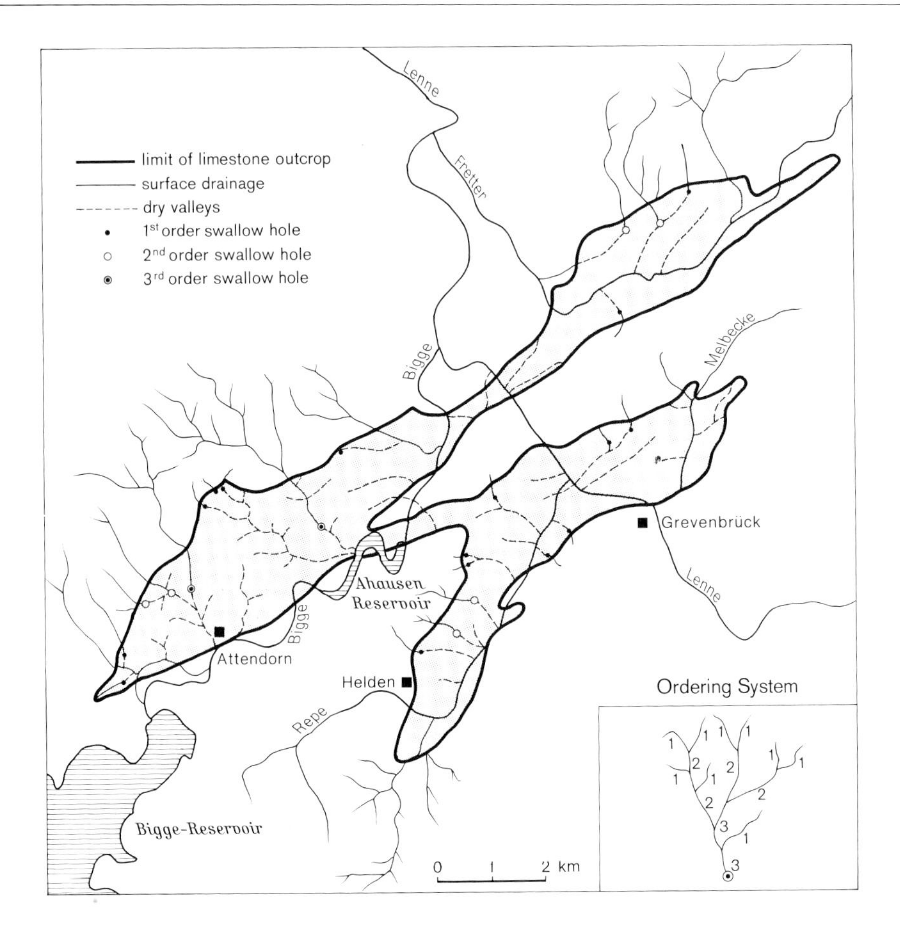

Fig. 4: *Surficial karst hydrography and swallow hole morphometry of the Attendorn/Helden karst depressions (after SCHMIDT 1977).*

Cretaceous and Tertiary periods. In the lower Oligocene, there was a major morphodynamically effective change from a humid tropical climate to a subtropical climate (HÜSER 1973, SCHMIDT 1975, ALBERS 1981). With this change in the climate, the formation of planation surfaces, which had incorporated rocks of different resistance, including the limestones (THOMÈ 1984), was terminated. A period of more differentiated weathering and erosion began, that made selective downwearing of the karst areas possible. Controlled, therefore, by a favourable palaeo-hydrologic and palaeo-climatic framework, the karst depressions were formed in the middle Tertiary before the Chattian transgression.

4 Present-day processes in the karst areas

The dissection of the Rhenish Slate Mountains was initiated with their uplift in the late Tertiary and Quaternary. The limestone areas are also crossed by deeply incised allogenous rivers, al-

Photo 4: *Collapse doline in the eastern part of the Wuppertal depression. The limestone is covered by loess loam. An artificial ditch had been constructed to divert a small stream from a swallow hole, which was located in an upstream direction. After a few years the doline collapsed into the ditch and a new swallow hole was formed.*

though the general appearance of undulating plateaus is not disturbed. In the shallow, saucer-shaped valleys on the plateaus, there is no surface runoff and the small streams that flow from higher areas of impermeable rocks end in swallow holes (fig.4). The drainage to the swallow holes follows the laws of fluvial morphometry (SCHMIDT 1977). The most well-developed swallow holes and blind valleys are found on the Brilon plateau. On approaching the main valleys, the tributaries have sharp knick points in their valley gradients and change to V-shaped cross-profiles. They hang above the local base-level. Fluvial processes were active in the tributary valley systems when there were permafrost conditions during periglacial periods. At the present time, runoff occurs in the dry valleys only during snow melt or heavy rainfall. Seepage water and swallow holes feed karst springs. The Alme spring on the northern margin of the Brilon plateau has an average discharge of about 1000 l/sec. Only a few tracing experiments have been made. The waters from the siliceous rocks are still very aggressive and have a high solution potential when they enter the swallow holes. There are numerous caves in the limestone areas, particularly on both sides of the Hönne river canyon. Present subsurface solution in the limestones is shown by collapse dolines (photo 4).

The rate of chemical denudation from the Hönne river catchment was calculated using the rating curve/flow duration curve method (SCHMIDT 1979). At the present time, about 29 $m^3/km^2 \cdot a$ of material in solution are transported by the Hönne out of the limestone area. This rate of chemical denudation also provides a guideline for the other limestone areas. If it is extrapolated over the entire Quaternary and Tertiary, an unrealistically large amount of carbonate removal results. This means that the Holocene is a period with a high rate of solution activity because of an effective karst hydrography and the enlargement of the vadose zone. Limestone solution was not possible during periods of covered karst conditions and there have been periods, such as the periglacial phases or the longer periods of high baselevel elevation, when subterranean corrosive processes were greatly reduced and solution of carbonates was restricted to the surface outcrops in the karst areas of the northeastern Rhenish Slate Mountains.

References

ALBERS, H.-J. (1981): Neue Daten zum Klima des nordwesteuropäischen Alttertiärs. Fortschr. Geol. Rheinld. u. Westf. **29**, 483–503, Krefeld.

BIBUS, E. (1980): Zur Relief-, Boden- und Sedimententwicklung am unteren Mittelrhein. Frankf. Geowiss. Abh. **D1**, 1–296, Frankfurt.

BURGER, D. (1987): Kalkmulden im Rheinischen Schiefergebirge, Strukturformen aus mikromorphologischer Sicht. Z. Geomorph. N.F., Suppl. Bd. **66**, 15–21, Berlin, Stuttgart.

CLAUSEN, C.-D. (1979): Über kreidezeitliche und pleistozäne Karstfüllungen im Warsteiner Massenkalk (Rheinisches Schiefergebirge). Aufschluß, Sonderband **29**, 113–124, Heidelberg.

CLAUSEN, C.-D., GREBE, H., LEUTERITZ, K. UFFENORDE, H & WIRTH, W. (1982): Zur Paläogeographie, Tektonik und Karstmorphologie der südlichen und östlichen Warsteiner Carbonatplattform (Warsteiner Sattel, Rheinisches Schiefergebirge). Fortschr. Geol. Rheinld. u. Westf. **30**, 241–319, Krefeld.

CLAUSEN, C.-D. & LEUTERITZ, K. (1984): Geologische Karte von Nordrhein-Westfalen 1:25,000. Erläuterungen zu Blatt 4516 Warstein. - 1–155, Krefeld.

DAVIS, W.M. (1912): Die erklärende Beschreibung der Landformen. Leipzig, Berlin, 565 pp.

FUCHS, K., v. GEHLEN, K., MÄLZER, H., MURAWSKI, H. & SEMMEL, A. (Ed.) (1983): Plateau Uplift. The Rhenish Shield - A Case History. Berlin, Heidelberg, New York. 411 pp.

GLATTHAAR, D. & LIEDTKE, H. (1984): Die tertiäre Reliefentwicklung zwischen Sieg und Lahn. Ber. z. dt. Landeskunde **58**, 129–146, Trier.

HELAL, A. H. (1958): Das Alter und die Verbrei/tung der tertiären Braunkohlen bei Bergisch-Gladbach östlich von Köln. Fortschr. Geol. Rheinld. u. Westf. **2**, 419–435, Krefeld.

HEMPEL. L. (1962): Das Großrelief am Südrand der westfälischen Bucht und im Nordsauerland. Spieker, 12, 3–44, Münster.

HOFFSTÄTTER-MÜNCHEBERG, J. & PFEFFER, K.-H. (1982): Petrographisch bedingte Oberflächenformen und Karsterscheinungen im östlichen Teil der Iserlohner Kalkmulde. Hölloch-Nachrichten, **5**, 27–39, Zürich.

HÜSER, K. (1973): Die tertiärmorphologische Erforschung des Rheinischen Schiefergebirges. Karlsruher Geogr. Hefte **5**, 1–135, Karlsruhe.

JUX, U. & STRAUCH, F. (1967): Zum marinen Oligozän am Bergischen Höhenrand. Decheniana **118**, 125–133, Bonn.

KÖRBER, H. (1956): Morphologie von Waldeck und Ostsauerland. Würzburger Geogr. Arb. **3**, 1–155, Würzburg.

MEYER, W., ALBERS, H.-J., BERNERS, H.P., v.GEHLEN, K., GLATTHAAR D., LÖHNERTZ, W., PFEFFER, K.-H., SCHNÜTGEN, A., WIENECKE, K. & ZAKOSEK, H. (1983): Pre-Quaternary Uplift in the Central Part of the Rhenish Massif. In: FUCHS, et al. (eds): Plateau Uplift. The Rhenish Shield. A Case History, Berlin, Heidelberg, New York, 39–46.

MNICH, J. (1979): Das mittlere Diemeltal bei Marsberg und seine angrenzenden Hochflächen. Dissertation, Bonn, 356 pp.

OEKENTORP, K. (1984): Die Saurierfundstelle Brilon-Nehden (Rheinisches Schiefergebirge) und das Alter der Verkarstung. Kölner Geogr. Arb. **45**, 293–315, Köln.

PAECKELMANN, W. (1938): Geologische Karte von Preußen und benachbarten deutschen Ländern. Erläuterungen zu Blatt 4613 Balve. Berlin, 70 pp.

PHILIPPSON, A. (1899): Entwicklungsgeschichte des Rheinischen Schiefergebirges. Sitz. Ber. niederrhein. Ges Natw.- u. Heilkde. 48–50, Bonn.

PFEFFER, K.-H., (1984): Zur Geomorphologie der Karstgebiete im Rheinischen Schiefergebirge. Kölner Geogr. Arb. **45**, 247–291, Köln.

QUITZOW, H. W. (1971): Tertiär. Der Niederrhein **38**, 101–103, Krefeld.

QUITZOW, H. W. (1978): Geomorphologie und Landschaftsgeschichte. In: ZIEGLER, W.: Geologische Karte von Nordrhein-Westfalen. Erläuterungen zu Blatt 4813 Attendorn, 155-160, Krefeld.

SCHMIDT, K.-H. (1975): Geomorphologische Untersuchungen in Karstgebieten des Bergisch-Sauerländischen Gebirges. Ein Beitrag zur Tertiärmorphologie des Rheinischen Schiefergebirges. Bochumer Geogr. Arb. **22**, 156 pp, Paderborn.

SCHMIDT, K.-H. (1976): Strukturbedingte tertiäre Reliefgestaltung am Beispiel von Kalkgebieten am Nordrand des rechtsrheinischen Schiefergebirges. Z. Geomorph. N.F., Suppl. Bd. **24**, 68–78, Berlin - Stuttgart.

SCHMIDT, K.-H. (1977): Morphometrie der Attendorner und Heldener Kalkmulden. Ein Beitrag zur morphometrischen Analyse von Karstgebieten des humid-temperierten Klimabereiches. Ber. z. dt. Landeskunde **51**, 11–28, Meisenheim/Glan.

SCHMIDT, K.-H. (1979): Karstmorphodynamik und ihre hydrologische Steuerung. Erdkunde **33**, 169–178, Bonn.

SCHRIEL, W. (1954): Neue Tertiärfunde im Sauerland als Zeitmarke für die junge Vererzung (Pb, Zn, Cu, Fe) im Massenkalk. N. Jb. F. Mineral. Mh. **10**, 226–230, Stuttgart.

SEMMEL, A. (1984): Geomorphologie der Bundesrepubik Deutschland. Stuttgart, 192 pp.

STEINMANN, H.G. (1953): Zur Morphologie des Bergischen Landes. Verh. Dt. Geographentag **29**, 173–177, Essen.

STORK, T. (1958): Der Flußtal der Hönne. Spieker **9**, 3–34, Münster.

THOMÈ, K. (1974): Grundwasserhöffigkeiten im Rheinischen Schiefergebirge in Abhängigkeit von Untergrund und Relief. Fortschr. Geol. Rheinld. u. Westf. **20**, 259–280, Krefeld.

WENZENS, G. (1974a): Morphogenese der Iserlohner Kalksenke. Decheniana **126**, 133–150, Bonn.

WENZENS, G. (1974b): Eine oligozäne Dolinenfüllung in der Iserlohner Kalksenke. Erdkunde **28**, 138–140, Bonn.

WENZENS, G. (1983): Ein Beitrag zur Morphogenese der Karstlandschaften im nördlichen Sauerland. Karst u. Höhle 1982/83, 7–13, München.

WIRTH, W. (1964): Über zwei Unterkreiderelikte im Nördlichen Sauerland. Fortschr. Geol. Rheinld. u. Westf. **7**, 403–420, Krefeld.

WIRTH, W. (1970): Eine Tertiärzeitliche Karstfüllung bei Eisborn im Sauerland. Fortschr. Geol. Rheinld. u. Westf. **17**, 577–588, Krefeld.

WIRTH, W. (1976): Bodenkarte des Kreises und der Stadt Iserlohn 1:50,000, Erläuterungen. Krefeld. 96 pp.

ZÖLLER, L. (1984): Reliefgenese und marines Tertiär im Ost-Hunsrück. Mainzer geowiss. Mitt. **13**, 97–114, Mainz.

Address of author:
Dr. Karl-Heinz Schmidt
Institut für Physische Geographie
Freie Universität Berlin
Altensteinstraße 19
D-1000 Berlin

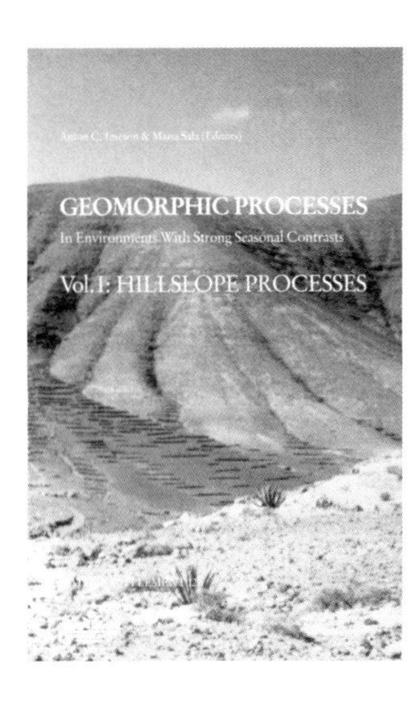

Anton C. Imeson & Maria Sala:

GEOMORPHIC PROCESSES

In Environments With Strong
Seasonal Contrasts
Vol. I: HILLSLOPE PROCESSES

CATENA SUPPLEMENT 12, 1988

Price: DM 149,– / US $88.–

ISSN 0722-0723 / ISBN 3-923881-12-3

CONTENTS

CATENA SUPPLEMENT 15 p. 179–188 Cremlingen 1989

THE IMPORTANCE OF LOESS IN THE INTERPRETATION OF GEOMORPHOLOGICAL PROCESSES AND FOR DATING IN THE FEDERAL REPUBLIC OF GERMANY

Arno **Semmel**, Frankfurt

1 Introduction

Loess can be used to interpret present and past processes in geomorphology, including various stages of landform development that occurred during the formation of the loess. Loess is also used frequently in the dating of landforms and as an indicator of the nature of paleoclimates. The value of loess research has been recognized since the turn of the century. Research has, however, been stimulated in recent years by a refined stratigraphic subdivision of loess which, in the Federal Republic, has been carried out especially by FREISING (1951), SCHÖNHALS (1951) and BRUNNACKER (1954).

Loess can generally be subdivided by means of fossil soils the characteristics of which vary depending upon the climatic conditions that prevailed at their time of origin. Frequently there are only remnants of such soils. Another means of subdivision are the volcanic layers that occur over large areas in the western part of the Federal Republic and that are derived from the Pleistocene volcanism in the Eifel. Using buried soils and volcanic layers a differentiated system has been developed that determines the structure and age of the loess cover of the last two Pleistocene glaciations. The loesses of the Würm or Weichsel glaciation are best preserved. Classifiable loess from the older Riss or Saale occurs less frequently and from earlier glaciations only rarely (fig.1).

It is generally assumed that a parabrown earth (Orthic Luvisol) or a strong pseudo-gley developed on loess during the interglacials while black earth soils (chernozems, humus zones) and soils with weaker brown horizons belong to the interstadials. Weak pseudogleys (Nassböden) cannot always be definitely associated with a particular climate. The suggestion that they represent warmer phases of climate is contradicted by the presence of a Columella fauna, which prefers low temperatures, in some deposits. In general, the interpretation of climate from loess profiles is problematic. This is particularly true for fossil soils in loess. The climatic characteristics that resulted in the formation of humus horizons and brownish horizons within the loess cannot be accurately interpreted. Also, the parabrown earth remnants might have

ISSN 0722-0723
ISBN 3-923381-18-2

3-923381-18-4/89/5011851/US$ 2.00 + 0.25

been formed in a climate colder than that of a genuine interglacial. For example, the Holocene derno-podzols on Siberian loesses, which correspond to the parabrown earths, originated in a much colder climate than that in which central European parabrown earths were developed in the Holocene. However, despite such uncertainties loess is of great value in the interpretation of the geomorphology of many areas.

2 The function of loess soils in the interpretation of recent and subrecent landform development

The loesses of the Federal Republic of Germany are commonly covered by parabrown earths. Only in some very dry areas are there loess chernozems (Rheinhessen) or degraded chernozems (Wetterau, Hildesheim). The profiles of parabrown earths can be used to estimate the extent of denudation by soil erosion. For example, in the Ap-Al-Bt-C profile of a parabrown earth on loess that is under cultivation the initiation of soil erosion is indicated by a diminished Al horizon that is poor in clay. Owing to its high silt content, this horizon is easily eroded, resulting eventually in an Ap-Bt-C profile (fig.2). The high clay content of the Bt horizon makes it more resistant to erosion, although where erosion is intensive this horizon also disappears. The remaining calcareous loess is then ploughed directly. The loess below the B horizon is often highly resistant because of a Cc horizon that has formed as a result of calcareous accumulation. Where this Cc horizon has been removed, it is not possible to estimate the depth of the erosion that has taken place. Recently, however, it has been shown that in many areas weakly-developed fossil pseudogleys (Nassböden) within loess are being cultivated on surfaces from which the entire recent soil profile has been eroded. This denser horizon is more resistant to erosion than normal loess because of calcareous ($CaCO_3$) infiltrations. The occurrence of this weak pseudo-gley can be used to estimate the approximate extent of erosion. This method is particularly useful where the thickness of the soil horizons and loess strata are relatively constant. When the decalcification depth and consequently the thickness of the B horizon vary greatly, estimating soil erosion is more difficult, especially on the coarser loesses of the northern and northwestern loess regions of Germany. The decalcification level in these loesses varies. Below the Bt horizon there is always a Bv horizon, poorer in clay, that normally has a thickness of 20 centimeters but in some areas reaches a thickness of over 400 centimeters. Clay-silt lenses (Lamellenflecken) are frequently present.

An estimate of soil erosion based on degraded chernozem profiles is often not possible because the original thickness of the soils, which are cultivated at the present time, is not known. It is not certain whether these soils are a degraded chernozem or whether they are "Ackerbraunerde" as they are defined on the Bavarian soil map. The latter are deeply eroded parabrown earths with humus-containing B horizons, the result of a long period of cultivation. Apart from these problems, however, it is possible to estimate the extent of recent soil erosion from an examination of loess soils.

It should be noted that loess soils that have been eroded require a long time to regenerate. On old fields that have been woodland for more than 200 years,

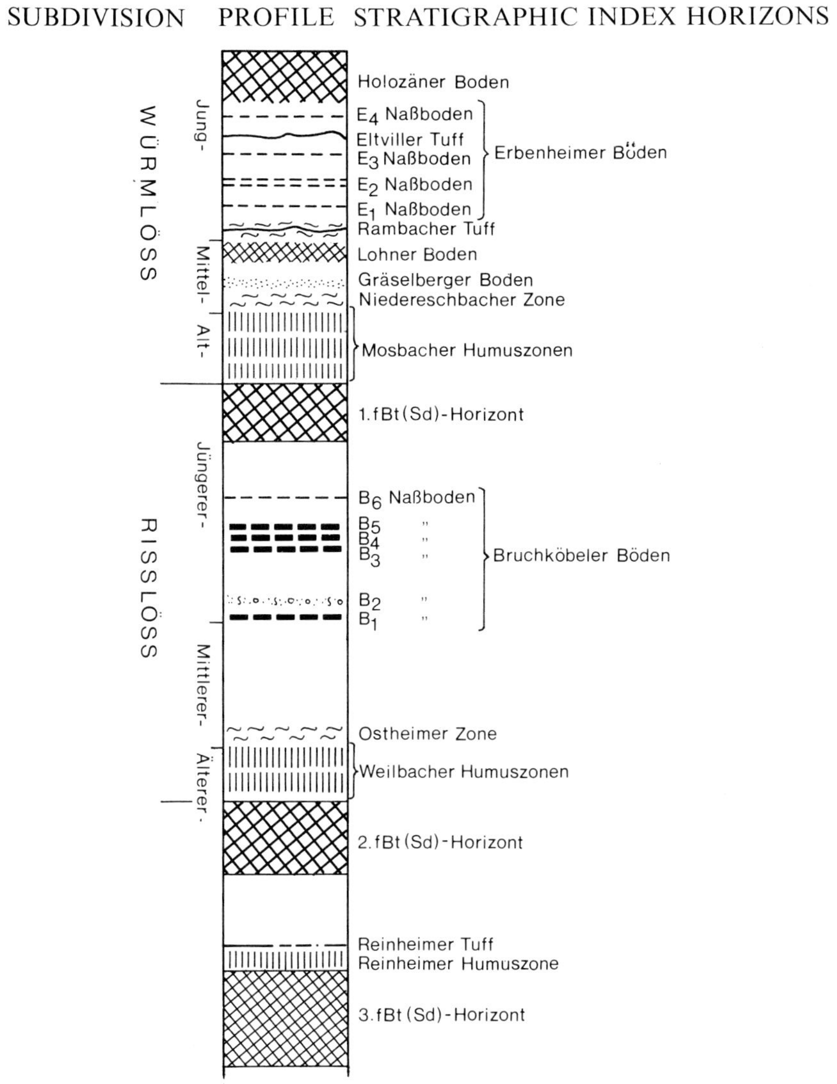

SUBDIVISION PROFILE STRATIGRAPHIC INDEX HORIZONS

WÜRMLÖSS
Jung-
Mittel-
Alt-
Holozäner Boden
E_4 Naßboden
Eltviller Tuff
E_3 Naßboden
E_2 Naßboden
E_1 Naßboden
Erbenheimer Böden
Rambacher Tuff
Lohner Boden
Gräselberger Boden
Niedereschbacher Zone
Mosbacher Humuszonen

RISSLÖSS
Jüngerer-
Mittlerer-
Älterer-
1. fBt(Sd)-Horizont
B_6 Naßboden
B_5 "
B_4 "
B_3 "
B_2 "
B_1 "
Bruchköbeler Böden
Ostheimer Zone
Weilbacher Humuszonen
2. fBt(Sd)-Horizont
Reinheimer Tuff
Reinheimer Humuszone
3. fBt(Sd)-Horizont

Fig. 1: *Classification of middle and late Pleistocene loesses in Hessen (Nassböden in Riss loess according to BIBUS 1974).*
Gliederung = Subdivision; Profil = Profile; stratigraphische Leithorizonte = stratigraphic index horizons

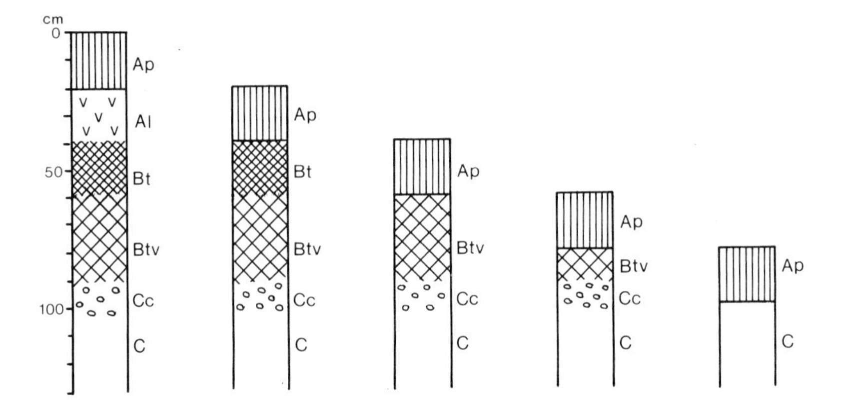

Fig. 2: *Loess-parabrown earth as indicator of the extent of soil erosion.*

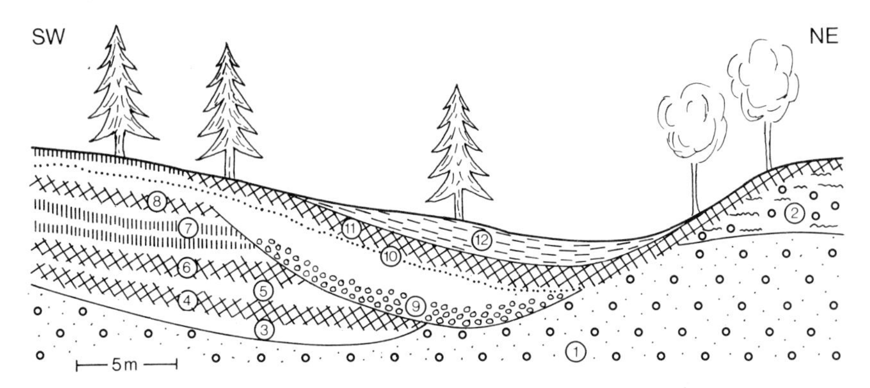

Fig. 3: *Reconstruction of the development of a dell in the Southern Taunus piedmont by means of loess classification.*
1 = Tertiary gravel; 2 = Early Pleistocene gravel; 3 = Earliest loess; 4 = Earliest buried Bt horizon; 5 = Middle loess; 6 = Latest buried Bt horizon; 7 = Early Würm humus zones; 8 = "Lohner Boden"; 9 = Solifluction debris; 10 = "Naßboden" in latest loess; 11 = Holocene parabrown earth; 12 = Colluvium.

there has been hardly any regeneration. There have been reports of newly-formed "brown earths" on eroded loess surfaces but upon examination it was found that the seemingly newly-formed Bv horizon was in fact the non-eroded remnant of an Ap-Al-Bt-Bv-C profile of a former parabrown earth.

3 Landform development in loess areas

At the time of their formation loess soils lie on the existing landforms. Renewed loess deposition preserves the previously developed soil and with it the shape of its surface. On the basis of such evidence the landform development in an area can be estimated. A dell on the southern edge of the Taunus provides an example (SEMMEL 1978).

The dell is incised in Tertiary and early Pleistocene gravels (1 and 2 in fig.3). The oldest recognizable surface after the incision was covered with loess, (3) on which a parabrown earth, (4) developed, probably in the penultimate interglacial. This soil occurs only in remnants, indicating a period of denudation before more loess (5) was deposited, on which parabrown earths were also subsequently developed. It is likely that several zones of humus and Nassböden were developed and then denuded during this latter period of loess deposition, since the parabrown earth (6) that developed on these deposits probably belongs to the last interglacial (fig.1).

The next younger loess deposit contains two humus zones (7) of the early Würm and the Lohner Boden (8) of the middle Würm. These soils indicate long periods of geomorphological quiescence which were, however, repeatedly interrupted by loess sedimentation and by phases of denudation, since several fossil soils found elsewhere are absent here. Above the Lohner Boden a pronounced discordance occurs in the lower late Würm loess. The older loesses and soils had all been eroded when the dell was shifted northeastward into the adjacent loess-free slope. Alluvial sediments and solifluction material (9) in the basal part of the dell at this new location indicate that a period of cold moist climate preceded the cold dry period of loess deposition (10) that subsequently partially filled the dell. During this latter period of deposition a short wetter period occurred, indicated by the presence of a humus-containing Nassboden.

The parabrown earth (11) developed on this younger loess in the Holocene was preserved under colluvium (12) in the low central part of the dell as well as on the gravel deposits on the steep northeastern flank which has remained uncultivated. On the less steep southeastern slope, this parabrown earth has been removed by soil erosion during a period of cultivation.

Similar patterns of development have been described on the basis of loess classifications in many areas of West Germany (ROHDENBURG 1968, SEMMEL 1968, GEILENKEUSER 1970, SEMMEL & STÄBLEIN 1971). The alternating periods of deposition and erosion in the Würm loess have been clearly identified and allow the reconstruction of the temporal sequence of landform development. For example, asymmetric periglacial dells and valleys shifted eastward during the Würm and also in earlier glacial periods. During each phase of denudation the west-facing loess-free slopes were undercut. On the gentler eastern slopes, loess deposition dominated rather than denudation. Present-day land use reflects these differences. The parabrown earth on the loess of the eastern slopes is generally arable land. The steeper west-facing slopes are used for viticulture, where the climate is suitable, or fruit growing or forestry. They also serve as building sites, despite the danger of sliding at some locations. Other economic and ecological aspects of these valleys are discussed in SEMMEL (1986).

The date and extent of young tectonic movement can be recognized from displaced loess strata. On the eastern edge of the northern Rhine Rift Valley, for example, fossil loess soils of different ages have been displaced by varying amounts on the "Bergsträsser diluvial terrace" (SCHMITT & SEMMEL 1971). On the western edge of the Rift Valley, movements have been continuous as indicated by displacements that

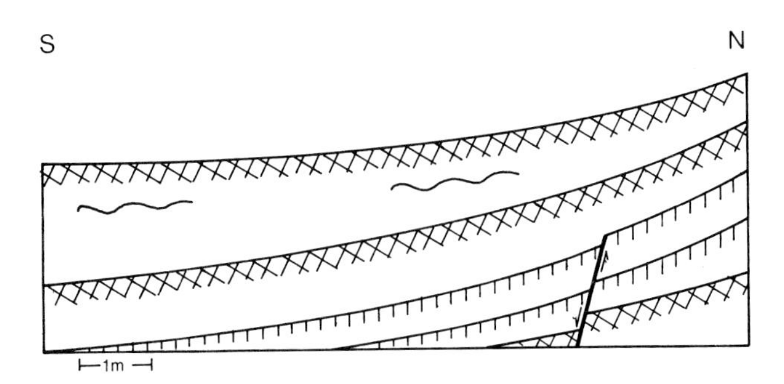

Fig. 4: *Tectonic displacements in loess profile (western boundary of the northern upper Rhine Rift Valley).*
The earliest buried soil (cross-hatching) is more displaced than the humus zones (vertical hatching). Above them the "Lohner Boden" is not disturbed.

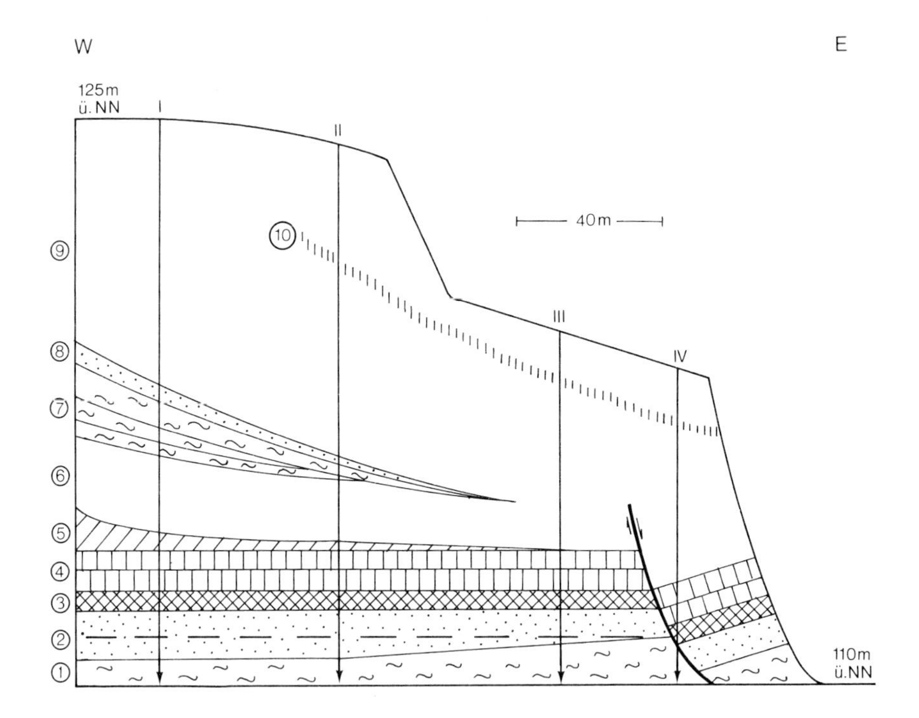

Fig. 5: *Slippage datings by means of loess classification (Rheinhessen).*
1 = Tertiary marl; 2 = Rhine sand and groundwater level (broken line); 3 = buried Bt horizon; 4 = Early Würm humus zone; 5 = Loess loam; 6 = Middle Würm loess; 7 = Solifluction marl, 8 = Rhine sand; 9 = Late Würm loess; 10 = "Nassboden"; I–IV = Drillings.

reach from the loesses into the present-day soil cover. Displacements in the late Würm loess have been smaller than in the middle Würm loess. Zones of humus in the early Würm and the Bt horizon of the last interglacial show the largest displacements (fig.4 and SEMMEL 1978, 1979, 1984).

Undisturbed loess horizons north of Flörsheim on the Main river suggest an absence of vertical movement during the last 20,000 years. Soil horizons in the pit of the brick works at Bad Soden am Taunus indicate that vertical movements ceased before the third Bt horizon, counted downward from the surface, was developed (SEMMEL 1978).

Loess deposits can also be used to date landslides. For example, on a vineyard slope north of Guntersblum (fig.5), undercutting by the Rhine during the middle Würm resulted in a major landslide. The soils of the last interglacial and the early Würm humus zones were affected by these movements. There was no undercutting during the late Würm, as shown by an undisturbed weakly-developed Nassboden ((10) in fig.5) that lies within the late Würm loess. Only since they have been used for viticulture have some parts of the slope been set in motion again (SEMMEL 1986).

4 Loess and the dating of landforms

Loess has often been used to date landforms in the Federal Republic of Germany. Most frequently a counting method is used for soils thought to have been developed during the interglacials. Using this method, glaciofluvial features in the German Alpine Foreland (LÖSCHER et al. 1979), thick loess deposits on gravel terraces in the southern Rhine Rift Valley (BRONGER 1966) or in the middle Rhine valley (BIBUS 1980) and volcanoes in the Eifel (WINDHEUSER & BRUNNACKER 1978) have been dated. A major disadvantage of dating by counting interglacial soils is that it is usually not clear whether one or more soils in a sequence are missing in denudational unconformities. In one instance, for example, I mistook a severely truncated Bt horizon for an interstadial form (SEMMEL 1961).

Only rarely do loess or loess soils in the Federal Republic contain absolutely or relatively datable material. Sometimes, molluscan faunas can be used for relative dating as, for example, in the loess of the Kärlich profile (BRUNNACKER et al. 1982). Pollen and charcoal remnants have not, so far, provided conclusive evidence. Absolute dates of volcanic tuffs within loess have been obtained at some locations in the middle Rhine area and in the Wetterau (LIPPOLT et al. 1986). Some dating has also been possible by thermoluminescence (TL) or paleomagnetism. For example, loesses were dated in the crater of the Plaidter Hummerich in the Neuwied Basin (SINGHVI et al. 1986) and in the Mainz basin and the Rhine Rift Valley (WINTL & BRUNNACKER 1982; WAGNER & ZÖLLER 1987) using the TL method. Datings have been possible as far back as the third oldest glaciation. By means of paleo-magnetism, loesses of up to about one million years old have been dated (fig.6; SEMMEL & FROMM 1976).

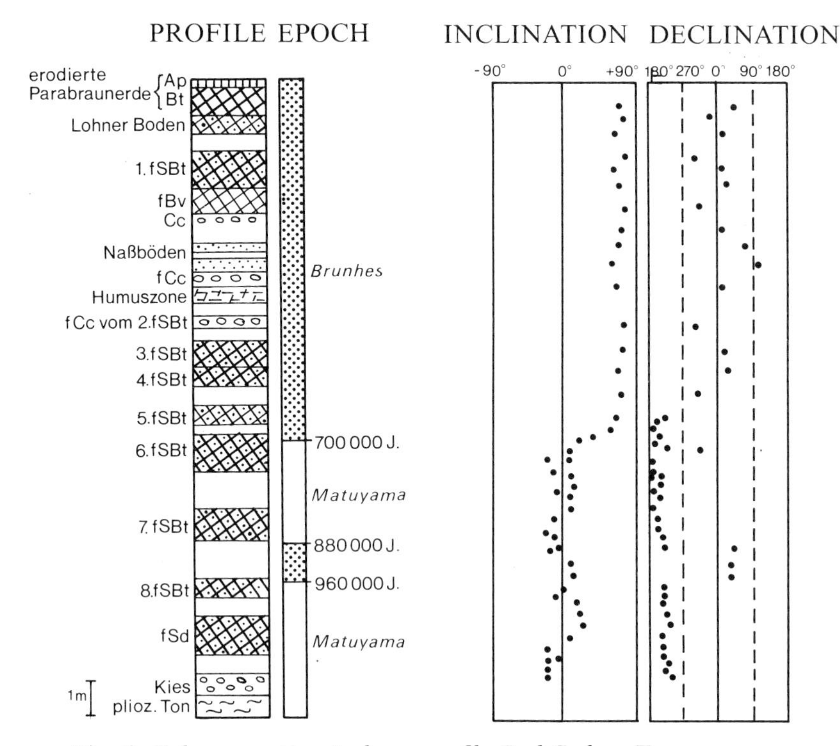

Fig. 6: *Paleomagnetism in loess profile Bad Soden, Taunus.*

5 Loess profiles as climatic indicators

The essential criterion for a subdivision of loesses is the presence of fossil soils developed under different climatic conditions from which it is possible to deduce ideal profiles for a cycle of interglacial and glacial periods. One example is the loess profile at Dreihausen in Hessen, described by SABELBERG et al. (1976) where 13 such cycles have been identified. Above each interglacial soil, or above its remnants slightly displaced by solifluction, there are wash-transported sediments that have become bleached or humus-containing. These are followed by soil materials that have been transported by solifluction, and above them lie solifluction covers composed of loess and basaltic material that was picked up by the movement on uncovered slopes. The succession of beds is completed with loess deposits. Based on this succession, the following climatic development has been suggested (SABELBERG et al. 1976):

"Intensive soil development occurred in the interglacial (parabrown earth on loess and clayey soils on weathered basalt).

During the transition to the glacial period, there was a short period of gelisolifluctional earth movement, although soil formation still dominated generally. Lessivage of clay or pseudogleyization took place at some sites. Short periods of pedogenesis accompanied by relatively weak soil movements occurred.

A period of predominantly fluvial downwearing followed at which time the vegetation must have changed considerably. Some topsoils were partially or wholly stripped and towards the end of this phase a sandy sediment accumulated in some areas, with the formation of a bleached horizon and the precipitation of iron and manganese.

After another phase of denudation and sedimentation, soils with humus horizons developed, or humus-containing sediments that had been formed earlier were removed and redeposited elsewhere. Clay was also lessivaged in non-calcareous substrates during short periods. In the subsequent interruptions of humus soil formation, strong geli-solifluction occurred and small amounts of aeolian loess sedimentation took place for short periods of time in the form of loess added to the upper layers of the humus zones.

During the transition to the middle glacial period, strong geli-solifluction occurred which initially affected the more clayey soil remnants. Ultimately, coarse basaltic material and loess were also transported. In the later part of the glacial period, geli-solifluction became weaker and the accumulation of loessial sediments dominated. On these sediments, a new parabrown earth developed at the beginning of the following interglacial".

References

BIBUS, E. (1974): Abtragungs- und Bodenbildungsphasen im Rißlös. Eiszeitalter und Gegenwart **25**, 166–182.

BIBUS, E. (1980): Zur Relief-, Boden- und Sedimententwicklung am unteren Mittelrhein. Frankfurter geowissenschaftliche Arbeiten **D 1**, 296 pp.

BRONGER, A. (1966): Lösse, ihre Verbraunungszonen und fossilen Böden. Schriften geogr. Inst. Univ. Kiel **24**, 114 pp.

BRUNNACKER, K. (1954): Löß und diluviale Bodenbildungen in Südbayern. Eiszeitalter und Gegenwart **4/5**, 83–86.

BRUNNACKER, K., LÖSCHER, M., TILLMANNS, W. & URBAN, B. (1982): Correlation of the Quaternary Sequence in the Lower Rhine Valley and Northern Alpine Foothills of Central Europe. Quaternary Research **18**, 152–173.

FREISING, H. (1951): Neue Ergebnisse der Lößforschung im nördlichen Württemberg. Jahrbuch der geologischen Abteilung des württembergischen statistischen Landesamtes, Stuttgart, 1. Jahrgang, 54–59.

GEILENKEUSER, H. (1970): Beiträge zur Morphogenese der Lößtäler im Kaiserstuhl. Freiburger geographische Hefte **9**, 111 S.

LIPPOLT, H.J., FUHRMANN, U. & HRADETZKY, H. (1986): Ar/ Ar Age Determination on Sanidins of the Eifel Volcanic Field (FRG): Constraints on Age and Duration of a Middle Pleistocene Cold Period. Chemical Geology, Amsterdam **59**, 187–204.

LÖSCHER, M., SCHIES, A., LEGER, M. & DABELSTEIN, H.-J. (1979): Pedologische Untersuchungen in den Deckschichten des oberen Hochterrassenschotters im Günztal und ihre Aussagen für die Schotterstratigraphie in der Iller-Lech-Platte. Heidelberger geographische Arbeiten **49**, 179–193.

ROHDENBURG, H. (1968): Jungpleistozäne Hangformung im Mittel-Europa. Göttinger Bodenkundliche Berichte **6**, 3–107.

SABELBERG, U., MAVROCORDAT, G., ROHDENBURG, H. & SCHÖNHALS, E. (1976): Quartärgliederung und Aufbau von Warmzeit-Kaltzeit-Zyklen in Bereichen mit Dominanz periglazialer Hangsedimente, dargestellt am Quartärprofil Dreihausen/Hessen. Eiszeitalter und Gegenwart **27**, 93–120.

SCHMITT, O. & SEMMEL, A. (1971): Zum Aufbau der Bergsträßer Diluvialterrasse südlich Bensheim. Notizblatt des hessischen Landesamtes für Bodenforschung, Wiesbaden **99**, 232–239.

SCHÖNHALS, E. (1951): Über fossile Böden im nicht vereisten Gebiet. Eiszeitalter und Gegenwart **1**, 109–130.

SEMMEL, A. (1961): Die pleistozäne Entwicklung des Weschnitztales im Odenwald. Frankfurter Geographische Hefte **37**, 425–492.

SEMMEL, A. (1968): Studien über den Verlauf jungpleistozäner Formung in Hessen. Frankfurter geographische Hefte **45**, 133 pp.

SEMMEL, A. (1978): Böden und Bodenkunde in der geomorphologischen Forschung. Beiträge zur Quartär- und Landschaftsforschung, Wien, 511–520.

SEMMEL, A. (1978a): Untersuchungen zur quartären Tektonik am Taunus-Südrand. Geologisches Jahrbuch Hessen **106**, 291–302.

SEMMEL, A. (1986): Angewandte konventionelle Geomorphologie. Frankfurter geowissenschaftliche Arbeiten **D 6**, 114 S.

SEMMEL, A. & Fromm, K. (1976): Ergebnisse paläomagnetischer Untersuchungen an quartären Sedimenten des Rhein-Main-Gebietes. Eiszeitalter und Gegenwart **27**, 18–25.

SINGHVI, A.K., SAUER, W. & WAGNER, G.A. (1986): Thermoluminescence Dating of Loess Deposits at Plaidter Hummerich and its Implications of the Chronology of Neandertal Man. Die Naturwissenschaften **73**, 205–207.

WAGNER, G.A. & ZÖLLER, L. (1987): Thermoluminescenz: Uhr für Artefakte und Sedimente. Physik in unserer Zeit **18**, 1–9.

WINDHEUSER, H. & BRUNNACKER, K. (1978): Zeitstellung und Tephrostratigraphie des quartären Osteifel-Vulkanismus. Geologisches Jahrbuch Hessen **106**, 261–271.

WINTL, M. & BRUNNACKER, K. (1982): Ages of Volcanic Tuff in Rheinhessen obtained by Thermoluminescence Dating of Loess. Die Naturwissenschaften **69**, 181–182.

Address of author:
Prof. Dr. Dr. h.c. Arno Semmel
Institut für Physische Geographie
Universität Frankfurt
Senckenberganlage 36
Postfach 111932
6000 Frankfurt am Main
Federal Republic of Germany

CATENA SUPPLEMENT 15 p. 189–197 Cremlingen 1989

PERIGLACIAL GLACIS (PEDIMENT) GENERATIONS AT THE WESTERN MARGIN OF THE RHINE HESSIAN PLATEAU

Nordwin **Beck**, Koblenz

1 Introduction

Pediments are commonly known from semiarid and subhumid tropical regions. It has been asked (SEMMEL 1972) whether pediments that occur in central Europe could have developed during the Pleistocene or whether they are relic forms dating from the periods of warmer climate in the Tertiary. An investigation of such landforms was made in the region of the Rhine Hessian Plateau, an uplifted area in the northern part of the Rhine Rift Valley, WSW of Mainz, with particular emphasis on the escarpment that forms the western margin of the plateau and on its dissected footslope that extends for several kilometers westward to the valley terrace of the Nahe river and its tributary, the Wiesbach (fig.1). In this paper, the term glacis is used to describe a pediment-like erosion surface developed on weakly-resistant Neogene sedimentary rocks or unconsolidated material.

ISSN 0722-0723
ISBN 3-923381-18-2

3-923381-18-4/89/5011851/US$ 2.00 + 0.25

2 Geological background

The Rhine Hessian Plateau is a tableland made up of Oligocene sands and marls, overlain by resistant Miocene limestones. The sea receded from the area during the Miocene and an initial fluvial system developed. A former course of the Rhine river can be traced by deposits of the Dinotherium sands that have been dated as upper Miocene (TOBIEN 1980, fig. 1). On the Steinberg near Sprendlingen, these Rhine deposits are mixed with gravels of the Nahe which contain Permian sedimentary rocks (Rotliegendes). Fe-concentrations derived from upper Miocene ferruginous clays and lower Pliocene pebbles and sands occur in the Pleistocene sedimentary caps on the Plateau. The upper Pliocene Arvernensis gravels (BARTZ 1950) which were deposited by the former Main river are also found on the plateau (fig.1). Their location indicates that the Rhine must have been flowing farther west before it shifted its course eastward to the Mainz-Bingen Rift, a local graben within the overall Rhine Rift Valley structure, along which it flows at the present time.

The uplift of the Rhine Hessian Plateau and simultaneous subsidence of the Rhine Rift Valley in the late Pliocene

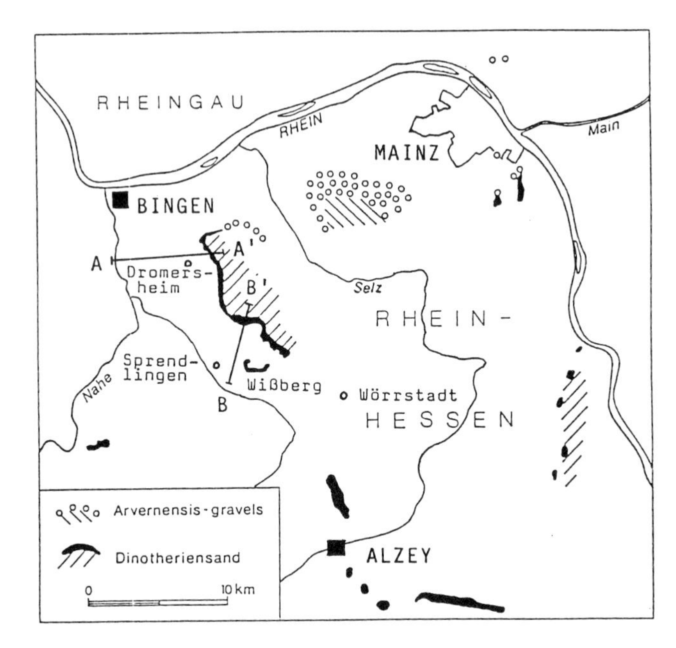

Fig. 1: *Dinotherien sands and Arvernensis gravels in the Mainz Basin (after BARTZ 1936, ROTHAUSEN & SONNE 1984, TOBIEN 1980, WAGNER 1970).*
A-A': Profile 1, near Sprendlingen, B-B': Profile 2, near Dromersheim

caused the rivers Nahe, Selz and Wiesbach to incise their courses. As a result the water tables in the limestones were lowered and karst developed. Pleistocene solifluction and runoff erosion were important factors in the further development of forms during the Quaternary.

3 The present landforms

The escarpment at the western edge of the plateau is between approximately 100 and 150 m high. Its steepest slope segments lie in the outcrop of the Miocene Corbicula limestone. The landforms on the footslope vary depending on the distance to the local baselevel. Where the stream channels lie far from the scarp, the footslopes are developed as broad and gently sloping glacis or pediments; where the stream channels lie close to the scarp, pediment-like forms are absent and the footslopes consist of steep limestone debris surfaces in their upper part and of dells and earth slide scars in their lower part.

The detailed investigation concentrated on four sites: the scarp-and-glacis profiles near Sprendlingen (B-B' in fig.1) and near Dromersheim (A-A' in fig.1) and the occurrences of older glacis remnants or "glacis generations" near Wörrstadt and the Alteberg glacis near Dromersheim.

4 The glacis near Sprendlingen

The profile (fig.2) shows that the steep scarp slope on Corbicula limestone grades upwards into a convex crest slope (walm) and downwards into a concave transition slope to the glacis, the angle

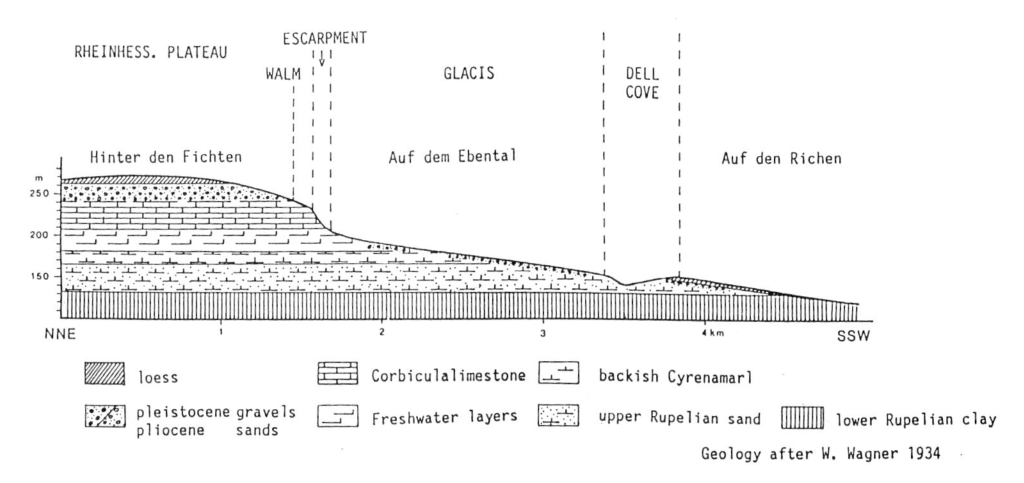

Fig. 2: *Longitudinal cross-section of a middle Pleistocene glacis extending NNE-SSW near Sprendlingen.*

of which decreases from 3° - 4° at the scarp foot to 2° farther away.

The glacis is, for the most part, covered with Pleistocene gravel lenses and sands which contain silty laminations and, locally, also ferrous hydroxide. The thicknesses of these deposits varies from 0.2 m to 1.8 m on weak Oligocene Cyrena Marls and Rupelian sands (fig.2). The coarse components of the gravel are oriented predominantly with their long axes transversally to the slope. 44% of the pebbles of Corbicula Limestone have a roundness index between 50 and 100 (CAILLEUX 1952). In addition to the limestone components, the gravels contain material from the upper Miocene and Pliocene sediments that occur on top of the plateau, and material from the Oligocene strata that underlie the lower scarp slope and the glacis.

The evidence indicates that the deposits on the glacis are of periglacial origin, transported by running water with a greatly varying runoff intensity. On the lower part of the glacis they are covered by loess up to 10 m thick. A stratigraphic section from a Sprendlingen brickyard shows the Schleichsand at the base is Rupelian in age. It is overlain by about 1.5 m of glacis gravels composed of 86% limestone and marl pebbles and 14% of other rocks. Many pebbles are frost-shattered. Above the gravels there is a similigley parabrown earth (10 YR 5/4) soil which is attributed to the Eem interglacial. Part of the soil has been destroyed by solifluction in the early Würm glacial.

The several soil profiles that occur higher in the column indicate pauses in the loess deposition during interstadials of the Würm. The Lohner Boden is an index paleosoil that marks-the interstadial between middle and late Würm (SEMMEL 1968). After deposition ceased, the Holocene chernozem at the present land surface developed.

The pedostratigraphical sequence in the loess, with the Eem soil at the base, and the periglacial character of the underlying gravels show that the latter have most probably been deposited during the Riss glacial.

Photo 1: *in the left foreground is the accumulation lobe of an earth slide from the northwestern slope of the Wißberg. In the middle and the background, gently sloping dissected glacis dated to the middle Pleistocene. Above is the escarpment of the Rhine Hessen Plateau.*

Photo 2: *In the background, glacis of the middle Pleistocene. Its flank is destroyed by dells. In the foreground there is a younger erosive plane sloping towards the Low Terrace of the Wiesbach.*

Photo 3: *Glacis G1 north of Dromersheim. Dells form the edge of the Rhine Hessen Plateau, inclined towards the periglacial glacis. On the left side, the older glacis level "Auf der Kreuzschanze" can be seen.*

Photo 4: *Glacis 1 near Dromersheim. In the background is the edge of the Rhine Hessen Plateau, formed by dells. On the right the older glacis level "Altberg" G4 near Aspisheim can be seen.*

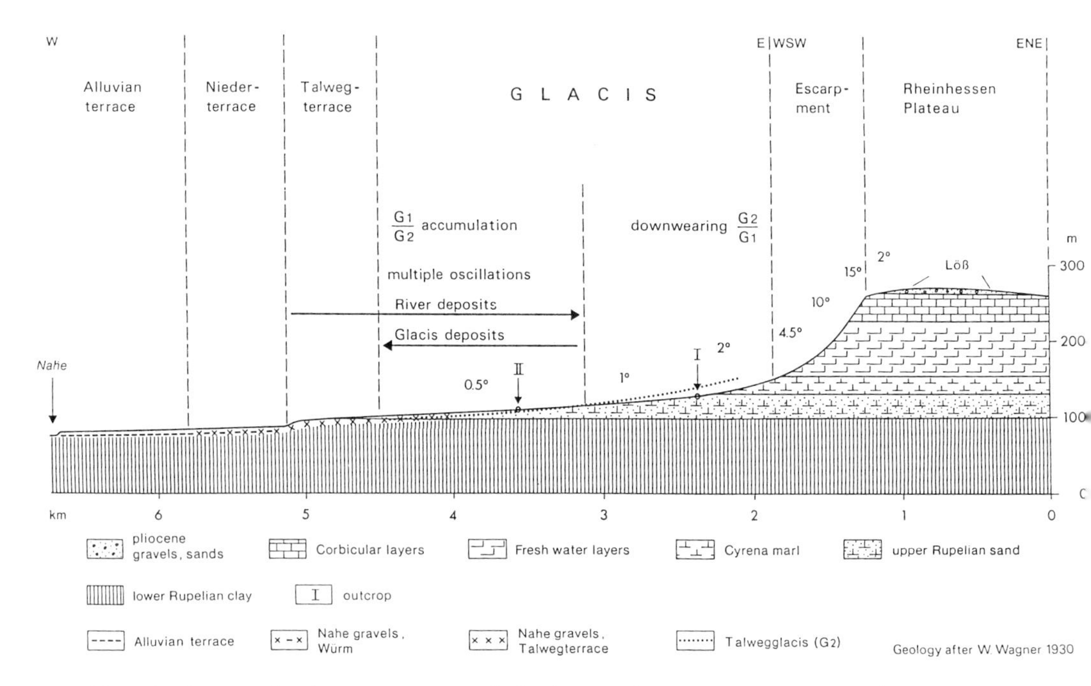

Fig. 3: *Longitudinal cross-section of the lower glacis generation extending into the Dromersheim area.*
The glacis G2 can be seen under the surface of the lower part of the glacis G1. The upper part, which once formed the glacis G2, is eroded.

5 The glacis near Dromersheim

This glacis extends from the foot of the steep cuesta scarp at about 145 m above sea level westward over a distance of about 3 km to where it grades into a terrace of the Nahe river, the Valley Road Terrace ("Talwegterrasse") at about 100 m above sea level. About 7 m below this terrace, and separated from it by a scarp, lies the Low Terrace (Niederterrasse). Clearly the glacis is similar in age to the Valley Road Terrace and younger than the Low Terrace. A section of the glacis deposits at the location I is shown in fig.3. Above the Cyrena Marl at the base lie 20–30 cm of debris composed of subangular fragments of limestone with Hydrobia inflata, Pliocene rounded quartz gravels, siliceous schists and oolitic iron concretions. This horizon is overlain locally by 10–15 cm of aeolian sands and 35 cm of humus-containing clay in which there are pieces of debris similar to the material at the base. On top lie 40 cm of colluvial loam and 20 cm of a loamy soil. The presence of Pliocene pebbles points to a post-Pliocene age of the deposits and the presence of the limestone fragments to transport in a cold period. The location of the glacis relative to the two terraces at its base indicates a pre-Holocene origin.

The lower part of the glacis at site II (fig.3) was analysed. Here the glacis debris is intercalated with deposits of the Nahe river. It is interpreted as an older glacis, termed Valley Road Glacis (G2 on fig.3). Towards the scarp, this older glacis

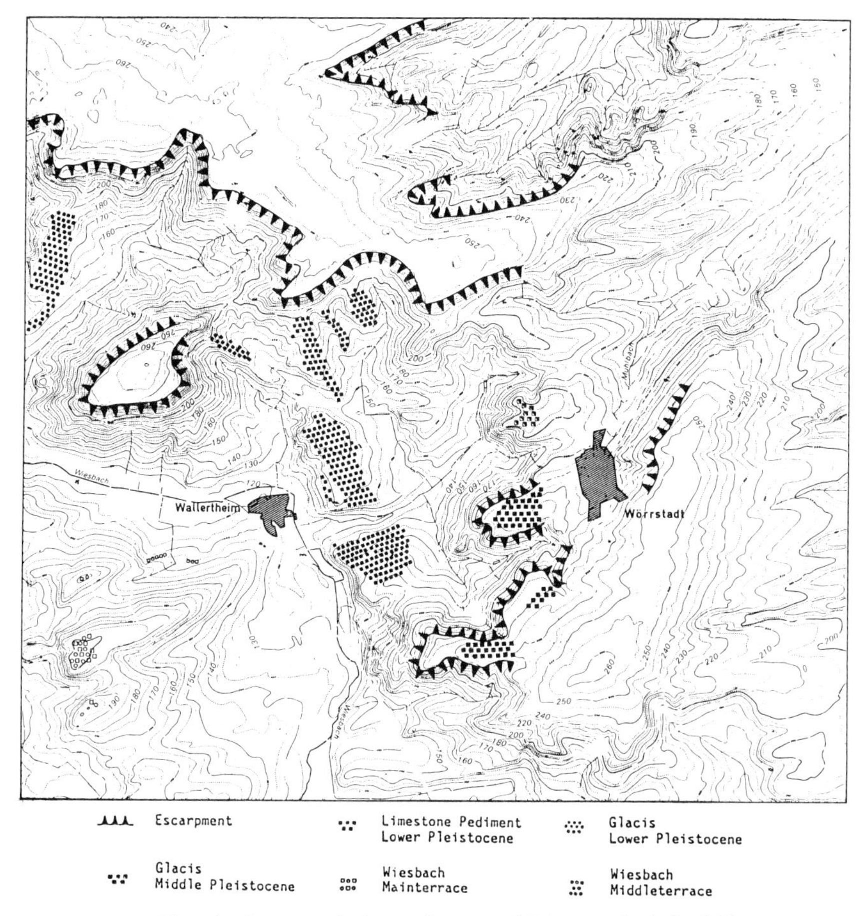

Fig. 4: *Geomorphological map of Wörrstadt 1:25,000.*

has been replaced by a younger one (G1 in fig.3) as a result of continued backwearing of the scarp and headward erosion of the footslope. The intercalation of the glacis material with Pleistocene Nahe deposits also proves the Pleistocene origin of the glacis.

6 Remnants of older glacis generations

Evidence of older pediment levels is found at higher elevations. Examples have been mapped in the Wörrstadt area (fig.4) and near Dromersheim (Alteberg glacis, fig.5). The glacis remnants are identified by their location below the level of the plateau and above the lower footslopes, by their gentle gradients and by their cover of Pleistocene deposits. In the area of Wörrstadt (fig.4), they lie at varying elevations and probably were formed at different times during the Pleistocene. Near Dromersheim two levels of glacis remnants can be distinguished (G3 and G4 in fig.5). The glacis G3 in fig.5 has its highest point at 167 m and the glacis G4 lies between 198 m and 190 m above sea level. These remnants are no longer connected to the plateau

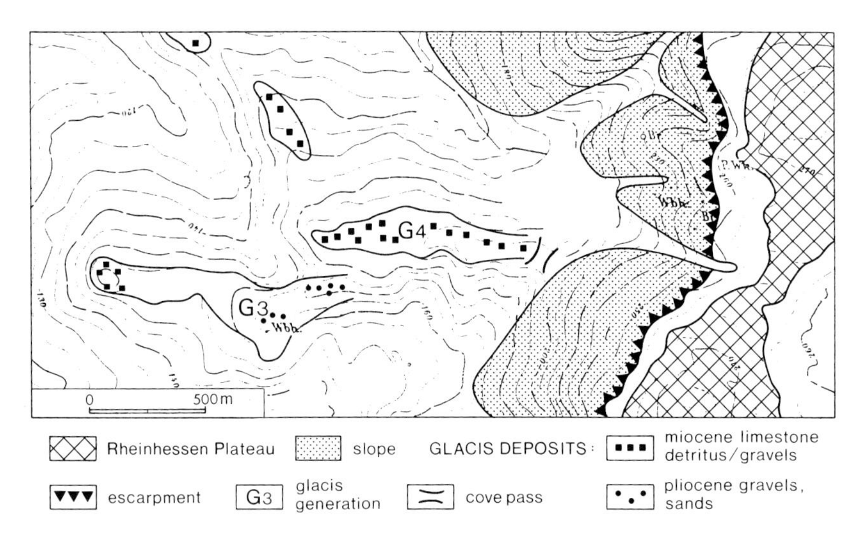

Fig. 5: *Geomorphological map of Dromersheim 1:25,000.*

by a unidirectional slope but are separated from it by lower passes (see "cove pass" in fig.5).

7 Glacis development

The fluvial late Pliocene Arvernensis gravels on the plateau and its outliers indicate that the development of the present escarpments and of their footslopes must have taken place during the Pleistocene. This development began with the uplift of the Rhine Hessian Plateau, as a structural block above its immediate surroundings (FALKE 1960). Fluvial dissection followed and the Nahe and Wiesbach valleys were established. Because the oldest pediments grade into the lower Pleistocene Main Terrace (WAGNER 1935), they must have been formed during this period. The Wiesbach had already cut through the limestone caprock of the plateau, as indicated by the gravel accumulation on the Streitberg WSW of Wallertheim (fig.4), and the weaker rocks below the limestone were exposed and subjected to glacis formation. Erosion was accentuated by more intensive uplift in the middle Pleistocene (FALKE 1960).

Retreat of the limestone scarp by intensive periglacial backwearing caused the glacis to develop as footslopes or replacement slopes which extended backwards as the scarp retreated and were at the same time regraded downward by periglacial wash denudation as their profile lengthened. The existence of disconnected remnants of older glacis at higher elevations suggests that glacis formation took place repeatedly during separate periods in the Pleistocene. The stratigraphical record and the morphological correlation to the Nahe terraces show that the latest glacis development took place during the Würm glacial and another during the Riss glacial. By analogy it was concluded that the older, higher glacis remnants were formed during the earlier Pleistocene glacials.

The western margin of the Rhine Hessian Plateau provides, therefore, evidence that glacis and, in a wider sense, pediments did form during the Pleistocene glacials.

References

BARTZ, J. (1950): Das Jungpliozän im nördlichen Rheinhessen. Notizblatt des hessischen Landesamtes für Bodenforschung VI, 1, Stuttgart, 201–243.

BECK, N. (1976): Untersuchungen zur Reliefentwicklung im nördlichen Rheinhessen. Erdkunde **30**, 73–83.

BECK, N. (1977): Fußflächen im unteren Nahegebiet als Glieder der quartären Reliefentwicklung im nördlichen Rheinhessen. Mainzer geogr. Studien **11**, Festschrift zum 41. Deutschen Geogaphentag in Mainz, 261–266.

CAILLEUX, A. (1952): Morphoskopische Analyse der Geschiebe und Sandkörner und ihre Bedeutung für die Paläoklimatologie. Geologische Rundschau **40**, 11–19.

FALKE, H. (1960): Rheinhessen und die Umgebung von Mainz. Sammlung geologische Führer, Bd 38, Berlin-Nikolasee, 156 pp.

MENSCHING, H. (1964): Die regionale und klimatisch-morphologische Differenzierung von Bergfußflächen auf der Iberischen Halbinsel (Ebrobecken-Nordmeseta-Küstenraum Iberiens). Würzburger Geogr. Arb. **12**, 141–158.

ROTHAUSEN K.H. & SONNE, V. (1984): Mainzer Becken. Sammlung geologischer Führer 79, Gebrüder Borntraeger, Stuttgart, 203 pp.

SEMMEL, A. (1968): Studien über den Verlauf jungpleistozäner Formung in Hessen. Frankfurter geogr. Hefte **45**.

SEMMEL, A. (1972): Geomorphologie der Bundesrepublik Deutschland. 2nd Edition, Wiesbaden, 149 pp.

TOBIEN, H. (1980): A Note on the Mastodont Taxa (Proboscidea Mammalia) of the "Dinotheriensande" (Upper Miocene, Rheinhessen, Federal Republic of Germany). Mainzer Geowissenschaftliche Mitt. **9**, 187–201.

WAGNER; W. (1935): Geologische Karte von Hessen, 1:25,000. Blatt Wörrstadt. Darmstadt.

WAGNER, W. (1972): Über Pleistozän in Rheinhessen (Mainzer Becken), Mainzer geowissenschaftliche Mitt.1 192–197.

WAGNER, W. & MICHELS, F. (1930): Geologische Karte von Hessen, 1:25,000, Blatt Bingen-Rüdesheim. Darmstadt.

Address of author:
Dr. Nordwin Beck
Seminar für Geographie,
Erziehungswissenschaftliche
Hochschule Rheinland-Pfalz
Rheinau 3–4
D-5400 Koblenz
Federal Republic of Germany

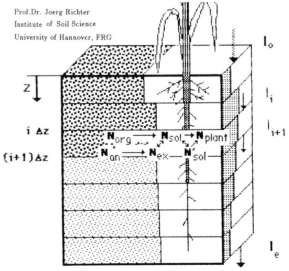

CATENA SUPPLEMENT 15 p. 199–210 Cremlingen 1989

GEOMORPHOLOGICAL ASPECTS OF THE ODENWALD

Adolf **Zienert**, Heidelberg

1 Introduction

The Odenwald is an area of upland that is bordered on the west by the eastern fault margin of the Rhine Rift Valley (Oberrheingraben) southeast of Darmstadt (fig.1). The western edge of the Odenwald is, therefore, a fault scarp with a maximum relative height of about 450 m above the sediment-filled rift valley. The maximum vertical displacement, near Heidelberg, is about 4000 m, two-thirds of which results from downfaulting in the graben itself.

The northern part of the fault scarp between Darmstadt and a point about 5 km north of Heidelberg is composed of the Paleozoic crystalline Hercynian basement (Grundgebirge). Near Heidelberg, the basement is exposed only on the lower slopes of the Neckar Valley, for example in the moat of Heidelberg Castle. In the southern part, the fault scarp, with some step faulting, is made up of lower Triassic Bunter Sandstone (Buntsandstein), which, apart from local Permian sediments, is the oldest rock of the sedimentary cover (Deckgebirge).

The boundary between the crystalline rocks and the Bunter Sandstone extends northeastward from the fault scarp north of Heidelberg into the Odenwald, dividing the region into a triangular area of crystalline rock in the north and northwest, the crystalline Odenwald, and a sandstone area in the south and southeast, the Bunter Sandstone Odenwald.

2 The crystalline Odenwald

The crystalline Odenwald is divided into several structural blocks as a result of Cenozoic tectonic movements. Some of these movements took place on existing old faults, particularly in the area west of the Otzberg fault (Otzbergspalte, fig.1). The cuesta scarp of the Bunter Sandstone, which forms the northwestern boundary of the south German scarplands, lies a few kilometers east of this fault zone. To the south and east of it lies the Bunter Sandstone Odenwald.

The tectonic and geomorphological development of the individual fault blocks in the crystalline Odenwald was varied. Fission track dates (WAGNER 1967) indicate that the "northern block" (ZIENERT 1957) between Darmstadt and Heppenheim has been uplifted above a particular temperature horizon of the crust about 20 million years earlier than the Weschnitz area immediately to the south. Sediments containing Eocene fossils at Messel, NE of Darmstadt, lie directly on the Permian and on crystalline rocks so that all the Mesozoic sed-

ISSN 0722-0723
ISBN 3-923381-18-2

D–3302 Cremlingen-Destedt, W. Germany
3-923381-18-4/89/5011851/US$ 2.00 + 0.25

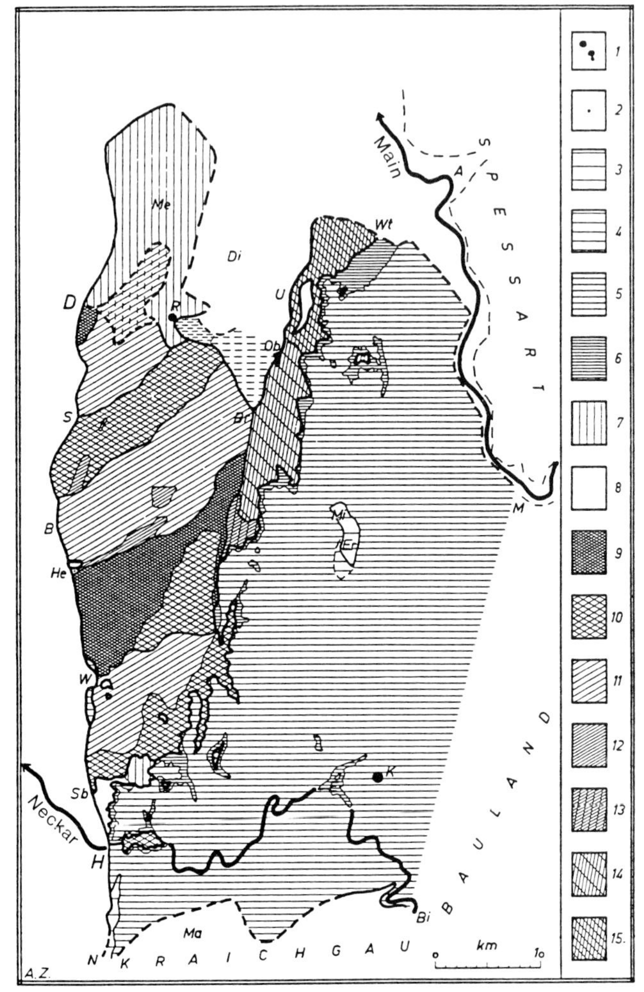

Fig. 1: *Geology of the Odenwald (modified after ZIENERT 1954).*

1 Basalt (Tertiary)
2 Oligocene sediments
3 Muschelkalk
4 Upper Bunter Sandstone
5 Main Bunter Sandstone (in the south including Upper Bunter Sandstone)
6 Lower Bunter Sandstone (5 and 6 dashed: Klingen basin)
7 Rotliegendes (Permian)
8 Porphyries, ignimbrites (Permian)
9 Granodiorites ("Hornblende-granit")
10 Predominantly biotite granites
11 Undifferentiated crystalline rocks
12 Predominantly schists
13 Gneisses and porphyries
14 Gneisses and granites
15 Gneisses and schists
Place names: A Aschaffenburg; B Bensheim; Bi Binau; Br Brensbach; D Darmstadt; Di Dieburg; E Eberbach; Er Erbach; H Heidelberg; He Heppenheim; K Katzenbuckel; M Miltenberg; M Mauer; Me Messel; Mi Michelstadt; N Nußloch; Ob Otzberg (Basalt); R Roßberg; S Seeheim; Sb Schauenburg; U Groß Umstadt; W Weinheim; Wt Wartturm near Pflaumheim.

iments, with a probable combined thickness of about 1000 m, had been worn away by the Eocene. At Heppenheim, 30 km to the south, the Bunter Sandstone was, however, present in the main fault scarp of the Rhine Rift Valley as late as the middle Oligocene. At this time the streams in this area were beginning to erode into the underlying crystalline rocks, as indicated by the composition of middle Oligocene coastal conglomerates. There was no faulting before the middle Oligocene in this part of the Rift Valley, although there was a "saxonic" upwarp in the northwest in the late Cretaceous/early Tertiary (STRIGEL 1949).

The present landforms of the northern block indicate that there has been uplifting in the south and southeast by at least 500 m of vertical displacement, but that the north of the block has remained in a low position since the Oligocene. Rem-

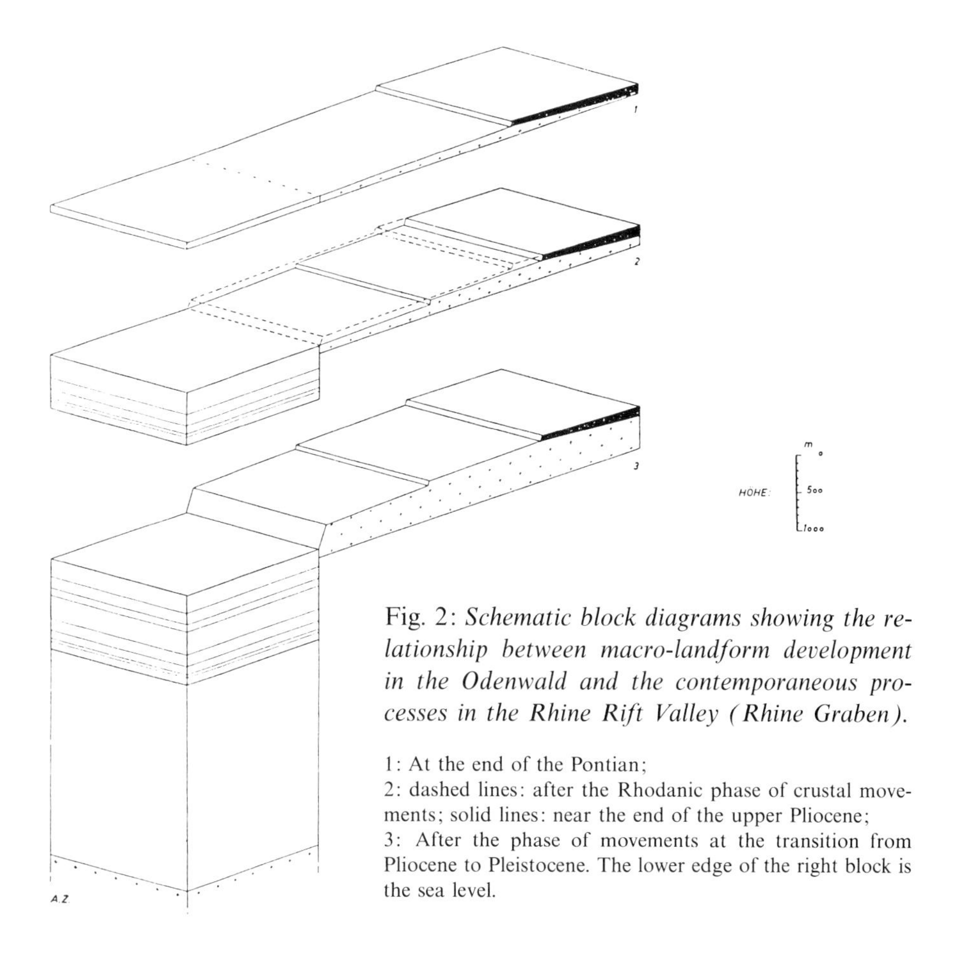

Fig. 2: *Schematic block diagrams showing the relationship between macro-landform development in the Odenwald and the contemporaneous processes in the Rhine Rift Valley (Rhine Graben).*

1: At the end of the Pontian;
2: dashed lines: after the Rhodanic phase of crustal movements; solid lines: near the end of the upper Pliocene;
3: After the phase of movements at the transition from Pliocene to Pleistocene. The lower edge of the right block is the sea level.

nants of the Permian denudation surface, the land surface on which the terrestrial sediments of the Bunter Sandstone were deposited, are present only in the northern part of the block. Elsewhere on the block neither Permian nor Tertiary deposits or paleosols are preserved.

A group of denudation surfaces extends largely undissected over the central part of the northern block. Above the lower of these surfaces there are remnants of a higher denudation level, including a large number of monadnocks on resistant rocks along the contact metamorphic zones that are aligned with the SW-NE strike of the Hercynian fold systems. The altitude difference between the denudation surface and the higher remnants increases from 10–15 m in the north to more than 100 m in the south and southeast. There are no sediments to indicate the age of the denudation surfaces but they extend across several lithological and structural units, such as the lower Permian Rotliegendes beds and the Eocene beds near Messel in the north, the remnants of the Permian denudation surface in the central area and the crystalline rocks further south.

The low relief denudation surface remnants have often been referred to as peneplains. Climatic conditions were suitable for peneplanation in the early part of the Tertiary but later the Tertiary paleoclimate seems to have been more conducive to the development of pediplains. During most of the Pliocene, the climate was probably too cool and too moist for either form of development. About 8 million years were, therefore, available from the end of the period of planation until the onset of the Pleistocene for the development of the present landforms. This would have been long enough for large areas to be lowered by denudation under widely varying climatic conditions, particularly on impervious rocks.

In the Weschnitz area where ridges predominate, there are no sediments from which past stages of form development can be dated. It is uncertain whether the approximate accordance of the ridge crests indicates the presence of old planation remnants or a young "upper denudation niveau" resulting from the intersection of valley side slopes at the interfluves in areas in which the valleys are more or less uniformly spaced. An argument against the latter interpretation is that the width of the ridges varies greatly and does not seem to have any recognizable influence on their heights. On the basis of either interpretation, an estimate of the minimum amounts of uplift of the different structural blocks is possible.

The "southern upland" (Südliches Bergland, CREDNER 1922) of the crystalline Odenwald provides additional evidence of uplift. In the east of this area there is an angle of up to 10° between the inclination of the Permian denudation surface underneath the Bunter Sandstone strata and the low-relief forms that occur on the crystalline rocks at 500 m above sea level in front of the scarp. To the west lies a denudational scarp of about 150 m relative height that crosses greatly varying crystalline rocks, with outliers up to two kilometers in length; in front of this scarp, numerous lower remnant hills lie above the lower level which ends at the edge of the Rhine Rift Valley. This lower level, is preserved over an area of nearly 20 square km and is dissected by younger valleys.

In both the northern block and the southern upland two denudational levels are undoubtedly preserved over relatively large areas; the lower level in the northern block may have been subdivided further by more recent tectonics. Neither level can be part of the exhumed Permian denudation surface, since both lie at elevations that are at least 100 m, and in some cases more than 200 m, lower than the estimated elevation of the Permian surface. They were formed after the removal of the Mesozoic rocks which must have been largely completed by the end of the Miocene (fig.2).

In the eastern crystalline Odenwald, at the base of the Bunter Sandstone scarp, termed the basal denudation level (basislandterrasse, SCHMITTHENNER 1956) of the south German scarplands, the size of the angle between the Permian denudation surface and the younger surface seems to vary regionally. East of the Tromm mountain massif (east of the Weschnitz basin) the younger surface also cuts across some remnants of Bunter Sandstone.

Two levels can, therefore, be distinguished in the crystalline Odenwald. It is uncertain whether they are peneplains, pediplains or upper denudation niveaus. They do, however, provide information about the tectonics, in particular the minimum relative uplift, because they occur on each of the tectonic blocks at different elevations. The tectonic movements of the crystalline area also had an influence on the morphological development of the adjacent cuesta scarplands to the east.

If it is assumed that the low-relief forms originated during periods of tectonic inactivity, the dates of their development can be estimated from the correlative deposits in the Rhine Rift Valley that were identified from borings drilled for a thermal well near Heidelberg (BARTZ 1951, ZIENERT 1957, 1981, 1986). The following sequence was found (fig.2),

from the surface to 230 m depth: Pleistocene gravels and sands (Mauer facies).

to 335 m depth: gravels and sands probably deposited during the transition from the Pliocene to the Pleistocene (Wiesenbach facies).

to 380 m depth: mainly quartz gravels, probably upper Pliocene.

These 380 m of Rift Valley sediments and an estimated uplift of the adjacent Odenwald by up to 300 m would account for the approximate 700 m total vertical displacement during the last major morphologically significant tectonic phase that has occurred since the upper Pliocene. Farther down the borehole revealed:

to 660 m depth : fine-grained material with high clay contents, formerly believed to be typical for the upper Pliocene.

to 850 m depth: transitional beds.

to 930 m depth: coarse material with blocks at the base.

These 550 m of beds and an estimated uplift of the Odenwald of 150 m would also amount to about 700 m of total vertical displacement. The tectonic phase (rhodanic) probably took place during the transition from the Miocene to the Pliocene.

Between a depth of 930 m and nearly 1200 m there were fine-grained sediments, perhaps of late Miocene (Pontian?) age. Such fine-grained materials indicate periods of tectonic inactivity of sufficient duration for the development of low relief denudational landforms. These were then uplifted in the last two phases of tectonic displacements, and further modified up to the present.

3 The Bunter Sandstone scarp and its valley gateways

The relative height of the Bunter Sandstone scarp varies greatly. Also, different segments of the scarp were developed at different times (ZIENERT 1957, 1986). The lowest and probably oldest segments of the scarp may date back to the late Miocene. They are cuesta scarps with relative heights of at least 50 m; the total thickness of the middle Bunter Sandstone is about 350 m. Other scarp segments are both younger and higher, especially along small streams, and have well-defined small denudation levels at the scarp foot. In addition, some segments are local fault scarps or fault line scarps. Only in the northeast of the crystalline Odenwald on the structural eastern margin of the Böllsteiner Odenwald (fig. 1) is the scarp absent for a straight line distance of about 10 km, probably the result of intensive local uplift of the crystalline rocks in the west and only minor uplift of the adjacent Bunter Sandstone in the east. Farther south, along this same margin, the Bunter Sandstone forms a cuesta scarp that attains a maximum height of 170 m above its crystalline foreland and has a broad denudational level, especially at its western foot, and clearly developed contact spring valley heads in the scarp slope. Beyond the northern scarp, there is a residual outlier the two summits of which have been lowered considerably by denudation. Wind gaps, consequent valleys that have been beheaded by the retreat of the scarp, occur in the southern part of this scarp, east of the Osterbach. There are no back scarps behind the scarp since no valleys have been cut through the Bunter Sandstone into the underlying crystalline rocks. In other areas, such as the northern part of the Böllstein Odenwald, there may have been back scarps in the past; the retreat of such back scarps updip could have accelerated the removal of the Bunter Sandstone in these areas.

The crest of the Bunter Sandstone cuesta, the crystalline Böllstein Odenwald and the Tromm form the local drainage divide over great distances. In the south, there are also smaller streams that flow into the scarp and form small gaps in its front. Above these gaps there are indications of old broader valleys that were more shallowly incised and the relative elevation of which would have corresponded to the old 50 m high scarps. In the southeast, the older valleys seem to merge gradually into the present dipslope, suggesting a genetic linkage between the basal denudation level, the cuesta scarp and the dipslope in the middle Bunter Sandstone.

4 The western Bunter Sandstone Odenwald

The western Bunter Sandstone Odenwald is a 15–20 km wide area between the Bunter Sandstone scarp in the west and a north-south fault zone in the east. It is characterized by ridges and flat-topped mountains separated by valleys that decrease in depth eastwards. Many of the valleys are aligned with faults, often with the N10°E strikes which also control the drainage directions but, in general, have no other morphological effects. Other valleys follow the southeastward dip of the Bunter Sandstone strata. This also applies to some valley systems that now lie in crystalline areas ahead of the scarp but that were also formerly covered by Bunter Sandstone. These valleys indicate, therefore, the past extent of the Bunter Sandstone and the probable location of its scarp before the last phases of uplift. Only in the crystalline area between the Otzberg fault and the Rhine Rift Valley farther to the west does the drainage system appear to be fully adapted to the morphotectonic development of a crystalline area (ZIENERT 1957, 1986).

5 The eastern Bunter Sandstone Odenwald and adjacent areas to southeast (Bauland, Hohenlohe Plain)

A west-facing fault scarp, with a maximum vertical displacement of 80–150 m, extends from near Obernburg on the Main river southward to the Neckar valley west of Eberbach. The upthrown block is characterized by high planation surface remnants on the middle and, in particular, on the upper Bunter Sandstone. Farther to the east, the surfaces become more extensive. They cross many local faults but are dissected by valleys that have developed largely independently of the faulting.

In large parts of the area, these surfaces, similar to dipslopes, extend from the middle Bunter Sandstone southeastward across the Muschelkalk (middle Triassic) of the Bauland to the lowest horizons of the middle Keuper at the southern margin of the Hohenlohe Plain and southward into the Kraichgau, without being interrupted by cuestas. This region has been termed the Great Dipslope (Große Landterrasse, ZIENERT 1986). In valleys oriented in the direction of the strike within the Muschelkalk, or at its base, vale-like widenings occur with structural benches on the more resistant beds that outcrop on the valley sides. Cuesta scarps in the Muschelkalk, parallel to valleys, occur in a few places, for example at Mosbach and northeast of Mauer. No Muschelkalk scarp has been traced over any great distance indicating that the short stretches of scarp were probably the downdip sides of shallow asymmetrical strike valleys which developed during an early phase of valley formation that occurred after a planation phase. Deep dissection and further accentuation of the local scarps followed later.

6 Summary of the major landform features

In the Odenwald and in the adjoining areas to the southeast both the strata and the planation surfaces descend, for the most part, in a southeasterly direction. Since, in general, the strata dip more steeply than the planation surfaces, the surfaces are cut across progressively younger formations in the downslope direction. The strata are truncated at an acute angle which tends to be larger in relatively resistant beds than in the weaker beds. More of the resistant Bunter Sandstone has, therefore, been worn away towards its western margin. The resistant middle Bunter Sandstone, for example, has been reduced in some places from a thickness of about 350 m to 50 m.

The planation surfaces also extend across many faults without forming scarps, although some of the faults have considerable displacements. This indicates that they are older than the surfaces in the areas in which the surfaces cross them. However, three major antithetic fault zones, the marginal fault of the Rhine Rift Valley in the west, the Otzberg fault and the fault zone from Obernburg/Main to Eberbach/Neckar, disrupt the general southeasterly descent of the planation surfaces. On the eastern upthrown side of each of these faults, the descent of the surfaces begins again. An exception to the antithetic structure is the northern block of the crystalline Odenwald which is uptilted in its southern and southeastern parts.

The geomorphological evidence and the fission track dates indicate that the main axis of uplift, which strikes SW-NE, has gradually shifted from north to south, so that in the last phase, since the Pliocene, the southern Odenwald was most strongly uplifted. The Neckar river flows from east to west into and across this southern area. It is, therefore, antecedent to these last phases of uplift.

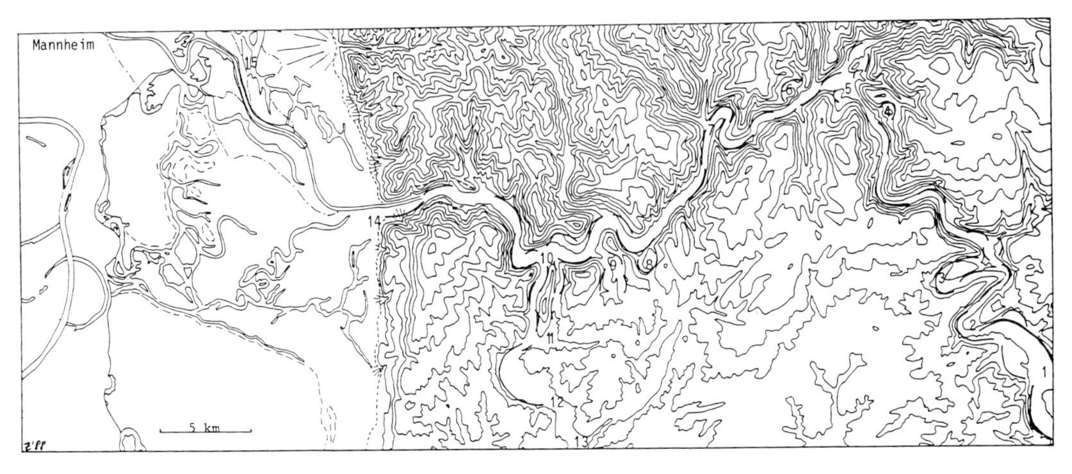

Fig. 3: *The valley of the Neckar from Mosbach (1) to Mannheim.*

Places identified by number: 1 Neckar west of Mosbach; 2 Binau; 3 Neckarkatzenbach; 4 Schollerbuckel; 5 Eberbach; 6 Igelsbach; 7 Hirschhorn; 8 Mückenloch; 9 Dilsberger Hof; 10 Neckargemünd; 11 Wiesenbach; 12 Mauer; 13 Elsenz valley; 14 Heidelberg; 15 Ladenburg.
The contour interval in the uplands is 50 m. In the Rhine Rift Valley the map shows the 100 m contour, the old meanders of the Neckar on its alluvial fan (note the distributary pattern) and the artificially straightened bed of the Rhine with former meander bends.

7 Valleys and valley slopes

The Neckar is the largest river in the Odenwald. It approaches from the south-southeast, reaching the eastern monoclinal margin of the present upland near Mosbach ((1) in fig.3), where it has formed two large, long and narrow meander bends, the meander bend near Binau (2) and the abandoned meander near Neckarkatzenbach (3). Such bends are common at young tectonic obstacles. To the northwest, against the dip of the strata, the valley becomes increasingly incised into the upland. Near Eberbach (5), where the planation surfaces north of the valley lie up to 580 m above sea level, the Neckar turns southwest and flows in a direction that is approximately along the strike of the strata to Neckargemünd (10). It continues diagonally updip and in part along the boundaries between fault blocks in a generally westerly direction to Heidelberg, 100 m above sea level. At Heidelberg (14), the granite basement outcrops on both sides of the valley. The unconformity between the granite and the overlying sedimentary rocks has resulted in the development of a structural bench on the westernmost spur of which Heidelberg castle has been built. Prior to the canalization of the Neckar, outcrops of resistant granite in the river bed caused the formation of rapids and potholes, known in Heidelberg as "Hackteufel" (chop devil) because of the danger to boats in the rapids. West of the old town of Heidelberg, there is a large, alluvial fan that extends into the alluvial plain of the Rhine Rift Valley and on which the river deposited its sediments until the late Pleistocene before it cut, successively, two shallow valleys during the Holocene. The first was cut in the direction of Wallstadt NE of Mannheim from where it swung back to continue north along the western foot of the Odenwald and the second westward directly to the Rhine.

Within the upland the Neckar has formed several incised meanders, most of which have been abandoned. A low-lying double bend near Eberbach (fig.3 (5)) and the three largest abandoned bends between Mückenloch (8) and Neckargemünd-Mauer (10–12) contain gravels and sands. These three bends were developed on the south side of the river and became particularly large because they were developed when the area was tilted to the south and formed the southern marginal flexure of the Odenwald.

A developmental sequence can be reconstructed from sands and gravels preserved in some of the old meanders and on adjacent terraces.

The Neckar valley, south of Neckargemünd, must have been incised at least 150 m below the level of the present plateau surfaces before the earliest phases of deposition, which probably occurred during the transition from the Pliocene to the Pleistocene (ZIENERT 1957, GRAUL 1977). The formation of the other meanders had begun by this time and they were also incised and enlarged.

On the east side of the former river valley between Neckargemünd and Mauer, (the dry valley of Wiesenbach, (11)) and in the two former meander bends that lie to the northeast at the Dilsberger Hof (9) and near Mückenloch (8), there is a terrace about 75 m above the present river level with predominantly limestone-free gravels (Wiesenbach facies, fig.4). Farther south at lower elevations deposits of the Mauer facies occur of which those deposited during Pleistocene cold periods contain a high proportion of limestone.

The highest of the Mauer terraces has a surface elevation of 45 m above the present river level. Its sands and gravels consist of a series, from top to bottom, of cold-period, warm-period and cold-period deposits (GRAUL 1977). The lowest 20 m have never been exposed and are known only from refraction-seismic evidence. In these Pleistocene deposits, which are preserved only in the southern part of the abandoned meander, the lower jaw of the Homo erectus heidelbergensis was found in the warm-period deposits together with partly "African", partly "European" fauna. After the deposition of the last cold period strata, the Neckar reduced the amplitude of its meander bend at this location by about one half (fig.4).

In the northern part of this old meander bend, there are two additional terraces, an upper terrace at about 30 m and a lower terrace at about 15 m above the present level of the river. The 30 m terrace also contains a cold-warm-cold sequence of deposits of typical Neckar material, namely, Bunter Sandstone and granite from the northern tributaries. After the formation of these terraces the river also abandoned this part of the meander. The maximum length of the meander had been 16 km.

The 15 m terrace contains predominantly material transported by the Elsenz river, a tributary that joins the Neckar from Kraichgau in the south and that flows in the western reach of the abondoned meander bend; the Elsenz gravels are mixed with reworked Neckar material derived from the gravels of the older terraces.

The valley bottom of the former

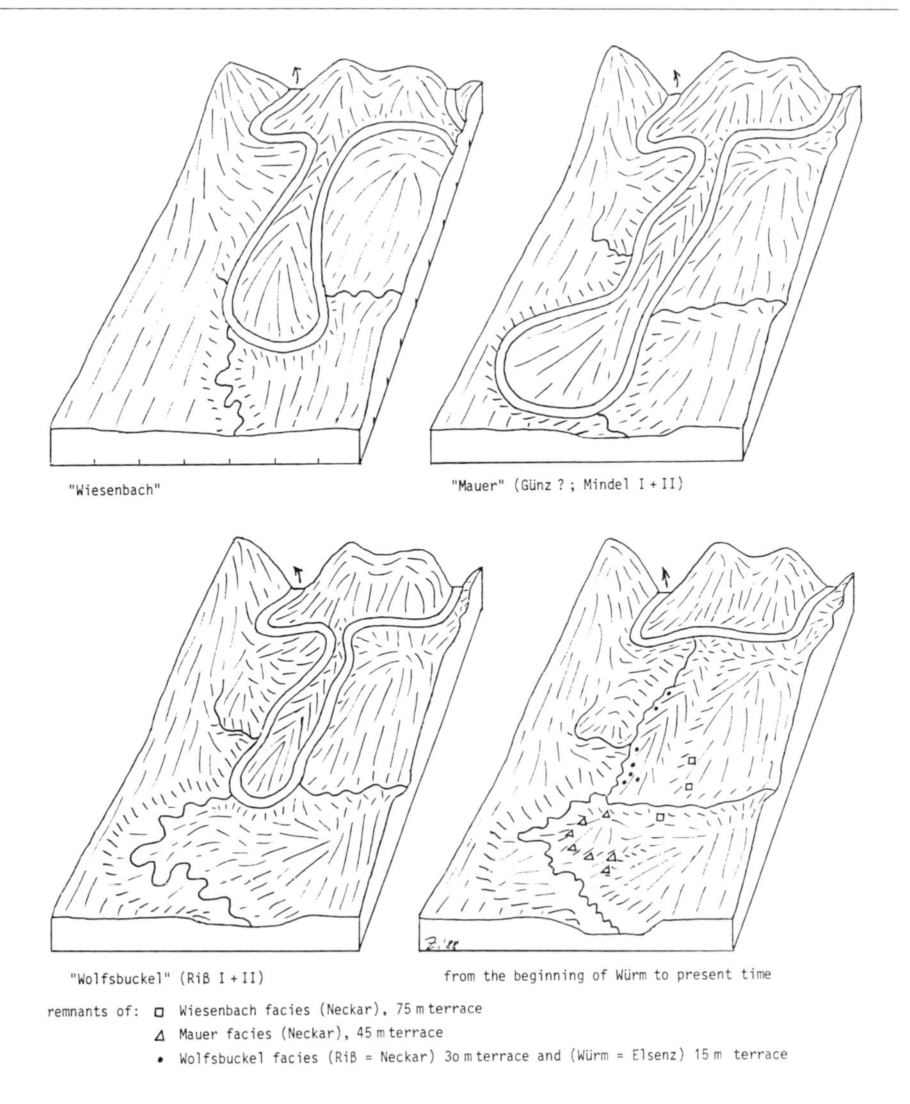

Fig. 4: *The evolution of the Neckar meander south of Neckargemünd.*

Neckar valley, subsequently the Elsenz valley, was formed during the Würm glacial (GRAUL 1977). The 15 m terrace was probably formed in the early Würm (ZIENERT 1981); the 30 m terrace would then have to be dated as Riss and the 45 m terrace would be dated at least as Mindel and perhaps as Günz. These are only estimates but they agree largely with the results that SOERGEL obtained (1928) for the Mauer Neckar gravels and which were based on the presence of separate loess deposits that overlie the gravels disconformably. The age of the 75 m terrace has not yet been determined.

Nearly all the smaller valleys in the Odenwald have a broad floor with a flood plain that has frequently been used as irrigated meadowland in the past. The low broad ridges and narrow furrows of the old irrigation systems can still be recognized in some places. The settlements were located on the lower slopes usually on the sunny side of the valley. The valley slopes in the Bunter Sandstone area are generally quite steep, with blocks derived from the strongly silicified strata of the sandstone. Many of these blocks have been removed and used as building stones. In hollows on the slope, there are often block streams, known locally as Felsenmeere (block seas). During field trips in conjunction with the German Geographic Congress in Heidelberg in 1963, GRAUL has pointed out the similarities of form between these block streams and glaciers, for example, the concave tributary area and convex depositional form with marginal channels. These block streams are of periglacial origin and occur particularly on Bunter Sandstone.

However, the best known Felsenmeer in the Odenwald, which consists of several block streams, lies on the crystalline rocks of the northern block: on the southeastern slope of the Felsberg (501 m) at Reichenbach near Bensheim. The often egg-shaped blocks, most with diameters of several meters, originated as core stones during past periods of deep chemical weathering and their edges and corners are rounded accordingly. Blocks have been removed for building purposes since Roman times; an almost completed Roman column, weighing more than 27 tons, can still be seen. On small outcrops next to the block streams, there are also corestones with onion-like concentric weathered layers still embedded in granodiorite grus. The deep weathering progressed along the joint planes into the rock. At the intersections of joints, the edges and corners were weathered on two or three sides and became rounded. The grus was removed from around the core stones mainly by periglacial wash denudation and they were left as tors, block heaps and block streams. After heavy rains and during snowmelt, the runoff flows beneath the blocks and removes the grus matrix at the present time (GRAUL 1977, BRAUN 1967, RÜGER 1928, ZIENERT 1981).

References

BARTZ, J. (1951): Revision des Bohrprofils der Heidelberger Radiumsoltherme. Jahresber. Mitt. d. oberrh. geol. Ver.,N.F. **33**, 101–125

BLUME, H.(1971): Probleme der Schichtstufenlandschaft. Erträge der Forschung **5**, Darmstadt.

BRAUN, U. (1969): Der Felsberg im Odenwald. Heidelberger geogr. Arb. **26**.

CREDNER, W. (1922): Die Oberflächengestalt der kristallinen Gebiete von Spessart und Odenwald. Dissertation (typewritten),Univ. of Heidelberg.

FLICK, H. (1986): Permokarboner Vulkanismus im südlichen Odenwald. Heidelberger geowiss. Abhandlungen **6**, 121–128.

FLICK, H. & FUCHS, K. (1986): Ein permokarboner Ignimbrit-Förderschlot im Wachenberg bei Weinheim/Bergstraße. Jh. geol. LA Bad.-Württ. **28**, 31–42.

FRIED, G. (1984): Gestein, Relief und Boden im Buntsandstein-Odenwald. Frankfurter geowiss. Arb. **D 4**.

GRAUL, H. (1977): Exkursionsführer zur Oberflächengestaltung des Odenwalds. Heidelberger geogr. Arb. **50**.

RÜGER, L. (1928): Geologischer Führer durch Heidelbergs Umgebung. Heidelberg.

SCHMITTHENNER, H. (1956): Probleme der Schichtstufenlandschaft. Marburger geogr. Schriften **3**.

SEMMEL, A. (1961): Die pleistozäne Entwicklung des Weschnitztales im Odenwald. Frankfurter geogr. Hefte **37**, 425–492.

SEMMEL, A. (1975): Schuttdecken im Odenwald. Aufschluß, Sonderband **27**, 321–329

SOERGEL, W. (1928): Das geologische Alter des Homo heidelbergensis. Paläontolog. Zeitschr. **10**, 217–233

WAGNER, G.A. (1967): Spuren der spontanen Kernspaltung des ^{238}Urans als Mittel zur Datierung von Apatiten und ein Beitrag zur Geochronologie des Odenwaldes. Dissertation, Heidelberg.

WAGNER, G.A. (1973): Anwendung anätzbarer Partikelspuren in Geochronologie und Geochemie. Max-Planck-Institut Heidelberg, vol. **22**

ZIENERT, A. (1954): Die Großformen des Odenwaldes. Dissertation Heidelberg (printed 1957, Heidelberger geogr. Arb. **2**).

ZIENERT, A. (1961): Die Großformen des Schwarzwaldes. Forschungen z. deutschen Landeskunde **128**.

ZIENERT, A. (1981): Geographische Einführung für Heidelberg und Umgebung. Heidelberg.

ZIENERT, A. (1986): Grundzüge der Großformenentwicklung Südwestdeutschlands zwischen Oberrhein und Alpenvorland. Heidelberg.

Geological maps 1:25 000, 1:100 000 (KLEMM Darmstadt 1929); 1:200 000: sheets CC 6318 Frankfurt a. Main - Ost, CC 7110 Mannheim, CC 7118 Stuttgart-Nord

Address of author:
Dr. Adolf Zienert
Gerbodoweg 7
D-6900 Heidelberg 1
Federal Republic of Germany

CATENA SUPPLEMENT 15 p. 211–227 Cremlingen 1989

HILLSLOPE HYDROLOGY DATA FROM THE HOLLMUTH TEST FIELD NEAR HEIDELBERG

Dietrich **Barsch**, Heidelberg
Wolfgang-Albert **Flügel**, Pretoria

1 Introduction

Valleys with broad depositional floodplains are common landforms in central Europe. Their stream channels are usually incised into the valley floor at some distance from the side slopes. Therefore, the water from the slopes reaches the channel indirectly and only by way of the groundwater in the valley floor (ATKINSON 1978). Under natural conditions, overland flow is infrequent; even on agricultural land it accounts for only 10% to 15% of total runoff (DIKAU 1986, SCHAAR 1989). Nevertheless, precipitation events which result in only minor overland flow can cause flooding in the valleys. This indicates that the subsurface flow, particularly the interflow (FLÜGEL 1979), must reach the stream channel and affect the discharge after a very short time.

Although research on interflow has increased in recent years (KIRKBY 1978, BARSCH & FLÜGEL 1988), there are few data on the magnitude of interflow under varying boundary conditions. This paper examines the relationship between interflow, groundwater recharge and stream discharge, based on empirical investigations at the Hollmuth test field.

ISSN 0722-0723
ISBN 3-923381-18-2

3-923381-18-4/89/5011851/US$ 2.00 + 0.25

2 The Hollmuth test field

A geomorphological-hydrological test field was established in 1976 on a west-facing slope with an inclination of 11° to 15° on Bunter Sandstone (lower Triassic). The slope borders the floodplain of the Elsenz river and is located about 20 kilometers southeast of Heidelberg (BARSCH & FLÜGEL 1978). The regolith on the slope consists of periglacial solifluction debris about 5 m thick, which is covered by 1–2 m of loess loam. The geomorphological and pedological characteristics of the site and the instrumentation are shown in fig.1. The test field is under grass.

Regular measurements have been made at several locations on the test field since 1976. The data used here are from the hydrological years 1977 (beginning Nov.1, 1976) to 1982. The following variables were measured:

air temperature
precipitation
soil moisture (at different depths)

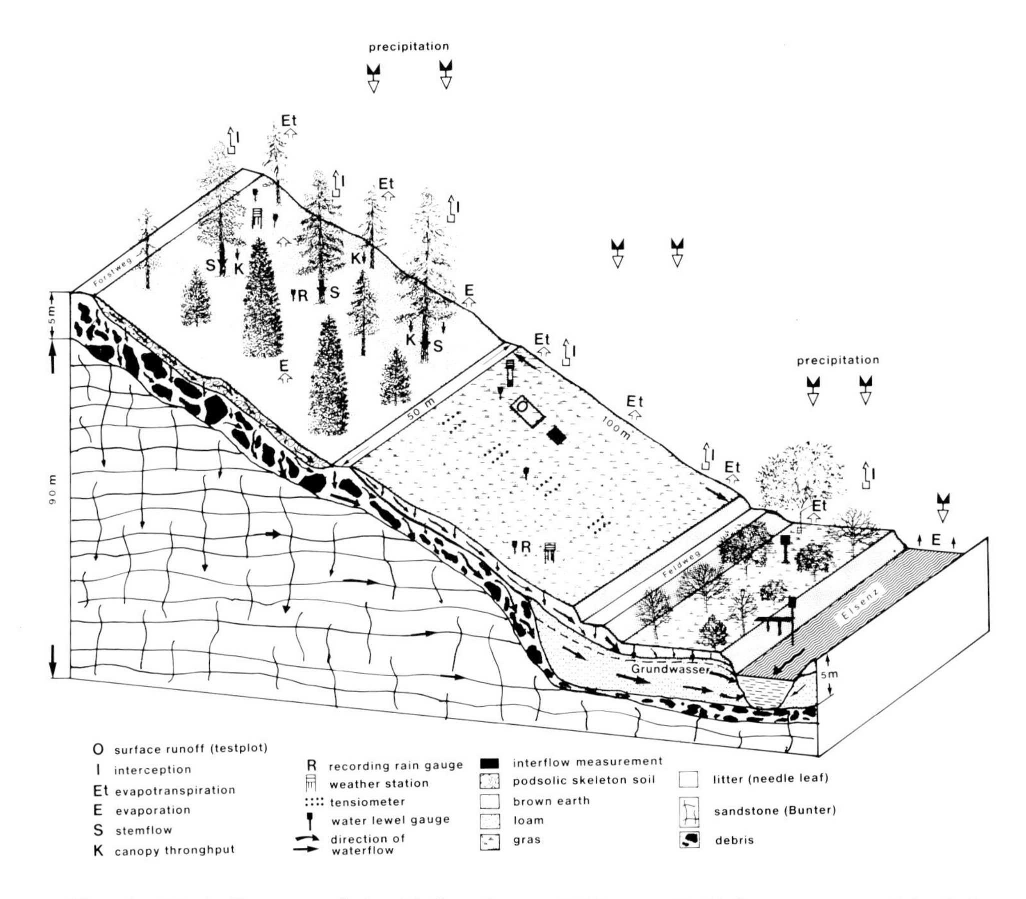

Fig. 1: *Block diagram of the Hollmuth test field near Heidelberg as a model of the hydrologic interaction between valley slope, valley bottom and channel.*

evaporation (Piche evaporimeter)
depth to the water table (at the slope foot)
discharge in the stream channel

In addition, static parameters such as the stratigraphic composition of the slope, the grain size distribution of the sediments and the pore volume were also determined. Despite the apparent uniformity of the slope, there is a considerable horizontal and vertical variability. The gradient of soil water tension parallel to the slope surface is, therefore, not uniform, although it has often been assumed to be (see ANDERSON & BURT 1977, HURLEY & PONTELIS 1985).

3 The occurrence of interflow

It was found that interflow is concentrated on the relatively dense surface of the periglacial waste cover beneath the loess loam and takes place during precipitation events in which the field capacity of the substrate (35–37 vol.% soil moisture) is exceeded. This occurs mainly in winter and in summer after a long period of rain during which the soil is

Precipitation (P) on testfield (mm):

WERT	Nov.	Dez.	Jan.	Feb.	März	Apr.	Mai	Juni	Juli	Aug.	Sept.	Okt.	Jahr
1977	87.6	91.6	86.9	81.9	33.0	65.3	18.1	88.7	41.5	95.2	29.2	24.6	743.6
1978	77.0	33.7	50.9	55.9	103.9	28.0	195.7	93.8	120.4	47.8	114.4	60.8	982.3
1979	16.9	97.1	53.1	69.1	138.1	82.3	54.3	37.4	80.4	59.9	32.5	35.1	756.2
1980	122.5	98.3	58.1	55.5	30.0	75.9	58.1	151.9	258.1	93.1	37.9	93.8	1133.2
1981	73.3	79.3	133.6	56.0	99.2	76.6	83.7	86.0	89.2	41.9	48.1	213.1	1080.0
1982	122.1	130.2	86.9	20.5	59.4	45.6	87.3	84.6	63.5	61.3	42.7	144.6	948.7
Mittel	83.2	88.4	78.3	56.5	77.3	62.3	82.9	90.4	108.8	66.5	50.8	95.3	940.7

Interception (I) on testfield (%):

	Nov.	Dez.	Jan.	Feb.	März	Apr.	Mai	Juni	Juli	Aug.	Sept.	Okt.	Jahr
	14.8	14.8	14.8	14.8	14.8	19.5	27.6	26.1	30.0	23.3	17.1	14.8	19.4

Reduced precipitation (P-I) in (mm):

	Nov.	Dez.	Jan.	Feb.	März	Apr.	Mai	Juni	Juli	Aug.	Sept.	Okt.	Jahr
1977	65.2	72.6	70.4	58.4	0.0	14.0	0.0	0.0	0.0	0.0	0.0	0.0	283.6
1978	47.8	23.3	39.8	36.2	52.4	0.0	71.0	0.0	0.0	0.0	50.7	24.0	346.1
1979	0.0	71.6	39.9	46.3	90.1	13.1	0.0	0.0	0.0	0.0	0.0	0.0	170.0
1980	92.4	70.6	41.7	30.5	0.0	13.0	0.0	45.9	130.0	0.0	0.0	58.3	482.7
1981	47.7	61.3	104.7	32.8	50.4	0.0	0.0	0.0	0.0	0.0	0.0	156.4	453.3
1982	87.5	105.2	67.4	3.3	17.9	0.0	0.0	0.0	0.0	0.0	0.0	107.4	388.7
Mittel	57.4	67.4	60.7	34.6	35.1	6.7	11.8	7.5	21.7	0.0	0.0	57.8	360.9

Potential evapotranspiration (pET) acc. to HAUDE (mm):

	Nov.	Dez.	Jan.	Feb.	März	Apr.	Mai	Juni	Juli	Aug.	Sept.	Okt.	Jahr
1977	9.4	5.4	3.6	11.4	36.1	38.6	84.1	75.4	111.0	78.0	54.1	32.2	539.3
1978	17.8	5.4	10.2	9.3	30.9	67.7	70.7	101.3	97.4	88.2	43.9	26.9	569.7
1979	10.9	11.1	5.3	12.6	27.6	53.2	102.2	105.6	81.2	72.1	64.8	40.2	586.8
1980	12.0	13.2	7.8	16.8	27.7	48.1	82.5	66.4	50.4	75.1	49.2	21.6	470.8
1981	14.8	6.3	9.1	14.9	34.1	64.1	84.0	79.7	75.4	90.7	63.4	25.2	561.7
1982	16.5	5.7	6.6	14.2	32.7	68.1	100.2	98.2	124.0	90.6	85.1	15.8	657.7
Mittel	13.6	7.8	7.1	13.2	31.5	56.6	87.3	87.8	89.9	82.5	60.1	27.0	564.3

Potential precipitation input into groundwater storage (mm):

	Nov.	Dez.	Jan.	Feb.	März	Apr.	Mai	Juni	Juli	Aug.	Sept.	Okt.	Jahr
1977	0.0	46.2	36.8	74.0	22.0	0.0	0.0	0.0	0.0	0.0	0.0	0.0	179.0
1978	33.8	0.0	48.8	63.5	22.8	0.0	0.0	0.0	0.0	0.0	12.5	0.0	181.4
1979	35.9	10.4	49.5	47.1	66.1	5.9	0.0	0.0	0.0	0.0	0.0	0.0	214.9
1980	60.0	146.2	0.0	0.0	43.5	8.2	0.0	18.3	118.3	0.0	0.0	42.7	437.2
1981	39.3	69.7	121.4	13.6	13.2	0.0	0.0	0.0	0.0	0.0	0.0	130.0	387.2
1982	87.5	69.2	98.6	0.0	27.5	0.0	0.0	0.0	0.0	0.0	0.0	53.4	336.2

Tab.1: *Precipitation (P), Interception (I), potential evapotranspiration according to HAUDE (pET), by interception reduced precipitation and the potential input by precipitation into the groundwater storage (P-I-pET) on the test field Hollmuth. Please note, that the potential input by precipitation into the deeper storages has been calculated on an hourly base. Thus, the monthly means given in the upper parts of this table cannot be used as a control.*

thoroughly wet. Precipitation that does not reach the stream as overland flow and that exceeds the concomitant interception and evaporation, infiltrates on the valley floor and on the slopes. If there is interflow on the slopes, such a rainfall event leads, in the valley plain near the foot of the slope, to

1. a rapid rise of the water table at the slope foot and
2. a slow fall of the groundwater hydrograph owing to delayed inflow from the slope and to seepage of water into the channel.

This corresponds to the differentiation into quick and delayed interflow (CHOW 1964). Recharge from the channel does exert a significant influence since the water table gradient is nearly always directed towards the channel.

4 Elements of the water balance

Table 1 shows data of monthly precipitation as input of the water balance. Hourly measurements were made and loss by interception calculated indirectly following SCHAAR (1989). The calculated values are similar to values that have appeared in the literature. Evaporation is assumed to be equal to potential evaporation calculated according to HAUDE (1955) . The possible error is probably small because the data are calculated on a daily basis. Also during the rainfall event, and for a short period thereafter, the actual evaporation is more or less equal to the potential evaporation. An overestimation of evaporation would cause an underestimation of the interflow.

An important component in the water balance is the valley floor aquifer. At the test site, the deposits are fine-grained so that the soil water budget above the water table must be included because the capillary rise and the infiltration of rainwater from the surface to the groundwater may affect the water balance under certain conditions. The soil moisture above the water table was, therefore, recorded at depths of 15, 30, 60, 90 and 120 cm. The data correlate very well down to depths of 90 cm because of a well-developed dry-weather moisture gradient between the surface and one meter depth and also because of the seepage of rainwater down to the groundwater in humid periods which causes the water table to rise. As a result, there is a close relationship between the position of the water table and the total stored volume of vadose water and groundwater above an arbitrary boundary plane located 3 meters below the land surface. This storage, comprising the zone of aeration and the upper part of the zone of saturation, is termed the total dynamic storage (GSI; fig.2).

Smoothing, using the means, reduced the 141 data pairs (water table height/storage volume) to 40. The water table values ranged from 79 cm to 181 cm below the surface. At a porosity of 48% this described a storage volume of between 1061 litres (for 79 cm) and 571 litres (for 181 cm) per square meter of surface area. The addition of the water stored in the vadose zone resulted in total dynamic storage volumes of 1423 litres and 1236 litres respectively for the two water table levels (fig.3). The regression equation for fig.3 is

$$GSI = 1442.6 \quad -1.90 \cdot 10^{-2} \cdot GWST \quad -1.96 \cdot 10^{-3} \cdot GWST^2 \quad -2.44 \cdot 10^{-5} \cdot GWST^3 \qquad (1)$$

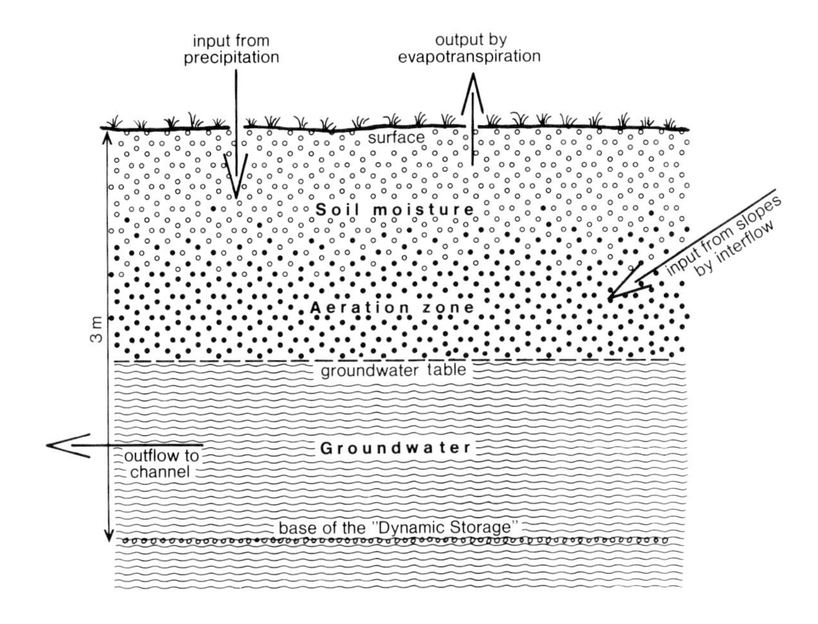

Fig. 2: *Model of the dynamic storage.*

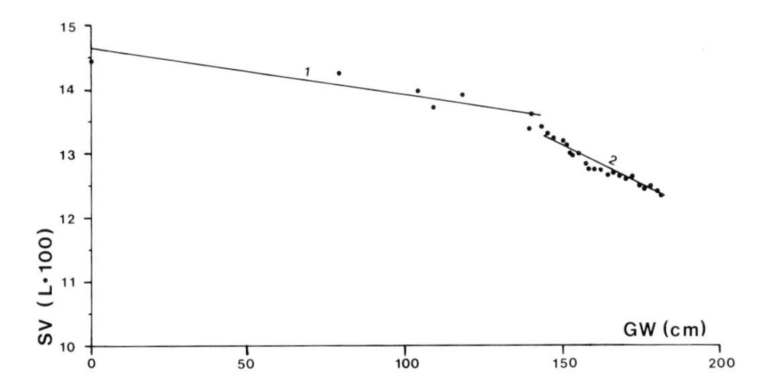

Fig. 3: *Relation between groundwater level GW (cm) below the surface and the volume SV (100 l) of the dynamic storage.*

where
GSI = total dynamic storage volume (l/m^2 = mm)
$GWST$ = Water table position below the surface (cm).

With this equation, which has a coefficient of determination B = 0.99, every water table change is converted into the corresponding change of storage which can then be correlated with changes in the slope water system. The storage change is the net result of addition by interflow from the slope, infiltration in the valley floor, loss by seepage into the channel and, to a minor extent, also evapotranspiration during the summer months (May to October).

Based on the hourly water table data, a balance model was designed (fig. 4)

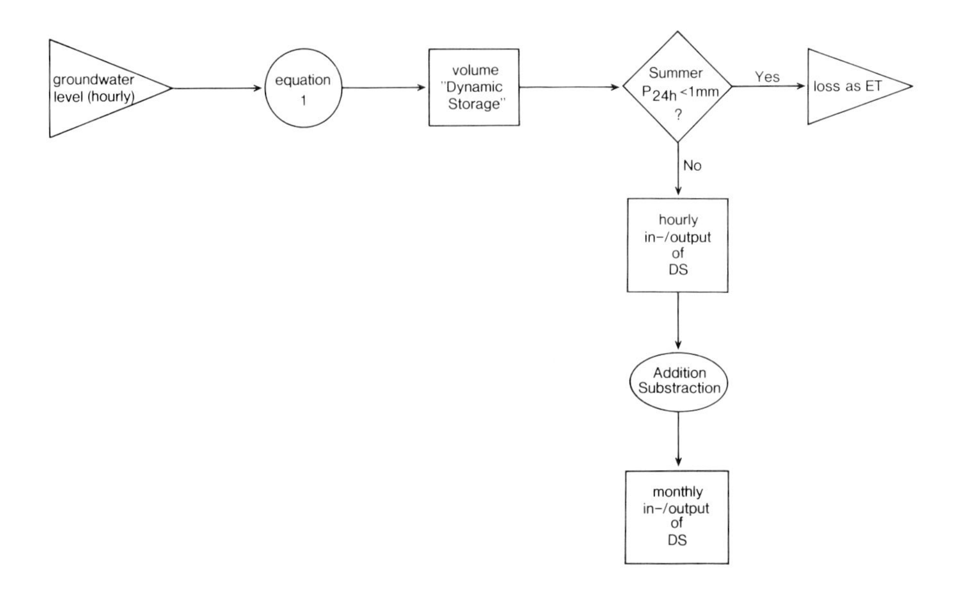

Fig. 4: *Schematic flow chart for the calculation of the balance of the dynamic storage.*

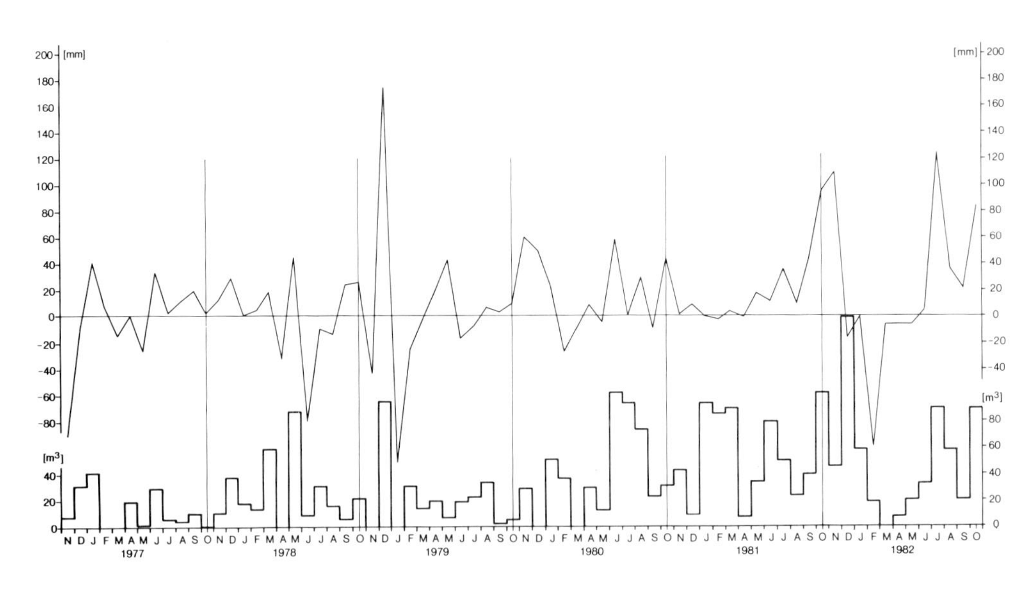

Fig. 5: *Monthly balance of the dynamic storage and the input by interflow for the hydrological years 1977–1982.*

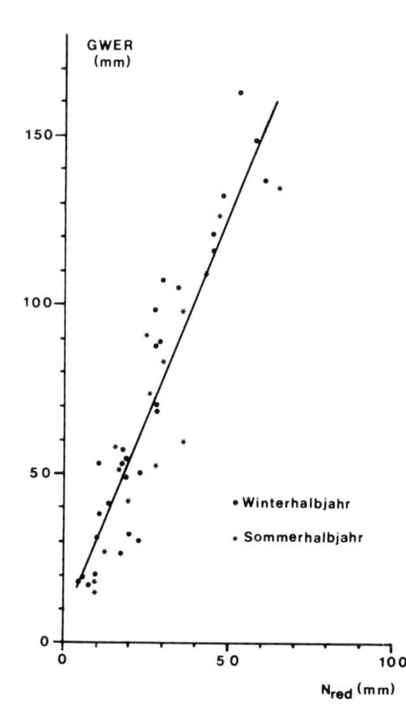

Fig. 6: *Correlation between the precipitation (P) reduced by interception ($N_{red} = P - I$) in (mm) and the input GWER into the dynamic storage ($mm = l/m^2$).*
Winter = 1. Nov. to 30. April;
Summer = 1. May to 31. October.

which calculates, for every hour, the volume of the total dynamic storage according to equation 1. On summer days with a large temperature range, diurnal water table fluctuations can occur that are caused solely by the capillary rise and the subsequent drop of the water table and that do not, therefore, affect the water balance. In order to identify such occurrences, the data are checked as to whether there was any precipitation during the 24 hours preceding a change of the water table. If there was none, or less than 1 mm, the corresponding water table fluctuation is disregarded. Addition of the hourly data yields daily and monthly values.

The plot of monthly values of the total dynamic storage and of the interflow (fig. 5) shows considerable fluctuations. These are due to recharges and outflows. The dryer years, 1977–1979, are clearly distinguishable from the wetter years, 1980–1982. The major outflows have been triggered mainly by inflows that caused the gradient of the water table towards the channel to increase. Generally, the storage surpluses of the winter half-year have disappeared by May. The high summer rainfalls that occurred in 1980 and 1981, for example, also produced considerable groundwater recharge.

A good linear correlation (r = 0.92) exists between precipitation, reduced by the interception (NRED), and the groundwater recharge (fig. 6):

$$GWER = 2.4 \cdot NRED + 6.3 (mm = l/m^2) \quad (2)$$

where
$GWER$ = groundwater recharge (mm) =
volume increase of total dynamic storage (l/m^2).

The correlation does not improve if the winter and summer data are treated separately.

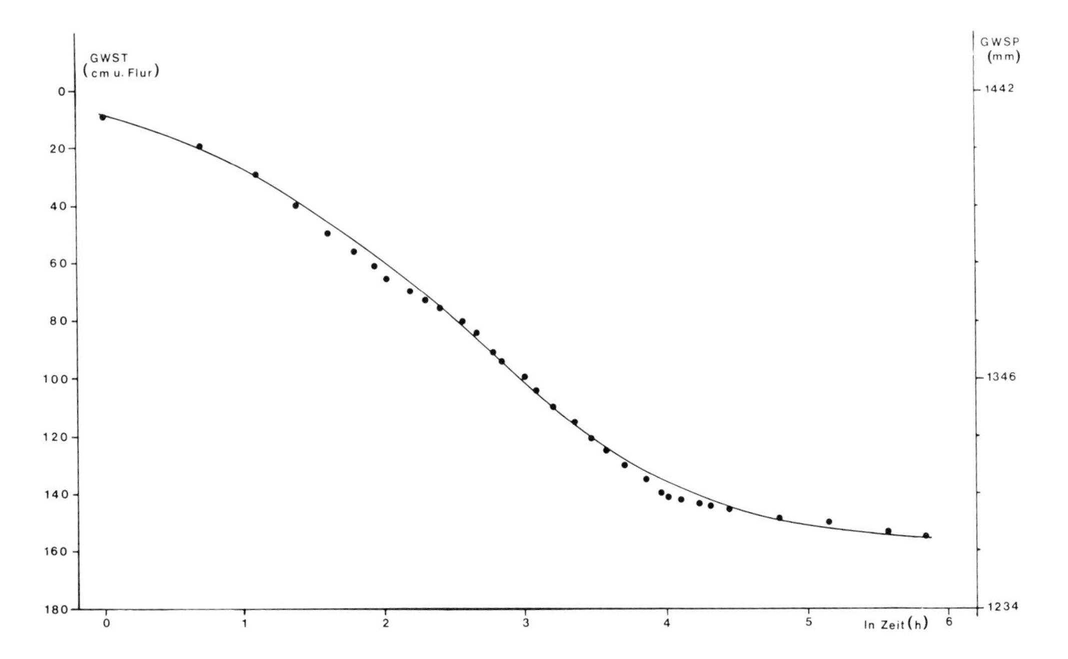

Fig. 7: *Output from the dynamic storage during dry weather: x-axis: time as ln(h); y-axis: groundwater below surface (GWST in cm) or groundwater storage (GWSP in mm).*

5 Examples of interflow

Increases in the total dynamic storage can take place within a few hours after a rainfall event. Decreases, on the other hand, are slow and often greatly delayed. Rapid increases are interpreted as short-term recharges, that is, as direct interflow. Slow increases and, especially also, slow decreases are the result of the delayed interflow that arrives in the storage after some retardation. The groundwater hydrograph describes the water supply and can be evaluated quantitatively if the decline of the hydrograph in dry weather is known. The curve of the dry weather hydrograph was determined following the example given by WUNDT (1953) for dry weather discharge in open channels. Between depths of 15 and 155 cm below the surface this curve is described by

$$GWST(cm) = 7.59 + 29.03 \cdot lnT - 18.01 + (lnT^2) + 13.39 \cdot (lnT^3) - 3.01 \cdot (lnT^4) + 0.21 \cdot (lnT^5) \quad (3)$$

where
$GWST$ = water table position in cm below the surface
$ln\ T$ = natural logarithm of time in hours after occurrence of the maximum.

This means that the water table falls by 20 cm in the first two hours, by 120 cm in the next 54 hours and by 15 cm in the following 344 hours, a total lowering of 155 cm. The polynomial describes the entire non-linear range of values. A subdivision of the fall line of the hydrograph as suggested by DRACOS (1980) is, therefore, unnecessary.

The time lag is between 7 and 12 hours for water table positions between 137 cm

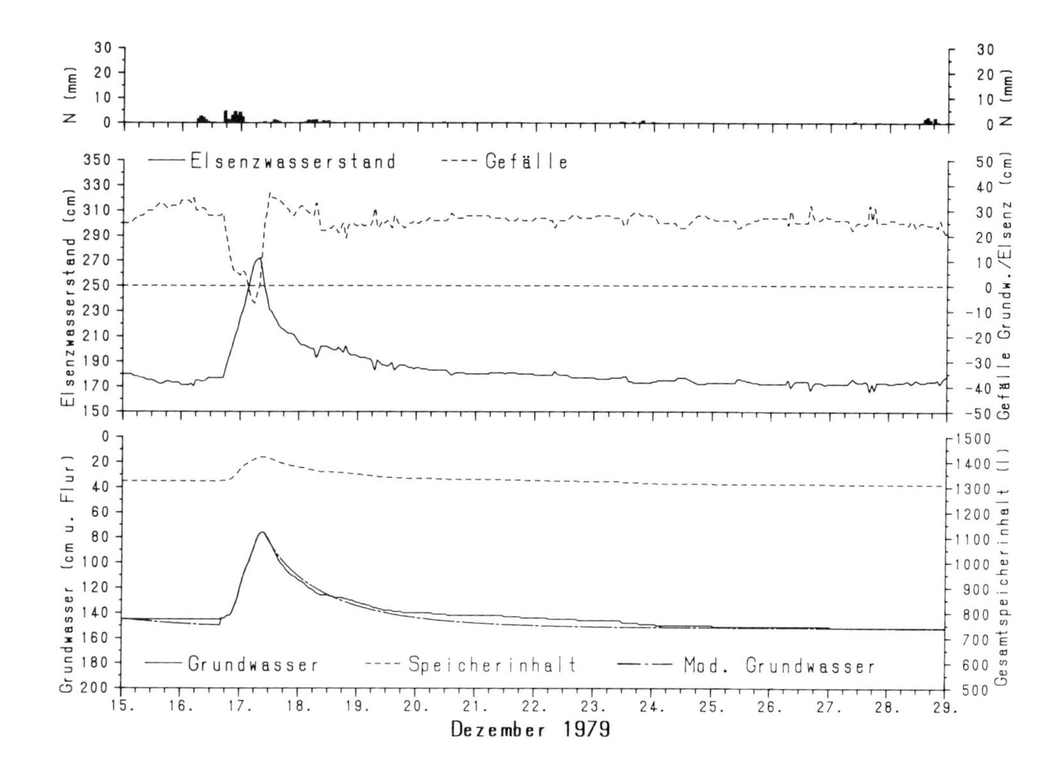

Fig. 8: *Precipitation (N in mm, water level in the Elsenz river (Elsenzwasserstand in cm), slope towards water level in the channel (Gefälle in cm), volume of the dynamic storage (Gesamtspeicherinhalt in $l=10^{-3}$ m^3), measured and calculated (broken line) groundwater level as hourly averages at the base of the Hollmuth test field during December 1979.*

and 145 cm hours below the surface. In some cases, the rise of the water table continued for up to 17 hours after the rainfall. Usually the drop of the water table begins directly after the maximum has been reached. However, when water is supplied from interflow for a long period, the hydrograph may stay at the maximum for up to 13 hours. High intensity summer rains of up to 25 mm/h cause rapid rises and exponential declines of the hydrograph. If the precipitation is combined with snow melt, recharge amounts of 100 mm (or l/m^2) can occur with precipitation of only 10 mm.

If the groundwater hydrograph is used to determine the water balance it is important to take into account that the drainage to the channel begins immediately after the water table has begun to rise. In dry weather the hydraulic conductivity was found to be

$$Kf = 1.36 \cdot 10^{-8} \tag{4}$$

At gradients of about 28 cm the loss to the channel of the Elsenz river is 0.16 mm/h. Allowing for the generally higher gradients, the outflow is set at 0.3 mm/h.

period	precipitation (mm)	reduced precipitation (mm)	interflow (mm)			d/q %	surchage	precipitation on testfield (m^3)	interflow from festfield						"loss"	
			quick	delayed	total				quick		delayed		total			
									m^3	%	m^3	%	m^3	%	m^3	%
15. - 29.12.79	57	48.6	102.2	33.1	135.3	32.4	2.8	197.8	51.1	25.8	16.6	8.4	67.7	34.2	130.1	65.8
30.06. - 14.07.80	167.1	117.9	259.7	114.4	374.1	44.1	3.2	479.9	129.9	27.1	57.2	11.9	187.1	39.0	292.8	61.0
13. - 27.07.80	101.2	76.7	153.3	138.8	292.1	90.5	3.8	311.0	76.7	24.7	69.4	22.3	146.1	47.0	164.9	53.0
31.12.80 - 14.01.81	89.5	76.3	171.4	39.2	210.6	22.9	2.8	310.5	85.7	27.6	19.6	6.3	105.3	33.9	205.2	66.1
01. - 15.02.81	52.5	44.1	120.8	96.7	217.5	80.0	4.9	179.5	60.4	33.7	48.4	27.0	108.8	60.6	70.7	39.4
07. - 21.03.81	55.0	45.4	177.2	72.7	249.9	41.0	5.5	184.8	88.6	47.9	36.4	19.7	125.0	67.6	59.8	32.4
01. - 15.06.81	53.7	39.7	138.8	21.3	160.1	15.3	4.0	161.6	69.4	43.0	10.7	6.6	80.1	49.6	81.5	50.4
28.11. - 12.12.81	134.8	109.1	452.2	63.0	515.2	13.9	4.7	444.0	226.1	50.9	31.5	7.1	257.6	58.0	186.4	42.0
27.01. - 10.02.82	46.6	39.7	167.1	111.7	278.8	66.8	7.0	161.6	83.6	51.7	55.9	34.6	139.5	86.3	22.1	13.7
average	84.2	66.4	193.6	76.8	270.4	45.2	4.3	270.1	96.8	36.9	38.4	16.0	135.2	52.9	134.8	47.1

Remarks: reduced precipitation = precipitation - interception
d/q = relation of delayed to quick interflow (%)
surcharge = relation of interflow to reduced precipitation testfield Hollmuth: 4.070 m^2
"loss" = the amount of the reduced precipitation on the testfield Hollmuth, which evaporated or infiltrated probably to deeper storages and from their to the groundwater and to discharge.

Tab.2: *Quick and delayed interflow during selected (storm) periods on the test field Hollmuth near Heidelberg.*

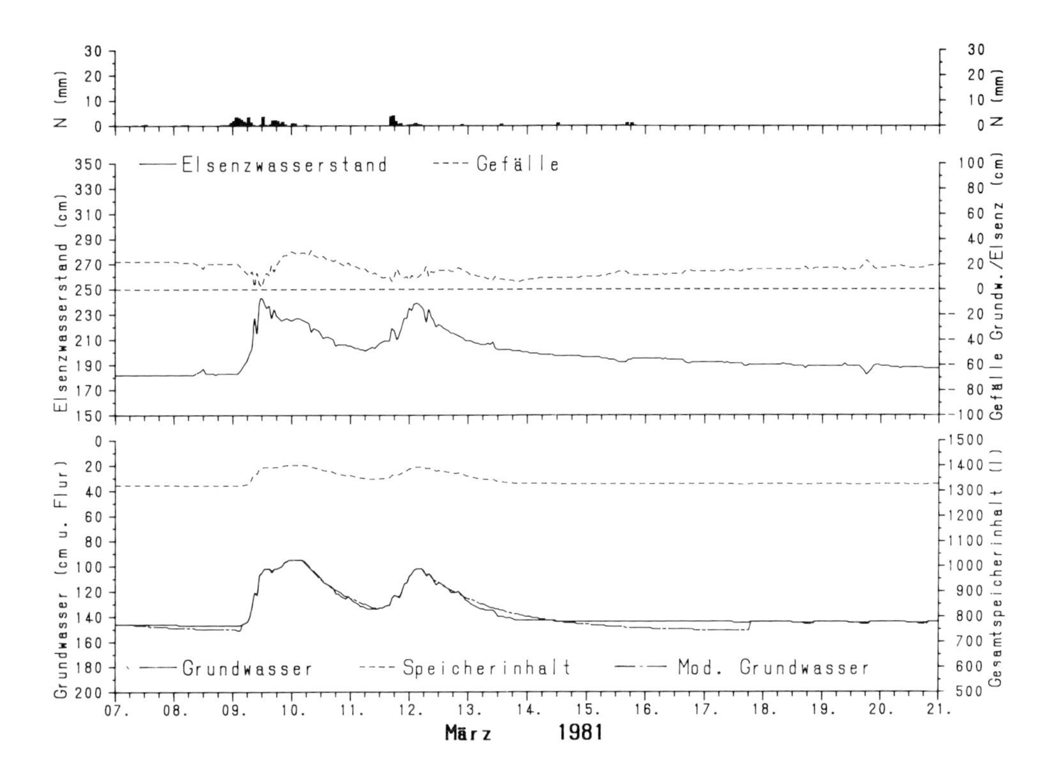

Fig. 9: *Precipitation, water level, groundwater, slope towards water level in the channel, volume of dynamic storage, measured and calculated groundwater level as hourly averages at the base of the Hollmuth test field March 1981. Symbols see fig.8.*

For individual rainfall events between 5.4 mm and 65.2 mm, the water table rises by direct infiltration and interflow ranges from 14.8 to 162.9 mm. Mean rainfall per event is 25.9 mm (= 32.1 mm - 19.2 % interception), causing a mean inflow into the total dynamic storage of 68.6 mm. If it is assumed that the reduced mean rainfall of 25.9 mm infiltrates directly from the valley floor into the storage, the interflow accounts for the remaining 42.7 mm. Therefore, 62% of the groundwater recharge is assumed to be the result of direct interflow.

Figs. 8–11 show hydrographs of several two week periods in different months that had significant interflows. The precipitation per period varies between 46.6 and 167.1 mm.

The major event of the summer of 1980 (June 30–July 14) resulted in an absolute maximum total which was, however, a below-average relative interflow of 39% of the reduced precipitation. By contrast, the event of November 28 - December 12, 1981 had a relative interflow value of 58%.

The delayed interflow is always considerably smaller than the direct interflow. The only exceptions are follow-up events

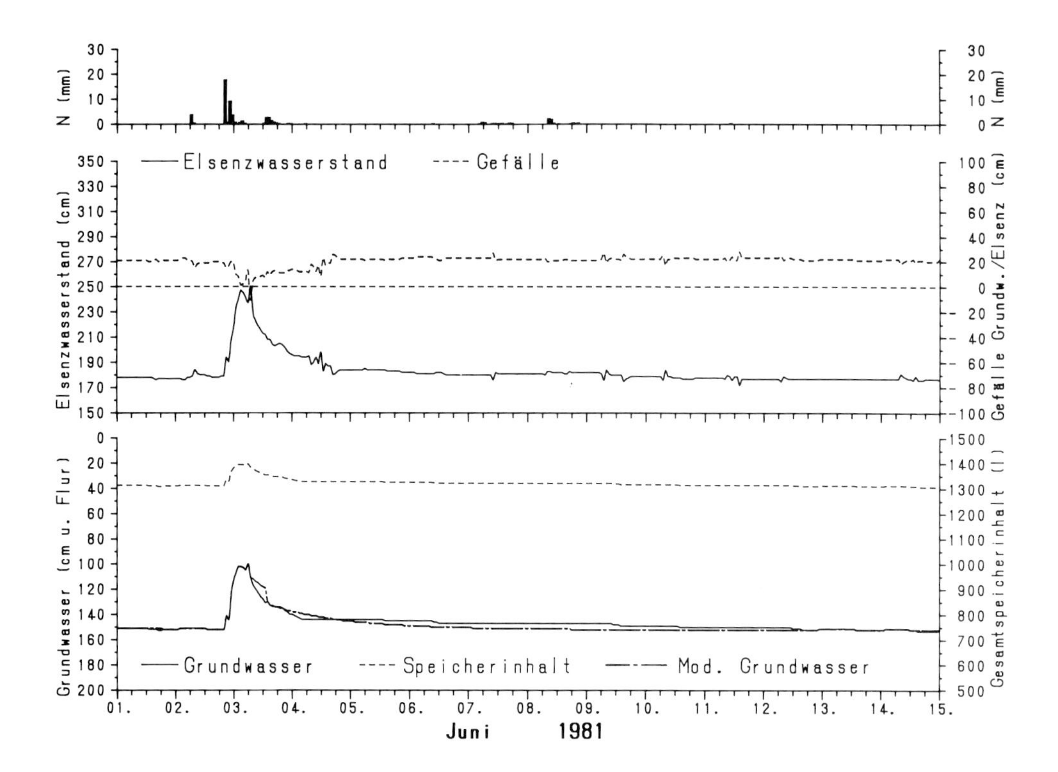

Fig. 10: *Precipitation, water level in the Elsenz river groundwater, slope towards water level in the channel, volume of the dynamic storage, measured and calculated groundwater level as hourly averages at the base of the Hollmuth test field June 1981. Symbols see fig.8.*

in summer or events in the late winter, for example, the events of July 13 - 27, 1980, February 1 - 15, 1981 and January 27 - February 10, 1982. The melting of snow was a contributing factor in the latter two events.

On average, nearly 50% of the reduced precipitation (NRED in equation (2)) that has been measured on the test field is not present in the balance data, the result probably of infiltration losses at depth and the consumption of water used to fill the soil storage up to field capacity. This consumption varies depending on the soil moisture contents before the precipitation event. On the test field of 4070 m^2, more than 2000 m^3 water are needed to fill the loam cover which has a mean thickness of 1.5 m and a field capacity of 35%. Unfortunately, the soil moisture data are not available in sufficient detail to allow a conclusive judgement to be made about this component of the water balance.

In order to estimate the total amount of direct and delayed interflow, the water input must be known. The precipitation data have, therefore, been reduced by the interception and evaporation. The resulting potential input, multiplied by

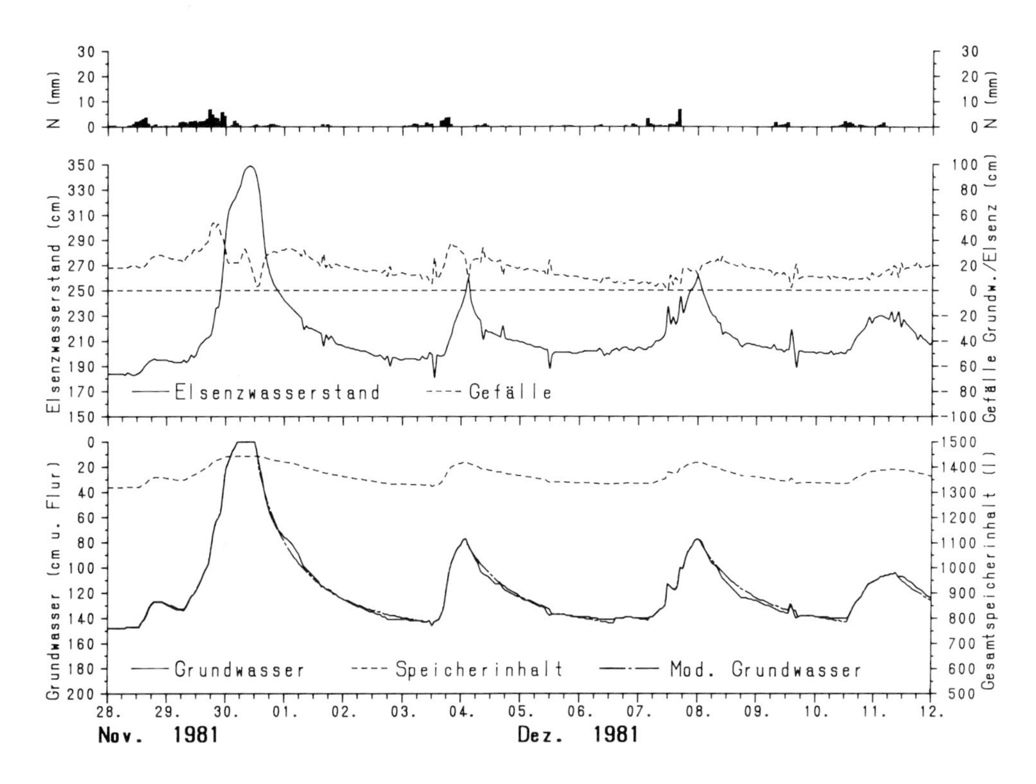

Fig. 11: *Precipitation, water level in the Elsenz river groundwater, slope towards water level in the channel, volume of the dynamic storage, measured and calculated groundwater level as hourly averages at the base of the Hollmuth test field November/December 1981. Symbols see fig.8.*

the area of the test field, represents the amount of water that infiltrates into the loam-covered slope and is then potentially available as supply to the total dynamic storage. In addition, there is the water that infiltrates directly into the valley floor.

Fig.5 shows these inputs at the bottom of the diagram, as the inflow that has been drained from the slope into the dynamic storage. The interflow has been set at zero for those months in which the potential evaporation was greater than the precipitation. Short term interflows during and after individual events have, nevertheless, occurred, and may have resulted in outflow. Interflow should, therefore, be analysed taking into consideration individual events and their dynamics rather than statistically on a monthly basis.

6 Slope water balance

Despite the limitations of the data, they can be used to estimate the approximate-mean water balance of a relatively dry year, 1977, a relatively wet year, 1980, and a "normal" year. Some of the values in the balances are reasonably accurate,

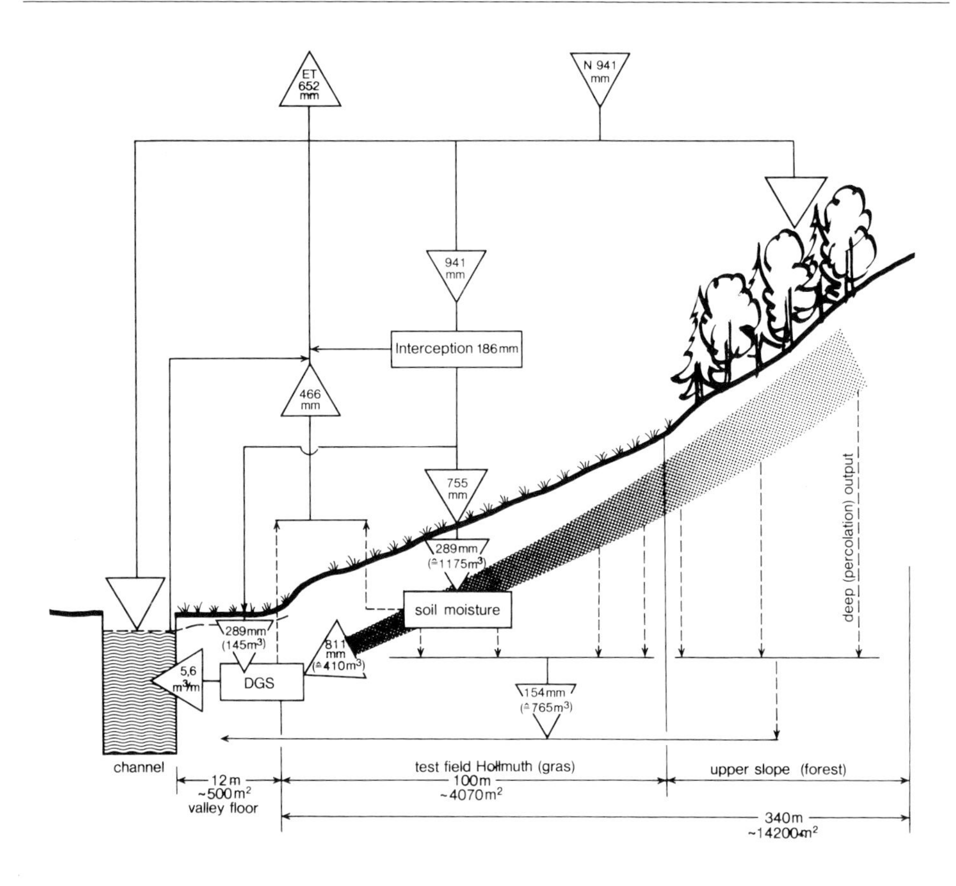

Fig. 12: *Water balance for a "normal" hydrological year on the Hollmuth test field.*

for example the changes in the storage volume. Others contain errors which are due either to unmeasured components, such as deep infiltration into the slope, or to the estimation methods used.

Tab.3 shows the data for 1977, 1980 and a "normal" year, whereby the "normal" year is defined as the mean of the six years of measurement. The six-year means represent a balancing-out of the large fluctuations that influence the storage and outflow terms beyond individual hydrological years.

For a given year, the measureable interflow amounts to about 20%–35% of the potential input. This is a relatively high proportion. Although the interflow is buffered and delayed, it is, nevertheless, drained to the storage relatively rapidly. The slope segments near the valley floor contribute a large proportion of the total groundwater recharge. Their areas are about 1.7 times, 2.2 times and 2.8 times as large as the area of the valley floor (500 m^2) in the test area. In the "normal" year, therefore, about one-third of the test area provides the entire potential input of the total dynamic storage. It is probable, however, that the proportion of the interflow that is contributed to the storage by different parts of the slope decreases exponentially with

	hydrological year					
	1977		1980		average	
	(mm)	(m^3)	(mm)	(m^3)	(mm)	(m^3)
precipitation (P)	744	-	1133	-	941	-
evaporation from interception (J)	(141)	-	(377)	-	(186)	-
potential evapotranspiration (pE) (1)	(539)	-	(471)	-	(564)	-
(2)	(424)	-	(319)	-	(466)	-
total evapotranspiration	565	-	696	-	652	-
potential input (PJ) (3)	179	729	437	1778	289	1177
total storage input (4)	486	243	1395	699	1100	550
storage input "interflow"	307	156	958	480	811	406
storage input "valley floor infiltration"	179	90	437	219	289	145
"loss" on testfield (5)	141	573	319	1298	189	771
"loss" from total slope (5)	168	2386	403	5725	206	2927
interflow in % of potential input		21 %		27 %		35 %
interflow in % of precipitation	5 %		10 %		11 %	

(1) Calculated from hourly data
(2) "Corrected" values according to the equation pE=P-J-PJ
(3) Potential input calculated on the base of hourly data of rainfall, interception and potential evapotranspiration horizontal surface of testfield: 4070 m^2
(4) Dynamic storage beneath the valley floor (500 m^2) in front of the testfield
(5) Loss means the deep percolation on the slope and the amount of water (temporarily) stored in the loam and other slope deposits

Tab.3: *The slope hydrology of the test field Hollmuth for a dry (1977), a wet (1981) and the average hydrological year calculated from hourly data for 1977–1982.*

the distance from the slope foot. Since the loam cover is absent on the slope above the test field, it can be assumed that the wooded upper parts of the slope deliver most of their infiltrated water into the deeper zones of the groundwater, the data for which are not included in these estimates.

Tab.3 contains deficits that are not accounted for. These occur particularly in months in which there are high inflows from the slopes and major rises in the water table. During these months the outflows from the groundwater were not measured but were undoubtedly higher. The loss from snow has also not been measured but based on HERRMANN (1974) and SCHAAR (1989), it is as-

	(mm)	(m^3)
Precipitation	941	—
Evapotranspiration (for all storages)	725	—
Potential input	216	878
Interflow	107	437
Deeper outflow	108	441
Interflow in % of potential input	49.8	
Discharge/river Elsenz at Meckesheim	(259.4 km^2)	
(1) in (m^3s^{-1}):	2.21	
(2) in (ls^{-1} km^2):	8.52	
(3) in (mm):	269	

Tab. 4: *Interflow from the test field Hollmuth under average conditions (c.f. text).*

sumed to be 0.3 mm/day for the days of snow precipitation with daily mean temperatures below 0°C. Interception is arbitrarily set at 30% for both summer and winter because of the dense grass cover of the test field. The values in Tab.4 are based on these assumptions. The most reliable data in Tab.4 are the precipitation data and, as inflow into the total storage, the interflow. The latter represents about 50% of the contribution of the precipitation to the groundwater recharge, or 11% of total precipitation. Since the discharge of the Elsenz river is equivalent to about 270 mm per year the interflow supplies about 40% of the discharge. This is a mean value. In the case of high intensity rainfall events which cause floods, the relative contribution of the interflow is far greater.

7 Conclusions

The slope water balance demonstrates the great importance of the interflow for groundwater recharge in the floodplain and for the river discharge. It should, therefore, receive greater weight in models of peak river flows. For the base flow, by contrast, the 50% of the potential recharge-contributing precipitation that was not taken into account in these estimates is more decisive. It has been assumed that this recharge reaches the groundwater after longer delays because it first infiltrates deeply into the slope. This component of the water supply is very important for the river discharge in dry weather.

Future empirical research should use tracers in order to identify the water movement of the deep infiltration and the interflow more precisely. An experimental approach using artificial sprinkling appears to be possible (FLÜGEL & SCHWARZ 1988).

Models for the description and calculation of interflow must be dynamic. They should proceed from individual events to an overall evaluation because mean values alone cannot correctly represent the rapid reaction of a slope system to a precipitation event.

Acknowledgement

We thank the Deutsche Forschungsgemeinschaft and the Kurt Hiehle foundation for their financial support of the research reported in this paper.

References

ANDERSON, M.G. & BURT, T.P. (1977): The role of topography in controlling throughflow generation. Earth Surface Processes **3**, 331–344.

ATKINSON, T.C. (1978): Techniques for measuring subsurface flow on hillslopes. In: Kirkby, M.J. (ed.): Hillslope Hydrology. Wiley, Chichester, 73–120.

BARSCH, D. & FLÜGEL, W.A. (1978): Das geomorphologisch-hydrologische Versuchsfeld "Hollmuth" des Geographischen Instituts der Universität Heidelberg. Erdkunde **32**, 61–70.

BARSCH, D. & FLÜGEL W.A. (1988): Untersuchungen zur Hanghydrologie und zur Grundwasserentwässerung am Hollmuth, Kleiner Odenwald. Heidelberger Geographische Arbeiten **66**, 1–82.

CHOW, V.T. (ed.) (1964): Handbook of applied hydrology. McGraw-Hill, New York.

DIKAU, R. (1986): Experimentelle Untersuchungen zu Oberflächenabfluß und Bodenabtrag von Meßparzellen und landwirtschaftlichen Nutzflächen. Heidelberger Geographische Arbeiten **81**, 195 pp.

DRACOS, TH. (198O): Hydrologie. Wien.

FLÜGEL, W.A. (1979): Untersuchungen zum Problem des Interflow. Heidelberger Geographische Arbeiten **56**, 170 pp.

FLÜGEL, W. A. & SCHWARZ, O. (1988): Beregnungsversuche zur Erzeugung von Oberflächenabfluß, Interflow und Grundwassererneuerung auf dem Hollmuth, Kleiner Odenwald. Heidelberger Geographische Arbeiten **66**, 196–200.

HAUDE, W. (1955): Zur Bestimmung der aktuellen und potentiellen Verdunstung auf möglichst einfache Weise. Mitt. d. Dt. Wetterdienstes Nr. **11**, 1–24.

HERRMANN, A. (1973): Entwicklung der winterlichen Schneedecke in einem nordalpinen Einzugsgebiet. Münchener Geographische Abh. **10**, 84 pp.

HURLEY, D.G. & PONTELIS, G. (1985): Unsaturated flow through a thin porous layer on a hillslope. Water Resources Research **21**, 821–824.

KIRKBY, M.J. (ed.) (1978): Hillslope Hydrology. Wiley, Chichester, 389 pp.

SCHAAR, J. (1989): Untersuchungen zum Wasserhaushalt kleiner Einzugsgebiete im Elsenztal/Kraichgau. Heidelberger Geographische Arbeiten (in press).

WUNDT, W. (1953): Gewässerkunde. Springer, Heidelberg, 320 pp.

Address of author:
Prof. Dr. Dietrich Barsch
Geographisches Institut der Universität Heidelberg
Im Neuenheimer Feld 348
Postfach 105760
D-6900 Heidelberg 1
Federal Republic of Germany
Dr. Wolfgang-Albert Flügel
Hydrological Research Institute
Private Bag X 313
Pretoria 001
Republic of South Africa

CATENA SUPPLEMENT 15 p. 229–242 Cremlingen 1989

A COMPARISON OF BUNTER SANDSTONE SCARPS IN THE BLACK FOREST AND THE VOSGES

Helmut **Blume**, Tübingen
Gerhard **Remmele**, Heilbronn

1 Introduction

Large areas of the Black Forest and the Vosges, located respectively on the eastern and western flanks of the Rhine Rift Valley, are composed of sandstones of the Bunter formation (Buntsandstein, Early Triassic). Because of the stronger tectonic uplift in the south of both the Black Forest and the Vosges, the sedimentary cover was removed and the crystalline basement exposed. In the north of both areas, the crystalline outcrops only in the valleys or in areas that have been uplifted as a result of locally differentiated tectonics.

The landforms developed on the Bunter sandstones differ from those developed in the crystalline basement rocks. Lithological variations in the gently inclined strata of the Bunter formation have resulted in the development of cuesta scarps in the northern Black Forest and the northern Vosges. Both areas belong to more extensive regions of scarps: the northern Black Forest to the scarplands system in southern Germany and the northern Vosges to the system of scarps surrounding the Paris Basin.

Since scarps in the Black Forest and Vosges have not been discussed in detail in the geomorphologic literature, this paper concentrates on a description of scarps in two structurally and lithologically comparable areas: the Kniebis area of the Black Forest and the Donon area of the Vosges.

2 Drainage patterns

The orohydrographic patterns of both areas are characterized by high plateaus or narrow divides and by deeply incised valleys that form dendritic valley systems. The valleys are, without exception, V-shaped (Kerbtal). Downstream they become flat-floored (Sohlenkerbtal). The high plateaus lie at an altitude of from 900 meters to more than a 1000 meters in the Black Forest and 700 meters to 900 meters in the Vosges. In both areas, the main divide is the watershed between the valleys that drain directly to the Rhine river and the valleys of those rivers that have a much longer course to the Rhine. As a result, not only the altitude of the local base level but also the local relief on the flanks of the main divides differ considerably in the Black Forest and in the Vosges.

ISSN 0722-0723
ISBN 3-923381-18-2

D–3302 Cremlingen-Destedt, W. Germany
3-923381-18-4/89/5011851/US$ 2.00 + 0.25

The base level of streams flowing directly to the Rhine is at an altitude of 260 meters in the Black Forest (the valley of the Rench near Oppenau) and of 304 meters in the Vosges (the valley of La Bruche at Schirmeck). The local base level of the streams flowing a greater distance to the Rhine is much higher. In the Black Forest it lies at 563 meters (the valley of the Murg near Mitteltal) and 450 meters (the valley of the Wolfach near Schapbach) and in the Vosges at 365 meters (the valley of La Plaine at Vexaincourt) and 435 meters (the valley of La Sarre Blanche near Malecôte). The Wolfach drains via the Kinzig, La Plaine via Meurthe/Moselle, and La Sarre Blanche via the Mosel to the Rhine.

These data agree with the observations of relief intensity made by KREBS (1922) for the Black Forest and by FREY (1965) for the Vosges.

3 Structural and lithological characteristics

Two major tectonic units determine the structural patterns in the area of Bunter Sandstone investigated in the Black Forest. In the north there is the Swabian-Franconian anticline with the uplifted Hornisgrinde block at its western end. In this area the boundary between the Bunter strata and the crystalline basement lies further to the east than in the area of the second tectonic unit which forms a structural syncline located south of the anticline. In the syncline the Mooskopf, west of Oppenau, forms the western edge of the Bunter Sandstones (GEYER & GWINNER 1986). The local differences in the distribution of the sedimentary rocks covering the basement and the altitude of the boundary between the basement and the sedimentary cover are, therefore, determined by tectonics. In the Vosges, the same type of structural dependency is apparent in the Bunter Sandstone. In the area of the Schirmeck massif and the adjacent massif of Champ-du-Feu, the Bunter scarp has retreated farther to the west and the crystalline basement is exposed. (MÉNILLET et al. 1978).

Series of faults extend through both areas investigated; the majority is oriented parallel to the Rhine Rift Valley (Rhenish direction), others lie in the Hercynian direction.

The dip of the strata is determined primarily by the tectonics in the Rhine Rift Valley and secondarily by local structural deformations. In the area investigated in the Black Forest, the Bunter strata dip towards the southeast at an average angle of 1°–2°. In the Vosges the strata generally dip towards the northwest at similar angles.

The northern Black Forest and the northern Vosges have, therefore, a similar structural pattern with the crystalline basement and the Permian sedimentary rocks outcropping only in the valleys, generally downstream from the headwaters.

The Bunter Sandstone attains a thickness of about 250 meters in the Black Forest and of about 300 meters in the Vosges. In both areas investigated, strata of the upper Bunter occur only locally as denudational remnants. In the Black Forest they are composed of "Plattensandstein" (so) and in the Vosges of the "Couches intermédiaires" (t2a). The middle and lower series of Bunter strata are characterized by a greatly varied lithology, not only in the stratigraphic sequence but also in the facies of the individual layers of rock (fig.1). The geo-

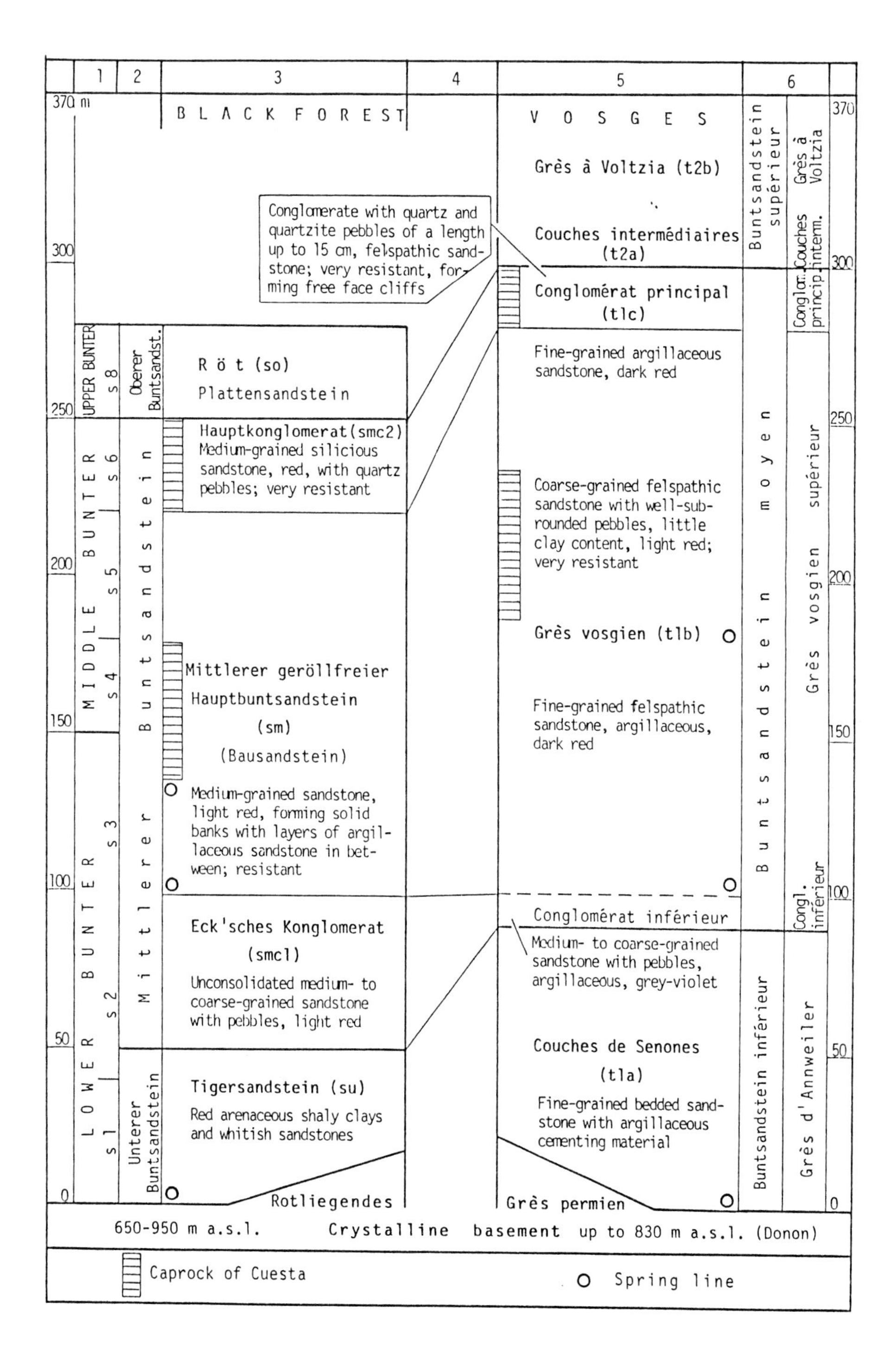

Fig. 1: *Stratigraphic and lithologic differentiation of the Bunter formation (Kniebis area of the Black Forest and Donon area of the Vosges).*
1 International Bunter (Scythian, Early Triassic) stratigraphy
2 Bunter stratigraphy according to H.v.ECK (1875)
3 Bunter stratigraphy and approximate thickness of strata, as given in: Geologische Karte von Baden-Württemberg 1:25000, sheet 7415 Seebach (with explanatory text by K. REGELMANN, 1907)
4 Correlation of Bunter stratigraphy in the Black Forest to that in the Vosges
5 Bunter stratigraphy and approximate thickness of strata, as given in: Carte géologique de la France à 1:50000, sheet XXXVI-16 Cirey-sur-Vezouze (with explanatory text by F. MÈNILLET et al. 1978)
6 Bunter stratigraphy in the Vosges according to J. PERRIAUX (1961)

morphology is also, therefore, very varied.

Strata that are strongly resistant to weathering and denudation form the caprocks of cuestas and determine landform development. In the Black Forest, these strata are the upper layers of the "Bausandstein" (sm) and the "Hauptkonglomerat" (smc2), in the Vosges, the "Grès vosgien supérieur" (t1b) and the "Conglomérat principal" (t1c).

Although in both areas of investigation the sequence of strata is almost alike, there are, nevertheless, two important differences:

1. the differences in facies and thickness of the "Hauptkonglomerat" and the "Conglomérat principal",
2. the differences in the thickness of the "Eck'sches Konglomerat" and of the "Conglomérat inférieur" and in the thickness of the "Tigersandstein" and of the "Couches de Senones" (fig.1).

Both these differences cause considerable variations in the geomorphodynamics and in the long term landform development. They are discussed in parts 4 and 6 of this paper.

4 Patterns of climate and recent geomorphodynamics

Both areas have a humid, oceanic temperate climate. The mean annual precipitation totals almost 2000 millimeters at windward locations on the western slopes of the Black Forest but decreases considerably towards the east on the leeward slope (tab.1). The total annual precipitation, which can be considered to be comparatively high, is distributed fairly evenly throughout the year with a maximum in summer (July) and a secondary maximum in winter (December or January). The Vosges has a similar precipitation regime but the total amounts tend to be lower at similar types of locations. During the summer, both areas experience heavy convectional rain events that may activate geomorphodynamic processes. Owing to their elevation, precipitation falls on a large number of days in the Black Forest and the Vosges and there are also a large number of days with frost. This is important in relation to weathering processes.

In both areas, recent geomorphodynamic activity is apparent from the headward erosion in the headwaters of ravines. On the side slopes of the ravines debris slides or block slides and, as a result of these, rock fall, all indicate active denudational processes. Where resistant sandstones outcrop as bare rocks on valley or cuesta slopes that are otherwise covered in debris, evidence of weathering is present in the form of pits, honeycombs or hour-glass structures, efflorences and pellicles, exfoliation and tafoni-type weathering. Congelifraction along joints and, particularly, the undercutting of resistant sandstone banks by means of slides in the underlying weak rock layers occasionally cause local rockfalls that contribute to the accumulation of boulders on the scree slopes.

With the exception of the headwater ravines even the steepest slopes in the Black Forest and the Vosges are stable. Nevertheless, one landform feature that indicates present-day geomorphodynamic activity and, therefore, slope instability, can be seen in the Bunter area of the Black Forest where it is termed "Schliff" (fig.2). A "Schliff" is a hollow that is sharply incised into the upper

Locality	Altitude (m)	Mean annual precipitation (mm)	Mean annual temperature (°C)	Number of days with frost	Number of days with at least 1 mm precipitation
BLACK FOREST (W-E)					
Hornisgrinde	1140	1975	4.9	139	172
Baiersbronn-Obertal	720	1793	n.a.	n.a.	n.a.
Baiersbronn	573	1503	7.2	126	155
Freudenstadt	710	1454	7.2	114	159
VOSGES (W-E)					
Col du Donon	725	1416	6.5	>100	>140
Rothau	349	1224	n.a.	n.a.	n.a.
Strasbourg	145	651	9.4	< 80	<130

Sources: TRENKLE, H. & H.V. RUDLOFF (1980) for German localities;
French data were collected locally.

Tab. 1: *Climatic data of selected localities in the areas of investigation.*

slope segment of the Bunter cuesta and has, therefore, not only extremely steep side slopes but an equally steep back slope as well. It is nearly oval in shape, narrowing downstream where it becomes a ravine. The form of the "Schliff" is caused by active headward erosion and by active denudation of the less resistant rock layers on both the back and the side slopes which are undercut. As a result, rockfalls from the sandstone layers above and sliding of the large boulders on the cuesta's scree slope may occur. These morphodynamic processes are active only after excessive rainfall. Following particularly extreme weather conditions there can be rapid mass movements of considerable dimensions, such as debris avalanches. On only one occasion has a settlement (Kalikutt, near Oppenau, see fig.2) been affected by rapid mass movements that took place in a "Schliff". However, in both the Black Forest and the Vosges slides may occur that have been induced by man, mostly the result of forest clearing, quarrying, building and the construction of forest roads and highways.

Despite the general similarity of structure, lithology and relief intensity in the Black Forest and the Vosges, "Schliffe" are absent in the Vosges because of differences in the lithology of the "Eck'sches Konglomerat" of the Black Forest and the "Conglomérat inférieur" of the Vosges (fig.1). The lower ends of all the "Schliffe" lie in the easily eroded and denuded "Eck'sches Konglomerat", which is of great thickness, while their upper ends lie in the "Bausandstein". In the Vosges on the other hand, the "Conglomérat inférieur" is too thin to cause geomorphodynamic processes of the same scale as those that occur in the formation of the "Schliffe" in the Black Forest.

5 Periglacial and glacial phenomena

Because of their elevation both periglacial and also glacial landforms were developed during the Pleistocene in the Black Forest and the Vosges. A large

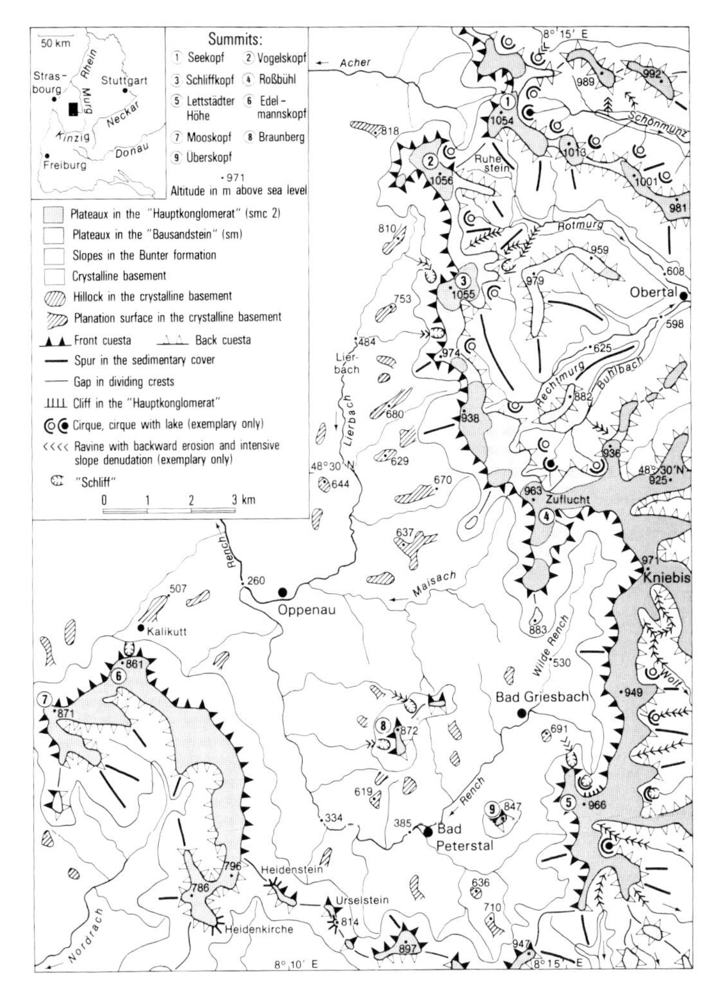

Fig. 2: *Landforms in the Bunter formation of the northern Black Forest (Kniebis area).*

number of publications deal with these phenomena. In the areas of investigation, they include DARMOIS-THÉOBALD (1972), FEZER (1957, 1971), FEZER et al. (1961), TRICART (1949, 1983), ZIENERT (1967).

Of the various periglacial features it can be stated, without describing them all and entering into the controversial discussion of their origin, that the accumulation of boulders on scree and debris slopes is the most conspicuous periglacial phenomenon in both areas. The boulders are of varying size but can attain several cubic meters. The scree, which is present on all slopes, varies from a superficial layer of rock fragments to a layer several meters in thickness. The boulders, which are composed of resistant sandstones or conglomerates, are embedded in an arenaceous-argillaceous matrix that in many locations has been, or is being, removed by the process of washing out and subrosion, with the result that on many parts of the slope only an accumulation of boulders is present. In spite of

Photo 1: *Ibacher "Schliff" at the front scarp of the Braunberg, an outlier of the Bunter near Bad Peterstal, Black Forest (5.6.1987).*

Photo 2: *Braunberg outlier of the Black Forest's Bunter cuesta, seen from the western slope of the Rench valley, west of Bad Peterstal. The "Bausandstein" cuesta rises above the dissected platform on the crystalline basement. The Bollenbacher "Schliff" at the front scarp slope is indicated by an arrow (30.7.1987).*

their steepness, these periglacially formed slopes are all stable at the present time.

Of the various glacial landforms, cirques are numerous in the northern Black Forest but occur less frequently in the Vosges (fig.2 and 3). The cirques of both areas and their characteristics were described by ZIENERT (1967). Based on these data, the number of cirques, their exposure, the mean altitude of cirque floors and the variation in their altitudes are given in tab.2. ZIENERT enumerates 35 cirques on the map sheets 7415 (Seebach) and 7515 (Oppenau) of the Topographische Karte 1:25,000 and 6 cirques in the Vosges on sheet 3616 (Cirey-sur- Vezouze, partie est et ouest) of the Carte topographique 1: 25,000. However, not all the cirques in the area under investigation are included in ZIENERT's enumeration, for example, those on the slope of the Chaume de Réquival in the valley of La Sarre Blanche (fig.3). On the other hand, some of the cirques enumerated by ZIENERT on the Black Forest sheets of Seebach

Photo 3: *Free face cliff at the front scarp of the "Conglomérat principal", Roche de l'Aigle, northern slope of the Plaine valley near Allarmont, Vosges (9.7.1987).*

Photo 4: *The Bunter cuesta forming the western slope of the Plaine valley, seen from the Roche de l'Aigle towards the southwest: the main cuesta is formed by the "Grès vosgien", the secondary cuesta above it by the "Conglomérat principal" (9.7.1987).*

and Oppenau lie outside the area of investigation.

From tab.2 it can be seen that most cirque floors in the Black Forest are located at a much higher altitude than those in the Vosges. The exposure of the cirques shows a marked maximum in the sector from NNE to ESE in the Black Forest. Owing to the smaller number in the Vosges, no pattern can be observed there. The floors of the cirques in both areas lie at a relatively low altitude. They have been developed in less resistant layers, in contrast to the steep back walls of the cirques which are formed in resistant sandstones that lie above the weaker layers. It is apparent, therefore, that in addition to the factors that favoured glaciation, such as altitude, reduced insolation due to exposure and snow accumulation in leeward sites, structure and lithology have also been of importance in the geomorphological development of the cirques. Moreover, in the Black Forest, some "Schliffe" also developed at the back walls of cirques because of the

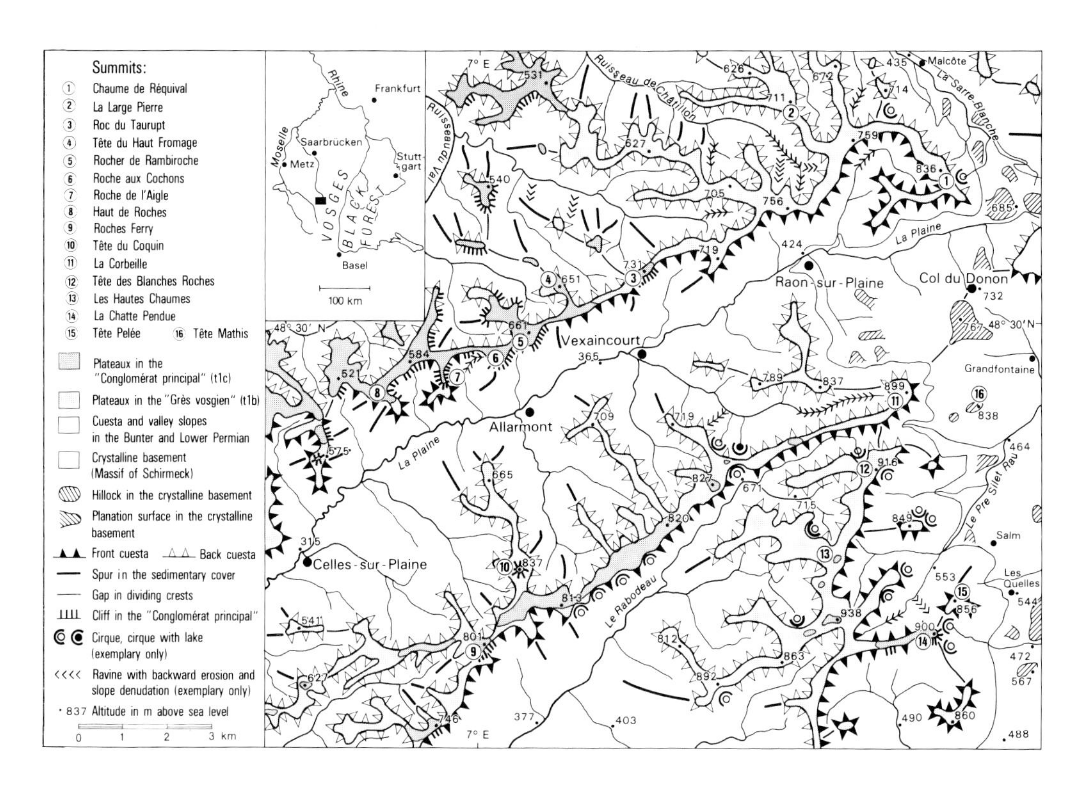

Fig. 3: *Landform features in the Bunter formation of the northern Vosges (Donon area).*

	Number of cirques (Mean altitude of cirque floors) (m)	Variation in altitude of cirque floors (m)	Number of cirques (and mean altitudes of cirque floors) with exposures towards								
			NNW	N	NNE	NE	ENE	E	ESE	SE	SSE
BLACK FOREST	35	1028–650	2	1	5	12	6	—	6	2	1
	(828)		(760)	(650)	(780)	(830)	(809)		(793)	(882)	(765)
VOSGES	6	780–615	—	1	1	1	—	1	—	2	—
	(686)			(675)	(700)	(635)		(615)		(745)	

Tab. 2: *Attributes of cirques in the areas of investigation.*

steepness of the slope as well as the structure and lithology.

6 Scarpland-type landforms

The variation in thickness of the gently inclined Bunter strata is the main factor determining the existence of scarpland-type landforms in the northern Black Forest and the Vosges (fig.1). Generally, two landform features characterize scarplands: cuestas and plateaus. Both are controlled by structure and lithology. The cuestas are limited to resistant rock layers that as caprocks overlying weaker strata constitute the upper segment of a cuesta slope; the underlying weaker strata outcrop on the lower segment of the cuesta slope. In the Black Forest and the Vosges, sandstones and conglomerates are the caprocks on the cuestas and form , consequently, the scarps.

The "Bausandstein" (sm) of the Bunter strata is generally the caprock of cuesta slopes in the Black Forest. It is a medium-grained sandstone, resistant to weathering, that forms relatively high faces. Thin argillaceous layers present within the sandstone do not influence the slope profile because of the accumulation of boulders on the slopes but they do cause a flattening of the longitudinal gradient in the ravines which are sharply incised into the slopes. Where there are remnants of the overlying "Hauptkonglomerat" (smc2), a previously much more extensive layer of strongly resistant rock, they form together with the "Bausandstein" the caprock of the Bunter strata in the Black Forest. Locally, bare outcrops of the "Hauptkonglomerat" form small cliffs on spurs at, for example, Heidenkirche and Heidenstein located on the edge of the divide separating the valleys of the Rench and the Kinzig southwest of Bad Peterstal. Other minor landforms, developed as a result of the resistance of the "Hauptkonglomerat", are present in the form of steep back walls of cirques. The Teufelskanzel southeast of Bad Griesbach is an example.

The lower segment of the cuesta slopes is composed of two weak layers: unconsolidated medium- to coarse-grained sandstones ("Eck'sches Konglomerat" smc1) and series of arenaceous shaly clays and argillaceous sandstones ("Tigersandstein" su). A spring line occurs at the contact between the resistant "Bausandstein" above and the weak layer of the "Eck'sches Konglomerat" below. It is also at this horizon that a large number of cirques have been formed and also all the "Schliffe" are developed. On the cuesta slopes, however, the contact between "Bausandstein" and "Eck'sches Konglomerat" is not visible in the morphology because of the accumulation of boulders. As a result, the "Bausandstein" cuesta, sometimes with the "Hauptkonglomerat" as an additional caprock, forms a concave slope above the crystalline basement. It attains its maximum steepness in the upper slope segment where the caprock, or caprocks, outcrop. Only along the valleys that are incised into the crystalline basement is the cuesta slope stepped, due to the basal platform developed at the contact between the crystalline basement and the sedimentary cover. It is also here that the major spring line occurs, first observed by SCHMITTHENNER (1913).

In contrast to the Black Forest, two strongly resistant layers form two separate cuesta caprocks in the Vosges. The more important is the "Grès vosgien" (t1b) which corresponds to the "Bausandstein" of the Black Forest (fig.1).

Above it, however, is a structural platform in the horizon of a fine-grained argillaceous sandstone. From this structural platform a secondary cuesta rises that is capped be the "Conglomérat principal" (t1c) a highly resistant layer that corresponds stratigraphically to the "Hauptkonglomerat". It is present only as remnants of a much more extensive stratum and appears everywhere in the form of a free face cliff, disintegrated locally into pinnacles. The main cuesta slopes, composed in the Vosges of "Grès vosgien" and in the Black Forest of "Bausandstein", are scree slopes in their upper segments. Isolated remnants of the resistant "Conglomérat principal" are usually totally surrounded by a free face cliff and have, therefore, a mesa-like shape. Some of the mesa-like outliers are so small, for example, the Tête du Coquin south of Allarmont, that it seems appropriate to term them buttes. They are quite different from the small cliffs that are locally present in the "Hauptkonglomerat" of the Black Forest.

As in most areas of cuestas, front and back scarps can be distinguished in both areas according to the direction of the cuesta slopes in relation to the dip of the strata. The front scarp rises in a direction that is opposite to the dip of the strata while the slope of the back scarp faces in the same direction as the dip. As a result of tectonics, the front scarps in both the Black Forest and the Vosges, face the Rhine Rift Valley, whether they form the forward edge of the sedimentary rocks or the slopes of valleys eroded into the rocks forming the cuesta. The back slopes are, therefore, found only along the valley sides and face in the opposite direction, that is, towards the southeast in the Black Forest and towards the northwest in the Vosges, apart from deviations due to local tectonics. The slopes of the front scarps are steeper than those of the back scarps. In the upper slope segment, formed in the caprock, the slope angles of the front scarp attain 30°–40° but on the back scarps only 20°–25°.

The spatial distribution of the front and back scarps also differs. As in many scarp areas, the front scarps of both the Black Forest and the Vosges extend in a straight line. By contrast, the back scarps are often curved as a result of the large number of embayments and spurs, formed by the headward erosion that dissects the plateaus (fig. 2 and 3). The strong dissection of the plateaus and the intensive backward erosion and denudation on the slopes of the ravines in the headwater reaches of back scarps are the result of the hydrological conditions because, owing to the dip of the strata, the drainage is directed mainly along the dip slope towards the back scarp. This is shown by the fact that the back scarp springs discharge more water and continue to flow much longer during dry spells than those on the front scarp. Moreover, the number of springs on the back scarp exceeds that on the front scarp.

Although the plateau surfaces cut the underlying strata at a slight angle, as is usual in areas of cuestas, they are structurally and lithologically controlled (SCHMITTHENNER 1954, BLUME 1971, 1987). Evidence for this is provided by the fact that wherever remnants of younger strata overlie the resistant caprock of a cuesta, the surface of the plateau is uneven with hillocks formed by the denudational remnants. Examples occur in the Black Forest where the structurally and lithologically controlled plateau above the "Bausandstein" is overlain by the "Hauptkonglomerat", and also on the plateau of the "Hauptkonglomerat" where it is overlain by the remnants of the "Plattensandstein". If the main divide has become very narrow as a result of denudation by periglacial processes, for example between the valleys of the Rench and Kinzig, the surface, underlain by the "Hauptkonglomerat", may consist of a few pinnacles and a large number of massive boulders nearby that have moved only short distances. The Heidenkirche and the Urselstein near Bad Peterstal are examples of this type of landform.

The structurally and lithologically controlled plateaus have been dissected intensively. Active headward erosion combined with denudational processes on the head slopes of ravines caused, and still continues to determine, the markedly embayed character of the back scarp and is also of major importance in the dissection of the plateaus. As a result of these processes the back scarp recedes progressively in the direction of the front scarp. In the Moos area of the Black Forest west of Oppenau, for example, the Bunter strata have become an outlier because of this and the cuesta and the plateau of the Braunberg north of Bad Peterstal have been separated completely from the continuous outer edge of the Bunter strata. In the Vosges, the widespread removal of the resistant caprock, also progressing from the back scarp, resulted in the development of divides with only a narrow strip of front scarp and broad area of back scarp. It is apparent, therefore, that in the areas investigated, as in many other areas of cuestas, the removal of sedimentary rock is caused not so much by the retreat of a front scarp but by erosion and denudation processes in the area of the back scarp.

7 Conclusions

In summary, the areas investigated in the Black Forest and the Vosges have certain similarities. Both have a plateau-like character because of the very low angle of dip of the Bunter strata. There is also a considerable local relief which is greater on the front scarps than on the back scarps. The intensity of headward erosion and of denudational processes attains its maximum, however, not on the front but on the back scarp, as a result of the drainage pattern. Apart from various types of weathering, washing-out and subrosion processes, there is little erosion or denudation on the slopes which are stable at the present time. Recent geomorphodynamic processes, such as headward erosion and denudation on ravine slopes are restricted to the areas of the headwaters.

Pleistocene landform development was also similar in both areas in the periglacial and glacial environments, indicated by the large accumulation of boulders on both the front and back scarp slopes and by the formation of cirques and other glacial, glacio-fluvial and periglacial minor landforms. Clearly in the Pleistocene cold periods, morphodynamics were intensive and also ubiquitous, as can be seen on front and back scarps slopes, on the plateaus and in the valleys. The morphodynamics were, in fact, much more intensive in the Pleistocene cold periods than they are at the present time.

Differences in the geomorphodynamics in the Black Forest and in the Vosges can be attributed to differences in the lithology of the two areas. These have resulted in two types of landform: the "Schliffe" present only in the Black Forest and the mesa-type landforms ("Steine") that are present only in the Vosges.

In both areas, landform development is similar to that in other cuesta areas in the humid mid-latitudes. Although no detailed evidence of pre-Pleistocene development is available, it can be assumed that the development of cuestas has continued for a long time. The subsidence of the Rhine Rift Valley during the Tertiary and, on its flanks, the uplift of both the Black Forest and the Vosges, which continued into the late Pleistocene, provided the impetus for the denudation of the sedimentary layers. In the area of the strongest uplift on the sides of the rift valley, progresssively older strata of the sedimentary cover have been removed so that eventually at the base of the Bunter Sandstone cuesta, the crystalline basement has been exposed.

Acknowledgement

The investigations on which this paper is based are part of a research project on scarpland-type landform features in southern Germany. The authors are indebted to the Deutsche Forschungsgemeinschaft for supporting this research by a generous grant.

References

BLUME, H. (1971): Probleme der Schichtstufenlandschaft. Wissenschaftliche Buchgesellschaft, Darmstadt.

BLUME, H. (1987): Schmitthenners Theorie der Schichtstufenlandschaft in heutiger Sicht. In: BLUME, H. and H. WILHELMY (eds.) (1987): Heinrich Schmitthenner Gedächtnisschrift. Erdkundliches Wissen **88**, 151–168.

DARMOIS-THÉOBALD, M. (1972): Cirques glaciaires et niches de nivation sur le versant Lorrain des Vosges a l'ouest de Donon. Revue Géographique de l'Est **12**, 55–67.

ECK, H: (1875): Über die Umgegend von Oppenau. Neues Jahrbuch für Mineralogie, Geologie und Paläontologie, 1875, 70–72.

FEZER, F. (1957): Eiszeitliche Erscheinungen im nördlichen Schwarzwald. Forschungen zur deutschen Landeskunde **87**.

FEZER, F. (197l): Zur quartären Formung des Nordschwarzwaldes. Jahresberichte und Mitteilungen des oberrheinischen geologischen Vereins, N.F. **53**, 183–194.

FEZER, F. et al. (1961): Plateauverfirnung und Talgletscher im Nordschwarzwald. Abhandlungen der Braunschweigischen Wissenschaftlichen Gesellschaft **13**, 66–72.

FREY, C. (1965): Morphometrische Untersuchung der Vogesen. Basler Beiträge zur Geographie und Ethnologie, Geographische Reihe, Heft 6.

Geologische Karte von Baden-Württemberg 1:25,000. Blatt 7415 Seebach (1907). Geologische Aufnahme von K. REGELMANN. Ed. by Geologisches Landesamt von Baden-Württemberg 1972 (Reprint).

Geologische Specialkarte des Grßoherzogthums Baden Blatt Gengenbach (1894). Geologische Aufnahme von A. SAUER. Ed. by Großherzogliche Badische Geologische Landesanstalt.

GEYER, O.F. & M.P. GWINNER (1986): Geologie von Baden-Württemberg. 3rd Edition. E. Schweizerbart'sche Verlagsbuchhandlung, Stuttgart.

KREBS, N. (1922): Eine Karte der Reliefenergie Süddeutschlands. Petermanns Mitteilungen **68**, 19–53.

MÈNILLET, F. et al. (1978): Carte Géologique de la France à 1/50,000, 270, Cirey-sur-Vezouze XXXVI -16. Notive Explicative. Orléans.

PERRIAUX, J. (1961): Contribution à la Géologie des Vosges gréseuses. Mémoires du Service de la Carte Géologique d'Alsace et de Lorraine 18.

SCHMITTHENNER, H. (1913): Die Oberflächengestaltung des nördlichen Schwarzwalds. Abhandlungen zur badischen Landeskunde 2.

SCHMITTHENNER, H. (1954): Die Regeln der morphologischen Gestaltung im Schichtstufenland. Petermanns Geographische Mitteilungen **98**, 3–10.

TRENKLE, H. & H. von RUDLOFF (1980): Das Klima im Schwarzwald. In: Der

Schwarzwald. Beiträge zur Landeskunde. Ed. by E. LIEHL and W.D.SICK. Veröffentlichung des Alemannischen Instituts Freiburg i. Br., Nr. **47**, 59–100.

TRICART, J. (1949): Les phénomènes périglaciaires dans les Vosges gréseuses. Compte Rendu sommaire des Séances de la Société Géologique de France, 1949, 351–353.

TRICART, J. (1983): Glacial forms and ice distribution in the northern Vosges during the last ice age. Geologie en Mijnbouw **62**, 557–561.

ZIENERT, A. (1967): Vogesen- und Schwarzwald-Kare. Eiszeitalter und Gegenwart **18**, 51–75.

ZIENERT, A. (1986): Grundzüge der Großformenentwicklung Südwestdeutschlands zwischen Oberrheinebene und Alpenvorland. Selbstverlag, Heidelberg.

Address of author:
Prof. Dr. Helmut Blume
Geographisches Institut der Universität Tübingen
Hölderlinstraße 12
D-7400 Tübingen
Federal Republic of Germany
Dr. Gerhard Remmele
Theodor-Heuss-Str. 154
D-7100 Heilbronn
Federal Republic of Germany

CATENA SUPPLEMENT 15 p. 243–252 Cremlingen 1989

FLUVIAL ACTION AND VALLEY DEVELOPMENT IN THE CENTRAL AND SOUTHERN BLACK FOREST DURING THE LATE QUATERNARY

Rüdiger **Mäckel**, Freiburg
Gaby **Zollinger**, Basel

1 Introduction

This paper examines three areas in the central and southern Black Forest that drain to the river Rhine: the valleys of the Schiltach and Gutach rivers in the catchment area of the Kinzig river near the main Rhine-Danube watershed, the upper reaches of the Dreisam river and the Brettenbach in the catchment of the Elz and the valleys of the Möhlin and the Sulzbach and their adjacent areas (fig.1).

Data have been collected on the geomorphology, hydrology and climatology of the Schiltach and Gutach rivers during field seminars of the Department of Physical Geography at the University of Freiburg since 1978. In addition, a research group has made detailed profiles and taken bore hole samples in the upper reaches and slope depressions of the Schiltach head valleys. In the valleys of the Dreisam river and the Brettenbach, RÖHRIG (1989) has examined Holocene slope and valley development and the morphological processes that result from heavy rainfall and flooding; the valleys of the Möhlin and the Sulzbach have been investigated by ZOLLINGER (1984).

ISSN 0722-0723
ISBN 3-923381-18-2

3-923381-18-4/89/5011851/US$ 2.00 + 0.25

2 The area of the upper Schiltach

There are four main groups of landforms in the areas of the upper Schiltach river: the plateaus of the upper Bunter Sandstone, the escarpment of the middle Bunter Sandstone, the re-exhumed subsurface of the Bunter Sandstone on the crystalline basement, for example, the granite massive of Triberg, and the moderately inclined valley heads of the Schiltach catchment area. The upper reaches of the upper Schiltach have incised into the crystalline basement and dissected the eastern edge of the sandstone areas, resulting in the formation of detached sandstone plateaus or outliers. These remnants, which indicate the existence of a formerly closed sandstone cover, include the Brunnholzer Höhe (945 m) in the north and the Hochwälder Höhe (968 m) in the south of the area investigated. The sandstone plateaus are accordant with the bedding planes (LIEHL 1934) and have very even surfaces. Examples are the plateaus of the

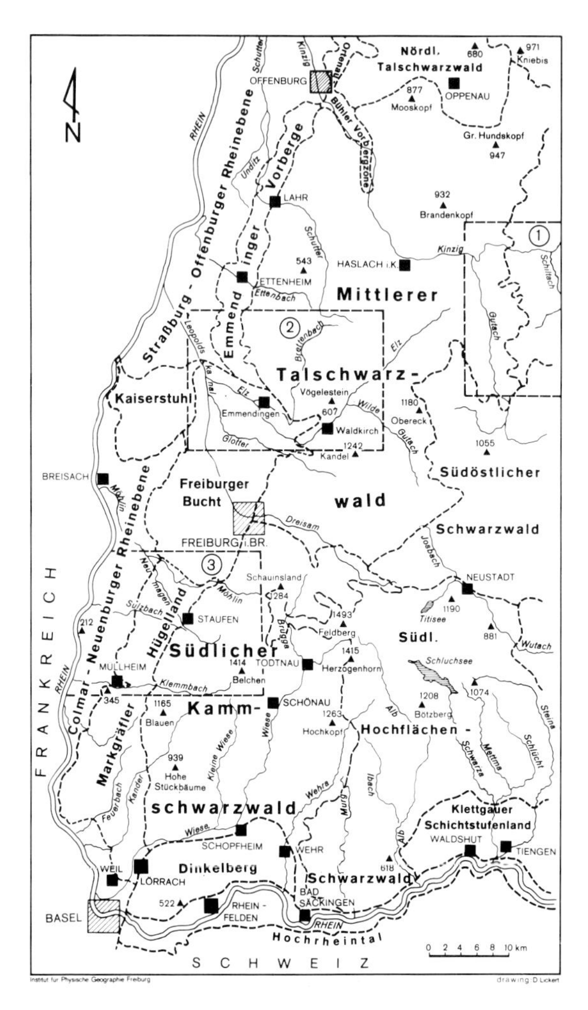

Fig. 1: *Map of the study areas in the black Forest based on the natural landscape units after REICHELT (1964).*

Hochwald and the Brunnholz (920 m) and the Benz plateau (between 880 and 920 m). The scarps have a relative height of 70 m. The height and form of the scarps depends mainly on the thickness and resistance of the sandstone layers. The foot of the sandstone scarp is covered by sandstone boulders and debris. On the basis of this, LIEHL (1934) assumed that the scarps do not retreat actively under present climate conditions. At the scarp foot and in the main gap between the plateaus, remnants of the exhumed pre-Triassic land surface occur between 870 and 970 m. These also form the watershed between the deeply incised upper reaches of the Gutach (Gremmelsbach, Rötenbach) and the moderately inclined upper slopes of the Schiltach.

In contrast to the permeable sandstone layers, the granite is impervious and there is a spring line along the contact between them. The valley heads at these springs are usually moderately inclined and deeply incised cuts occur only rarely.

2.1 The valley heads of the Schiltach

Weakly or moderately sloping valley heads, followed downstream by trough-shaped valleys and floodplain valleys, are typical of the relief in areas drained by the Danube and contrast with the steep valleys of the Rhine tributaries. This contrast can be clearly seen from the road between the V-shaped Nußbach Valley (Vordertal) and the weakly inclined valley head of the Sommerau, a tributary of the Brigach which is a headwater stream of the Danube.

The valley heads of the Schiltach river face northeast or east and are also only weakly to moderately inclined but do not belong to the Danubian drainage system. The eastward-flowing Schiltach turns to the north after 5 kilometers and joins the Kinzig river, a tributary of the Rhine (fig.1). The change in direction and the related river terraces indicate that the Schiltach was captured by the Rhine during the Würm glaciation. In the middle and lower course the valley is steep-sided but the area of the upper course still has a Danubian type of relief.

2.2 Valley development during the late Pleistocene

The valley floors of the trough-shaped valleys consist of periglacial and alluvial fill, suggesting that the valley form before the accumulation was more V-shaped. However, information obtained from bore holes indicates that the valley shape was similar to that of the present-day. The bedrock was found at about 1.5 meters below the surface and there was no evidence of a more deeply incised river bed. An exception is the Schachenbronn valley in which periglacial accumulation has led a V-shaped valley. It is probable that the Schachenbronn river cut its channel during a pre-Würm erosive phase.

Apart from the Schachenbronn valley, the upper Schiltach basin shows no evidence of glacial morphodynamics. Glacial relief formation, or at least modification, in the Schachenbronn valley can be assumed for the following reasons. First, the valley head shows typical glacial forms, such as a deep cirque-like back wall, a flat valley floor with peat formation and a rock bar downvalley. Second, the height of the area ranges from 850 to 910 m, that is, above the height of the glacial forms described by FEZER (1957) in the Bunter Sandstone area of the northern Black Forest. Third, the short tributary valleys which could be supply areas for the ice or firn cover face east or northeast. Fourth, the material found on the floor and the adjacent slopes is unsorted and has a high proportion of rocks and a loamy component which is derived from a weathered loess mixture.

2.3 Morphologically active phases during the Holocene

A phase of vertical erosion seems to have influenced the valley development during the early Holocene. Erosional remnants in the form of scars and steps along the valley slopes indicate a base level lowering of 6 m. In the Schachenbronn valley, downcutting of about 5 m took place during the Holocene in the periglacial material. The characteristic form of the slope scars developed subsequently as a result of ploughing. In the Schiltach area there is a striking contrast between a south facing ploughing terrace and a convex ramp, although both slopes are formed of periglacial debris (fig.2).

Radiocarbon dating yielded an age of

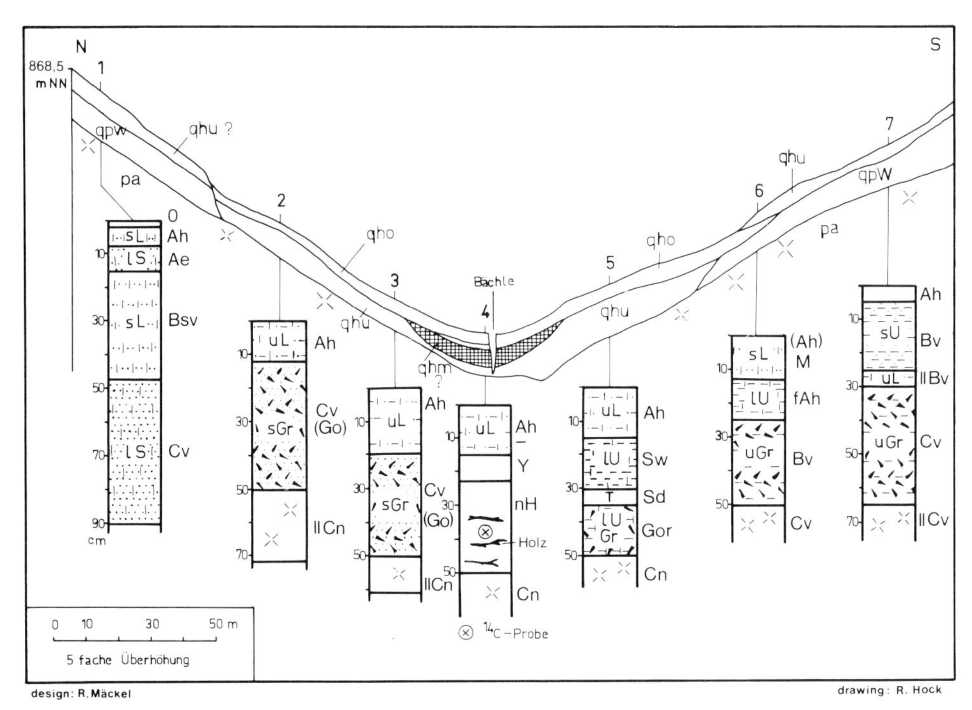

Fig. 2: *Cross profile of an upper Schiltach tributary east of Langenschiltach (study area 1).*

5660±75 years B.P. for a tree trunk embedded in peat on the valley floor and of 7760±90 years B.P. for wooded material in slope peat above the valley scar. This indicates an erosional phase between the two dates that is, during the transition from the Boreal to the Atlanticum; the erosion was probably caused by an increase in rainfall during a warmer period.

Another date of 4120±70 years B.P., found for tree remnants in the peat floor of the neighbouring Vogte valley, indicates that the peat development continued throughout the Atlanticum until the early Subboreal during a morphologically stable phase.

Late medieval settlement in the central Black Forest, forest clearing and ploughing resulted in the erosion of material from the slopes which was deposited on the valley floors as loam (Auelehm), 40 to 100 centimeters thick. Below this sediment cover, remnants of trees which occur mostly in an organic horizon, have yielded a radiocarbon age of 850±41 years B.P.. Some slope profiles show a buried A-horizon or charcoal layers, an indication of an active phase of soil erosion and accumulation in the sloping valley sections as a result of human activity. The recent organic horizon is preserved in only a few places. It has been largely destroyed when the land was drained in recent decades.

In the sandstone areas the rocks are weathered to a depth of several meters. Several ravines have developed along roads in the last few hundred years in the unconsolidated material. High rainfall converts these highways into temporary streams which remove the loose sand material as well as the pebbles of the sandstone conglomerate. An example of such a steep-sided ravine is the Hohe Straße (high level road), a road developed in medieval times for trade (KLEPPER 1984).

3 The Möhlin and Sulzbach catchment and adjacent areas

The southern Black Forest is characterized by a variety of drainage features. The largest eastern tributaries of the Rhine river in this area, the Möhlin with Neumagen (Münstertal) and the Sulzbach have their sources on the western slopes of the Black Forest summit ridges (Kammschwarzwald) in several trough-like valley heads that lie between 800 and 1080 m. Owing to the high gradient (20% to 30%), V-shaped valleys are characteristic for the upper courses below the valley heads. Between 500 and 700 m the gradient ranges from 5% to 10%, resulting in a narrow valley floor filled with cobbles and boulders through which the present-day river has cut down several decimeters into the parent rock.

A broad valley of 250 m with two terrace levels at 340 m and 440 m occurs near the exit of the valley into the Rhine basin. The upper terrace level corresponds with the Würm glaciation terrace (Niederterrasse), the lower terrace was formed during the Holocene. Because the gradient is low, about 0.3%, the channel is not well-developed. Only on the edge of the main bank of the Rhine river (Hochgestade) is there a noticeable channel incision. The courses of the Möhlin and the Sulzbach valleys were developed during the Riss glaciation, as indicated by the occurrence of the "Older Gravels". The courses of the rivers frequently coincide with major Hercynian alignments.

The rivers' discharges vary during the year and may range from high flood after heavy rain or snow melt to a total absence of flow in the late summer or early autumn (ZOLLINGER & BUCHER 1988).

3.1 The Sulzbach valley

In the transition area from the western part of the Black Forest to the Rhine Basin near the main Black Forest border fault against the Rhine Rift Valley, morphodynamic changes took place during the Quaternary. In the Sulzbach valley, there are Holocene series of sediments several meters thick on the valley floor between Sulzburg and Dottingen. Cut and fill phases were interrupted by two stable phases during which soils developed (ZOLLINGER & MÄCKEL 1988). According to radiocarbon dating of charcoal, one buried soil was formed during the Roman occupation between 40 BC and 350 AD and another during the late Middle Ages between 1050 and 1265. Only a few remnants of gravels exist from the Würm glaciation. These are underlain by Tertiary marl (Pechelbronner Schichten) (fig.3). Charcoal in the Würm gravel deposits has been dated at 12,600±240 years BP. The gravels were, therefore, accumulated not earlier than the end of the Würm glaciation (ZOLLINGER & MÄCKEL 1988). The Older Gravels (SCHREINER 1981, ZOLLINGER 1987) occur 10 to 20 m above the present-day floodplain level (fig.4). The loess/soil sequences that cover them at Oberdottingen indicate that the grav-

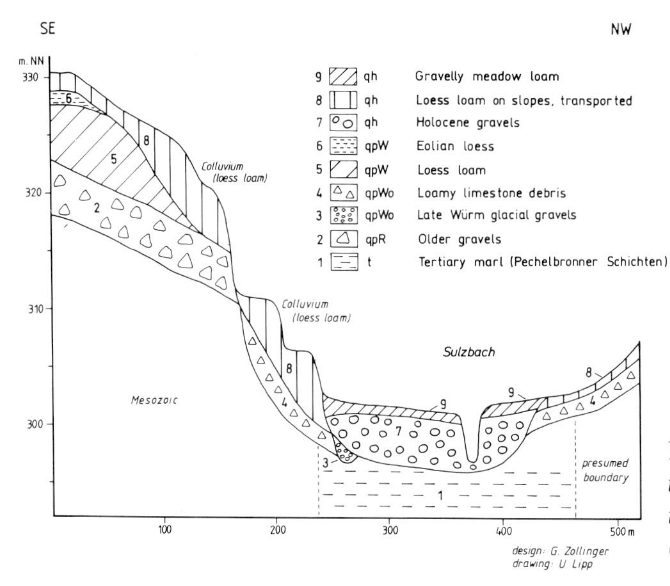

Fig. 3: *Cross profile of the Sulzbach at Oberdottingen west of Sulzburg (study area 3).*

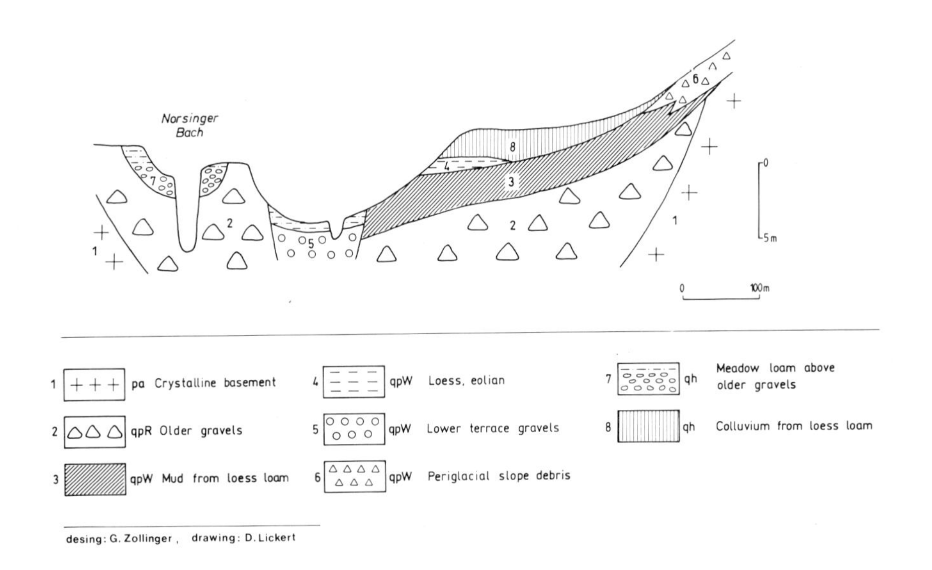

Fig. 4: *Cross profile of the Norsinger Grund near Ehrenstetten (study area 3).*

els were deposited in the Riss glaciation (ZOLLINGER & MÄCKEL 1988), in contrast to the older Pleistocene gravels found near Buggingen (BRONGER 1969) and Hügelheim (ZOLLINGER 1985), which are covered by loess/soil sequences that are up to 20 m thick.

3.2 The catchment area of the Möhlin

The valley cross-section of the Möhlin where the river leaves the western ridge of the Black Forest is similar to that of the Sulzbach. The valley fill consists of valley-floor loam and, locally, of a gravelly loam underlain by a sandy, gravelly sediment below which are gravels. The entire sequence was formed during the Holocene. The higher ground above the floodplain is made up of transported loams and weathered parent rock (gneiss). It is not a terrace from the Würm glacial. South of the Wittnau saddle, which separates the Möhlin and the Dreisam, both drainage areas have gently to moderately sloping valley heads. In the northern part of the Hexental, (Dreisam catchment), below the valley heads, trough-like valleys occur that cross the Würm terrace (MÄCKEL 1981). The topography and the periglacial debris cover indicate that these valley heads are solifluctional forms from the Würm glaciation. On the Möhlin side of the saddle, these valley heads are not connected to the present river systems because of the rapid downcutting by the Möhlin during the Holocene. The channels of the Heidenbächle and the Rainbächle, which join the Eckbach, a tributary of the Möhlin, have dissected the former solifluctional valley head. The latter can be reconstructed from the higher level relics of the former Pleistocene trough. On the basis of this, the total rate of downcutting of the Möhlin river is estimated to be 40 to 60 m during the late Quaternary.

A feature of the western flank of the southern Black Forest are the relatively short V-shaped valleys that are termed dobels. On reaching the Würm terrace on the plain of the Rhine valley, their channels becomes indistinct because of the eveness of the plain and the permeability of the material, loess underlain by gravel. The dobels are of Holocene age and were formed as a result of the large difference in the baselevel between the uplifted western ridge and the downwarped Rhine valley. Other reasons for their occurrence are the deeply weathered gneiss, the joints near the main Black Forest rift, and the high precipitation on the western side, up to 1300 millimeters annually, with high rainfall concentrations during thunderstorms in July and November and a high overland flow during snow melt.

3.3 Valleys with misfit gorges

Some rivers have cut a gorge or gully-like channel in the middle of their course. An example of this valley form is found along the Norsinger Ahbach (MÄCKEL 1981, ZOLLINGER 1984). The deepest part of the valley which belongs to the Würm terrace level, is drained only by a small ditch, the main channel lies on the higher terrace along the northern side of the valley (fig.4). The bottom of the river bed had already reached the basement rocks in some parts of its course. An unstratified body of Older Gravels of Riss age, 3 to 5 meters thick, lies on the bedrock. These gravels are covered by a thick loess loam so that the relief of the terrace is no longer apparent. During the Holocene, a river cut a channel into the loess-covered terrace; the channel was later filled with gravels. The composition of these gravels shows that they were transported from the basal Riss Older Gravels. Pollen analysis and radiocarbon dates indicate that the transportation and redeposition took place during a period that ranged from the Subboreal to the early Subatlanticum (ZOLLINGER 1987). After the channel was filled completely with valley floor loam (Auelehm), an incision phase began which resulted in the present-day gorge. A cut and fill phase and redeposition of Older Gravels account, therefore, for the sequence of deposits along the banks of the channel.

Various reasons have been suggested for the recent downcutting and gorge development on the slope above the valley bottom.

1. The river had to find a new course following landslides from the opposite slopes.

2. The channel incision occurred along joints that were close to the Main Black Forest Fault line where there were deeply weathered rocks which could be easily eroded by overland flow.

3. The headwater flow was restricted to the opposite slope by heavy lateral overland flow causing incision on the valley side (ZOLLINGER 1984).

4. The Norsinger Ahbach was redirected artificially in order to drain the wet grasslands in the valley bottom.

5. The Norsinger Ahbach was dammed in order to allow the rafting of timber.

6. The deep gorge was, in part, a former ravine that was deepened by concentrated overland flow.

However, investigations of the gorge by ZOLLINGER & BUCHER (1988) show that the upper 3 to 4 meters of the incision began in the pre-Alemannian period.The more recent phase of incision of an additional 3 to 4 meters has taken place during the past 150 years as a result of the correction of the Rhine river channel by Tulla, which caused downcutting by the Rhine and also headward erosion by the Norsinger Ahbach. This change, combined with other natural and man-made causes listed above, resulted in the development of the gorge at a morphologically unusual location. The channel continues to be altered by undercutting of the river banks and incision at the present time. The annual rate of incision was measured by ZOLLINGER & BUCHER (1988) and estimated to be 2 to 3 centimeters per year.

4 Distribution and importance of the Older Gravels

The term "Older Gravels" is commonly applied to the pre-Würm gravels which occur along nearly all the rivers that drain the western ridges of the Black Forest (SCHREINER 1981). The gravels are fan-like deposits which extend from the valley exits in front of the western ridge and coalesce laterally to form a continuous cover (ZOLLINGER 1986). During the late Pleistocene and early Holocene these Older Gravel fans were dissected so that the present-day deposits are relics of a more extensive distribution. As in the case of the Norsinger

Ahbach gorge, the terrace character cannot be seen in the field because the Older Gravels are covered by loess, transported loess loam and other sediments of varying ages.

The Older Gravels can be dated relatively using fossil soil horizons. For example, in the interfluve area adjacent to the Sulzbach (fig.3) the Older Gravels are covered by a loess series with a Bt horizon that was formed in the Eem (Riss-Würm Interglacial) and which indicates that the Older Gravels belong to the Riss or an earlier period (ZOLLINGER & MÄCKEL 1988).

The lower foothills of the Black Forest (330 m) are covered by thick loess sequences containing soil horizons. The valleys in this area have a box-shaped form or, at least, a flat floor drained only by a minor stream. In the Markgräflerland between Sulzbach and Klemmbach, there are several small periodical drainage systems in which the upper valley heads have a semicircular form that adjoins a box-shaped valley downstream. These valleys dissect the hilly loess landscape before reaching the Würm terrace of the Rhine valley. Bore holes in the floors of the valleys show that the loess cover is underlain by a gravelly- sandy material that has been derived from weathered Older Gravels and which was probably deposited in the late Pleistocene. The age of the deposits and the shape of the valley heads indicates that the valleys developed under periglacial conditions. In recent centuries the use of the land for vineyards and arable farming has resulted in a high rate of erosion from the slopes and the accumulation of fine material on the valley floors. The present-day form is, therefore, the result of man's activity.

Acknowledgements

The authors wish to thank the German Research Council (DFG) for financial support of this resesearch project since 1985. The manuscript was commented on by A. Röhrig and A. Laur. The radiocarbon dating was carried out in the 14C laboratory at the University of Kiel under the direction of Professor H. Willkomm.

References

BRONGER, A. (1969): Zur Klimageschichte des Quartärs von Südbaden auf bodengeographischer Grundlage. Peterm. Geogr. Mitt. **113**, 112–124.

FEZER, F. (1957): Eiszeitliche Erscheinungen im nördlichen Schwarzwald. Forsch. dt. Landeskunde **87**, 86 pp.

KLEPPER, D. (1984): St. Georgen den Hauptpässen nahe gelegen. Ein Buch von alten Straßen. Verein für Heimatgeschichte St. Georgen i. Schw. (Ed.), 254 pp.

LIEHL, E. (1934): Morphologische Untersuchungen zwischen Elz und Brigach (Mittelschwarzwald). Ber. Naturf. Ges. Freiburg i. Br. **34**, 94–212.

MÄCKEL, R. (1981): Staufener Bucht und Hexental. Eine Fußexkursion entlang der Schwarzwaldrandverwerfung südlich von Freiburg im Breisgau. Freiburger Geogr. Mitt. **1980/2**, 85–106.

REICHELT; G: (1984): Die naturräumlichen Einheiten auf Blatt 186 Freiburg im Breisgau. Naturräuml. Glied. Deutschland; Bad Godesberg.

RÖHRIG, A. (1989): Das Hochwasserereignis im Einzugsgebiet des Brettenbaches (Mittlerer Schwarzwald) am 8. Juli 1987. Frankfurter Geowiss. Arb. (in press).

SCHREINER, A. (1981): Quartär und Tektonik der Vorbergzone und der Oberrheinebene. In: Erl. Geol. Karte Freiburg im Breisgau, 174–198.

ZOLLINGER, G. (1984): Die Landschaftsentwicklung am Schwarzwaldrand zwischen Freiburg und Müllheim. Diss. Geowiss. Fak. Freiburg, 192 pp.

ZOLLINGER,G. (1985): Löß-Boden-Sequenzen am südlichen Oberrhein (Markgräflerland) und ihre Interpretation. Jh. geol. Landesamt Bad.-Württ. **27**, 113–143.

ZOLLINGER, G. (1986): Die Entwicklungsphasen des Quartärs am Engebächle (Aufschluß Pfefferlessandgrube) in Au südlich Freiburg im Breisgau. Ber. Naturf. Ges. Freiburg i. Br. **76**, 125–134.

ZOLLINGER, G. (1987): Die Älteren Schotter am Schwarzwaldwestrand: ihre Verbreitung, Genese und stratigraphische Stellung. Eiszeitalter und Gegenwart **37**, 57–66.

ZOLLINGER, G. & BUCHER, B. (1988): Erosionsmessungen im Norsinger Ahbach südlich von Freiburg im Breisgau und ihre geomorphologische und hydrologische Interpretation. Ber. Naturf. Ges. Freiburg i. Br. **77/78**, 67–79.

ZOLLINGER, G. & MÄCKEL (1988): Quartäre Geomorphodynamik im Einzugsgebiet des Sulzbaches und der Möhlin, Südbaden. Ber. Naturf. Ges. Freiburg i. Br. **77/78**, 81–98.

Addresses of authors:
Prof. Dr. Rüdiger Mäckel
Institut für Physische Geographie der Universität Freiburg
Werderring 4
D-7800 Freiburg i.Br.
Federal Republic of Germany
Dr. Gaby Zollinger
Geographisches Institut der Universität Basel
Klingelbergstraße 16
CH 4056 Basel
Switzerland

CATENA SUPPLEMENT 15 p. 253–260 Cremlingen 1989

THE KARST LANDFORMS OF THE NORTHERN FRANCONIAN JURA BETWEEN THE RIVERS PEGNITZ AND VILS

Karl-Heinz **Pfeffer**, Tübingen

1 Introduction

The Franconian Jura (Fränkische Alb) is a hilly upland of flat-lying Jurassic limestone (Malm: Oxford-Tithonian) and, locally, upper Cretaceous sediments, in eastern Bavaria east and southeast of a line from Bamberg to Nürnberg and Weissenburg. The west and northwest facing margin is a cuesta scarp that rises up to 300 m above its foreland. On its western end, the Franconian Jura is separated from the Swabian Jura by the large meteorite crater of the Nördlinger Ries. The Danube forms the southern and southeastern boundary of the Franconian Jura as far as Regensburg. The eastern boundary is the Naab basin and the northern boundary, the river Main.

Fluviokarst landforms dominate in the area. In the south there are dry valleys with dolines and shallow karst depressions that lie in an otherwise flat plateau. In the north there are also dome-shaped hills that rise above the general plateau surface. Allogenic rivers from the non-karstic areas farther east have cut deep valleys, some of them canyon-like, through the karst landscape.

ISSN 0722-0723
ISBN 3-923381-18-2

3-923381-18-4/89/5011851/US$ 2.00 + 0.25

A number of interpretations have been suggested for karst landforms of the Franconian Jura, particularly for the dome-shaped hills. These domes have usually been explained as exhumed early Cretaceous paleo-karst forms (CRAMER 1928, SPÖCKER 1952, BÜDEL 1977, GUDDEN & TREIBS 1961, TILLMANN & TREIBS 1967, GÖTZE & MEYER 1977). However, HÖHL (1963) and LIPPERT (1973) concluded that the domes developed during the Tertiary. Similar forms in the Swabian Jura have been explained lithologically because they consist of more resistant reef limestone; they have been dated as Quaternary (GWINNER 1968, DONGUS 1977). A more detailed review of the extensive literature is given by PFEFFER (1986).

The region between the rivers Pegnitz and Vils was chosen to investigate the karst landforms because its rock formation, landforms and landform associations are typical of the northern Franconian Jura. The geomorphological map (sheet 7, International Atlas of Karst Phenomena, PFEFFER 1986) shows a large number of dry valleys and domes. In addition, there are planation surface areas on the divides and between domes.

The Pegnitz, an allogenic river, flows through the upland in a steep-sided valley between 375 m and 345 m above sea level. It is the local baselevel for the greater part of the research area.

2 Geology and paleo-karst evolution

The karst in this area evolved in two phases. After the deposition of marine limestones in the Jurassic, the region became a land area in the lower Cretaceous in which a karst relief developed that had a dense vegetation cover (GUDDEN 1984), steep karst towers and karst depressions, some of which were 200 m deep and reached into the lower Malm and, in some cases, the Dogger (fig.1). Small steep-sided dolines formed, and also some larger depressions. A polje, about 5 km long and 1 km wide, developed in the vicinity of Königstein. Some karst hollows appear to have been connected by canyon-like valleys (TILLMANN & TREIBS 1967).

A network of underground water courses exists. Two drainage horizons of the Cretaceous karstification were found at depths of 35–40 m and 118 m below the present level of the Pegnitz, with channels that have diameters from 2 m to 6 m. The total height of the paleo-karst relief indicates that the region had been uplifted by several hundred meters above its regional baselevel.

In the Cenomanian (lowest upper Cretaceous) this landscape was progressively buried under a cover of, initially, fluvial and later marine sediments.

The second phase of karst development began with the renewal of erosion during or after the Campanian and continued from the upper Cretaceous through the Tertiary and Quaternary until the present.

A factor in this landform development are the local tectonic deformations which began in the lower Cretaceous and became more intense later. The Jurassic and Cretaceous beds were arched up in the area of Eschenfelden and Königstein and structural basins developed near Ranna north of Neuhaus and southwest of the Eschenfeld (FREYBERG 1969, GUDDEN 1984). These deformations caused spatially differentiated erosion of the Cretaceous beds. In regions of upwarps they are only preserved as fill in dolines and other depressions but in the structural basins, the Cretaceous sediments extend over large areas.

3 Dry valleys

Dry valleys occur in all parts of the area (fig.2). They are usually broad, have shallow trough-like cross-sections and very low gradients. There are also steep V-shaped valleys near the main trunk valleys, the Pegnitz and the Hirschbach. In the Jurassic limestones the valleys are narrow but where they cross Cretaceous sediments erosion has widened them into broad basins. The erosion of Cretaceous fill from paleo-karst depressions that have been laterally exposed in the valley sides or at the valley heads has left embayments in the valley sides and semicircular valley heads.

The dry valleys frequently have asymmetrical cross sections. Their slopes and floors are often covered with loess loam and solifluction debris. The valley floors also have up to 1 m of colluvial deposits. Shallow karst depressions and dolines in the valley floor have, for the most part, been filled in artificially to improve the agricultural usability of the land.

Open ponors have been found in exhumed Jurassic rock surfaces on the valley floors and cave entrances on the val-

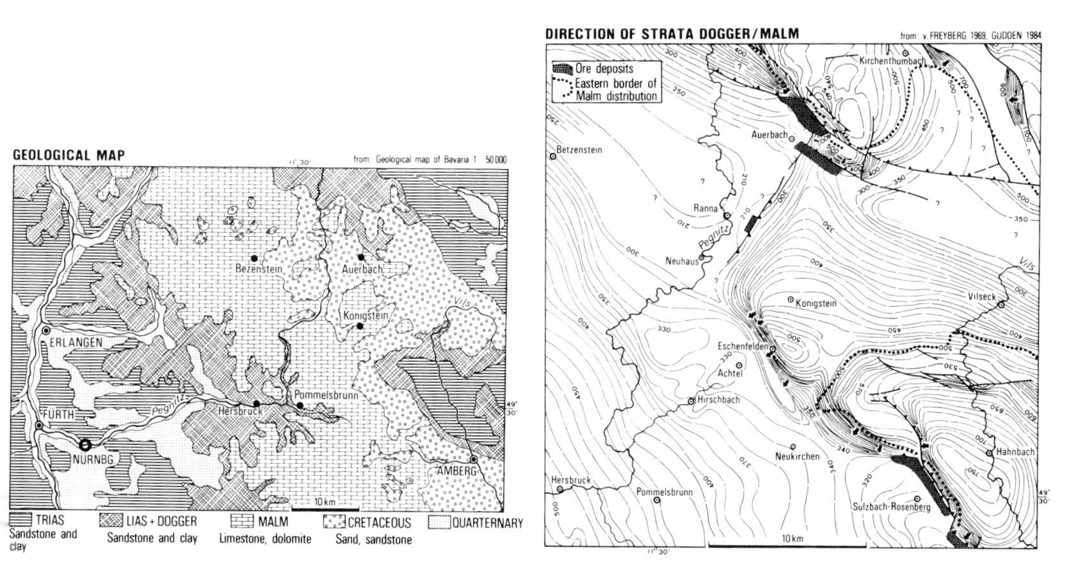

Fig. 1: *a: Geological Map of the northern Franconian Jura; b: Strike contours of the stratigraphical boundary between Dogger and Malm.*

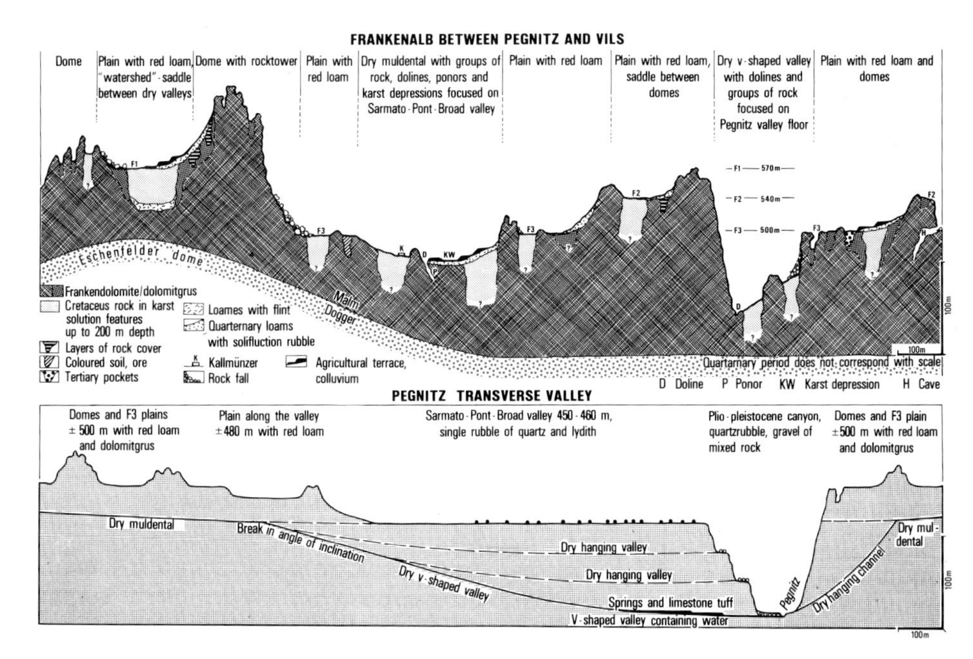

Fig. 2: *Valley development, domes and planation surfaces.*

ley sides. Runoff occurs on the valley floors during snow melt but remains on the surface for only short distances before it disappears in ponors and open joints. Karst springs and permanent surface runoff are confined to elevations close to the present baselevel, that is, to the elevation of the Pegnitz and Hirschbach channels. Precipitation of calcium carbonate from the spring waters has caused travertine terraces to form on the valley floor. In the V- shaped tributary valleys, the floors of which are close to the level of the Pegnitz and the Hirschbach, pinnacle shaped rocks have been exhumed on the valley sides following the removal of Cretaceous sediments and dolomite grus. There are some broad, usually shallow, dry valleys, the lower ends of which hang above the Pegnitz valley at 450 m to 460 m above sea level. Some have been incised down to the Pegnitz valley about 100 m below. The present valley of the Pegnitz has been incised into an old broad valley at 450 m to 460 m. This elevation appears to be a former regional baselevel.

4 Stages of drainage development and the age of valley incision

The development of the drainage pattern can be reconstructed and the valley incision dated on the basis of field evidence. Two stages of valley formation can be distinguished in the transversal valley of the Pegnitz, downstream from Neuhaus: the old flat-floored, broad valley at 450 m to 460 m above sea level, which is about 1 km wide, and the younger incision of the present valley into the old valley floor. Allochthonous fluvial sediments, particularly quartz pebbles and, very occasionally, lydite pieces (communication from Prof. BRUNNACKER and Dr SPÄTH), in the broad valley indicate that transport from the Fichtelgebirge and adjacent areas north of the Franconian Jura has occurred. The lower end of the broad valley lies in the cuesta scarp above the foreland. The old broad tributary valley floors at 450 m to 460 m were adjusted to the old, broad valley of the Pegnitz.

A pre-Quaternary river system flowed from the Fichtelgebirge to the Danube (BRUNNACKER 1973, TILLMANNS 1977, KALOGIANDIS 1981 and BRUNNACKER 1983). TILLMANNS (1977) has reconstructed the courses of these old valleys across the Franconian Jura. North of this region, SCHIRMER (1984) found deposits containing lydite on old valley floors at high elevations. The southward direction of transport seems, therefore, certain.

LOUIS (1984), who also found lydites in old drainage ways at these elevations, compared the drainage development in the Franconian Jura with that in the adjacent regions, in particular the Naab and Danube systems, in which debris from the meteorite impact crater of the Nördlinger Ries provide geological time markers. LOUIS concluded that the old drainage system developed in the Sarmatian/Pontian (upper Miocene). This dating is supported by the occurrence of lignite deposits in a Miocene valley system and the Miocene damming by debris from the Nördlinger Ries event (Miocene Altmühl-Rezat lake) of a north- south flowing river (BAYERISCHES GEOLOGISCHES LANDESAMT 1981).

It is not certain for how long the broad valley system in the karst region has been active. The north to south drainage was progressively intercepted and directed westward owing to captures by the headward extending Main river system, which was tributary to the Rhine, from the middle of the Pliocene (BOENIGK 1978). The oldest lydite-bearing deposits of the lower Main date from the Tegelen (earliest Quaternary) (SEMMEL 1974).

The incision of the Pegnitz canyon into the old, broad valley probably began during a Plio/Pleistocene period of uplift which has not yet been dated more precisely. In the narrow Pegnitz valley, terraces with gravels of mixed origins occur at elevations 430–440 m and 390–450 m. The water tables were lowered because of downcutting and the old, broad valleys became dry. Subsequently dolines and other karst depressions developed on the floors of these valleys.

The tributary valleys have been incised into their old, broad valley floors by headward erosion for only short distances upstream, although periglacial conditions prevailed during the Pleistocene cold periods.

5 Domes and planation surfaces

The domes and planation surfaces of the Franconian Jura are associated with one another. Planation surfaces have developed at several levels. The lowest, at 480 m above sea level, occurs along the Pegnitz valley and in narrow strips along the broad valley south of Eschenfeld. Three higher levels, at 500 m, 540 m and 570 m occur independently of rock types and geological structures throughout the region.

The planation surfaces extend across Jurassic as well as Cretaceous rocks. They have been preserved on the divides between dry valleys and in the saddles between domes. In many areas the dolomite substrate of the planation surfaces has been weathered into dolomite sand (dolomite grus or dolomite ash) to great depth along vertical solution funnels within solid rock. Borings in the planation surfaces have revealed a great heterogeneity of material and irregular bedrock subsurfaces. Pillars and domes with nearly vertical walls occur under the weathered mantle, many of which are made of solid rock. Others show all transitions from hard rock to completely weathered material. Between the prominent subsurface forms there are a large number of steep-sided hollows that are filled with Cretaceous material. The planation surfaces cut across all these forms and materials. Above the weathered dolomite and Cretaceous quartz sands, residual red loams occur. These have been interpreted as relicts of former weathering processes and of soil formation on the planation levels. The highest domes on the 570 m surface have a relative height of about 80 m. They are found at the margin of the former Königstein polje and south of Hirschbach. The domes on the lower surfaces generally have relative heights of 10–30 m. Their summits are all at or below the elevation of the next higher level. This indicates that the domes originated as resistant residuals of deep weathering which then remained when the planation surface was being lowered from one level to the next.

The shape of the domes seems to be affected by the weathering of the dolomite. If the dome consists of weathered dolomite, it is a gently rounded hill that does not rise much above the planation surface. Domes of solid dolomite are, however, steep-sided and rocky. Differential weathering of parts of the dome before exhumation has resulted in the de-

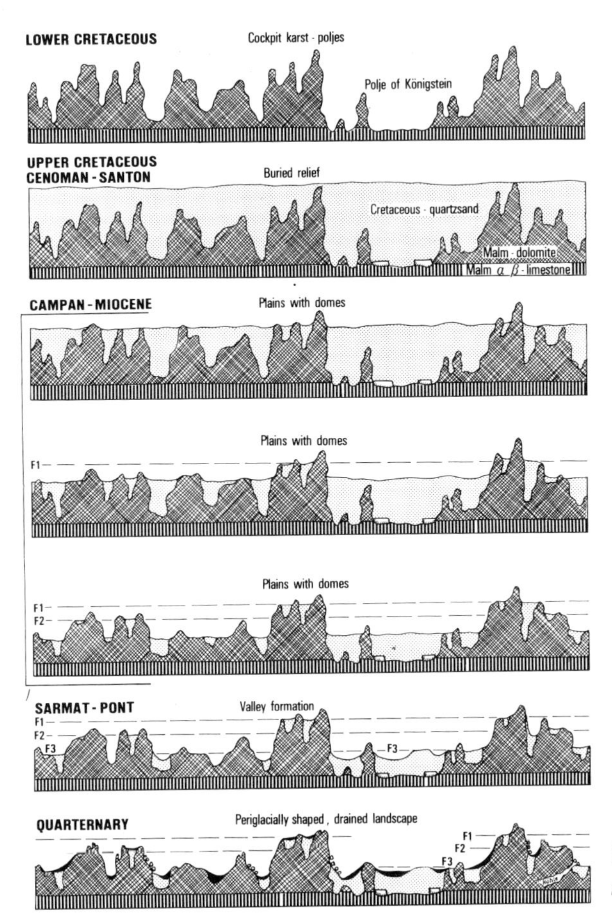

Fig. 3: *Evolution of the landscape.*

velopment of rock pinnacles and arches when the weathered residue was removed. Sections show, however, that the domes are composed of several cockpit karst forms of the lower Cretaceous and include hollows filled with Cretaceous quartz sand. At some locations, the transition from dome to surrounding planation level coincides with the change from dolomite to Cretaceous material, in others it lies within the dolomite. On the slopes of some domes there are hollows filled with Cretaceous material. Where the side of such a hollow has been breached by erosion and the Cretaceous material has been removed, a more or less semi-circular depression with dolomite walls and a floor of Cretaceous rock remains.

Lobes of Pleistocene solifluction material, made up of loamy material, dolomite grus and small pieces of solid rock extend from the slopes of the domes to the planation surface levels. The solifluction material is intercalated with loess loam. On domes with steep slopes, rockfalls have created nearly vertical rock faces and towers. Some of the fallen blocks lie in the transition zone to the adjacent planation surface.

6 Interpretation of domes and planation surfaces

The landform association of domes and planation surfaces originated after the lower Cretaceous karst landscape had been buried under later Cretaceous sediments. An inselberg relief developed similar to those in subtropical and tropical limestone regions. The development of each planation surface, with its domes, was controlled by the major trunk streams which were the regional baselevels. Deep weathering changed the dolomite to grus which, together with the Cretaceous quartz sands, could be easily removed. Red loamy soils, of which there are only remnants, developed on the surface.

Three separate periods of tectonic uplift resulted in the development of three separate planation surface levels. The inselberg relief has, therefore, developed after the burial in the upper Cretaceous of the paleo-karst relief and between the Campanian and the initial formation of the broad valleys in the upper Miocene.

7 Evolution of the landscape

At the end of the Jurassic the area was above sea level for a period of about 45 million years during which intensive development of cockpit karst with poljes took place. In the relatively short time between the upper Cenomanian and the Santonian, this karst relief was entirely buried under a sedimentary fill, which in the area of Auerbach-Amberg was up to 250 m thick. Between the Cenomanian and upper Miocene, planation surfaces and low dome-shaped inselbergs developed at elevations close to baselevel and under conditions of deep weathering of the dolomite. Red loam soils were also formed at this time (fig.3).

In the Sarmatian/Pontian (upper Miocene), systems of broad valleys developed that drained to the Danube. The Pegnitz broad valley, the remnant of such a system, served as a regional baselevel for the broad valley network in the research area.

Renewed tectonic uplift and the reorientation of the drainage from the Danube system to the Rhine system during the transition from the Pliocene to the Pleistocene caused the incision of the narrow Pegnitz valley, lowered the regional water table and resulted in the development of new karst landforms. There were several phases of dissection in the Quaternary, indicated by hanging valleys at various levels (fig.2) and by river terraces with terrace-related single cave systems. The landforms were also modified by periglacial processes.

References

Bayerisches Geologisches Landesamt (Hrsg.) (1981): Erläuterungen zur Geologischen Karte von Bayern 1:500,000. 3. neubearb. Aufl. München.

BOENIGK, W: (1978): Zur Ausbildung und Entstehung der jungtertiären Sedimente in der Niederrheinischen Bucht. Kölner geogr. Arb. **36**, 59–68.

BRUNNACKER, K. (1973): Gesichtspunkte zur jüngeren Landschaftsgeschichte und zur Flußentwicklung in Franken. Z.Geomorph. N.F. Supp. Bd. **17**, 72–90.

BÜDEL, J. (1977): Reliefgenerationen der Frankenalb. Klimamorphologie, 224–228, Berlin.

BURGER, D. (1982): Zur Dolomitverwitterung der Fränkischen Alb. Abh. z. Karst- u. Höhlenkunde, Reihe A, 17, 105–109.

CRAMER, H. (1928): Untersuchungen über die morphologische Entwicklung des fränkischen Karstgebietes. Abh. naturhist. Ges. Nürnberg, **22, 7**, 243–326.

DONGUS, H.J. (1977): Die Oberflächenformen der Schwäbischen Alb und ihres Vorlandes. Marburger geogr. Schriften, **72**.

FREYBERG, B. VON (1969): Tektonische Karte der Fränkischen Alb und ihrer Umgebung. Erlanger geol. Abh. **77**.

GÖTZE, F., MEYER, R.K.F. & TREIBS, W. (1975): Erläuterungen zur geol. Karte von Bayern 1:25,000, Blatt 6334 Betzenstein. Bayer. Geol. Landesamt, München.

GUDDEN, H. (1984): Zur Entstehung der nordostbayerischen Kreide- Erzlagerstätten. Geol. Jb., Reihe d, **66**, 3–49.

GUDDEN, H. & TREIBS, W. (1961): Erläuterungen zur Geologischen Karte von Bayern 1:25,000, Blatt 6436 Sulzbach-Rosenberg Nord, Bayer. Geol. Landesamt, München.

GWINNER, M.P. (1968): Paläogeographie und Landschaftsentwicklung im Weißen (Oberen) Jura der Schwäbischen Alb (Baden-Württemberg). Geologische Rundschau **58**, 32–40.

HAARLÄNDER, W. (1961): Erläuterungen zur Geologischen Karte von Bayern 1:25,000, Blatt 6436 Hersbruck, Bayer. Geol. Landesamt, München.

HÖHL, G. (1963): Die Siegritz-Voigendorfer Kuppenlandschaft. Festschr. O. Berninger, Erlanger geogr. Arb. 221–223.

KALOGIANIDIS, K. (1981): Geologische Untersuchungen zur Flußgeschichte der Naab (NO-Bayern). Köln.

KALOGIANDIS, K. & BRUNNACKER, K. (1983): Der Albenreuther Schotter und seine Bedeutung für die Landschaftsgeschichte von Nordost-Nayern. Z.Geomorph. NF **27**, 65–91.

LIPPERT, H. (1973): Die Oberflächenformung des Karstes der mittleren Frankenalb unter besonderer Berücksichtigung der Kuppenalb. Erlangen.

LOUIS, H. (1984): Zur Reliefentwicklung der Oberpfalz. Relief, Boden, Paläoklima **3**, 1–66.

PFEFFER, K.-H. (1986): Das Karstgebiet der nördlichen Frankenalb zwischen Pegnitz und Vils. Z. Geomorph. Supp.Bd. **59**, 67–85.

PFEFFER, K.-H. (1986a): Zur Genese von Karstformen auf den Westindischen Inseln. Festschr. 150 Jahre Frankfurter Geographische Gesellschaft.

SCHIRMER, W. (1984): Moenodanuvius - ein uralter Fluß auf der Frankenalb. Hollfelder Blätter, Nr **2**, 29–32.

SEMMEL, A. (1974): Der Stand der Eiszeitforschung im Rhein-Main-Gebiet. Rhein-Mainische Forschung, Heft **78**, 9–56.

SPÖCKER, R.G. (1952): Zur Landschaftsentwicklung im Karst des oberen und mittleren Pegnitzgebietes. Forsch. dt. Landeskunde, **58**. .

TILLMANN, H. & TREIBS, W. (1967): Erläuterung zur Geol. Karte von Bayern 1:25,000, Blatt 6335 Auerbach. München.

TILLMANNS, W. (1977): Zur Geschichte von Urmain und Urdonau zwischen Bamberg und Neuburg/Donau. Sonderveröff. Geol. Inst. Univ. Köln, **30**.

TREIBS, W., GOETZE, F. & MEYER, R.K.F. (1977): Erläuterungen zur Geol. Karte von Bayern 1:25,000, Blatt 6435 Pommelsbrunn. München.

Address of author:
Prof. Dr. Karl-Heinz Pfeffer
Geographisches Institut der Universität Tübingen
Hölderlinstr. 12
D-7400 Tübingen
Federal Republic of Germany

CATENA SUPPLEMENT 15 p. 261–267 Cremlingen 1989

DOLOMITE WEATHERING AND MICROMORPHOLOGY OF PALEOSOILS IN THE FRANCONIAN JURA

Dieter **Burger**, Tübingen

1 Introduction

The weathering of the Jurassic dolomite in the Franconian Jura is of great importance for the development of the relief (PFEFFER 1986). This type of weathering has been studied in the laboratory using chemical and micromorphological analysis. It has been demonstrated on the basis of field work that paleosoil development and disintegration of dolomite can be correlated with each other. The micromorphology of paleosoils has, therefore, also been examined.

2 Weathering of dolomite

In the karst area of the Franconian Jura between the Pegnitz and Vils rivers (sheet 7, International Atlas of karst phenomena, PFEFFER 1986) the dolomite is disintegrated far below the surface. The dome-shaped hills that have developed between the planation surfaces are also influenced by the weathering of dolomite. This process changes the solid rock into an unconsolidated, loose material known as dolomitic grus, grit, sand or ash. The fresh dolomite is white, 10 YR 8/2 (MUNSELL soil colour chart) and generally changes into pale yellow, 2.5 Y 8/4 and brownish yellow, 10 YR 6/8 (tab.1).

ISSN 0722-0723
ISBN 3-923381-18-2

3-923381-18-4/89/5011851/US$ 2.00 + 0.25

3 The age of the disintegration of dolomite

The disintegration is correlated with the planation surfaces, or etch plains, which have developed at elevations close to the former baselevel. Based on the morphology it has been established that this type of weathering took place in the Tertiary. In addition, SCHNITZER (1963) has shown that the disintegration of dolomite in the southern Franconian Jura is also of Tertiary age.

In quarries, isolated towers of dolomite have been found buried in Cretaceous quartz sands. The contact is often vertical. The boundary between the Jurassic dolomite and the Cretaceous sand is distinct and there are no layers of mixed material at the boundary. At the present time the dolomite is an unconsolidated, loose material, not able to form towers with vertical walls. The towers were formed when the dolomite was unweathered and resistant to the marine transgression and sedimentation that took place during the lower Cretaceous. The landform development in the Cretaceous had not yet been influenced

	$CaCO_3$ %	$MgCO_3$ %	Other %	Fe_2O_3 %	$CaCO_3$ in dolomite %	Other $CaCO_3$ %	Degree of Consolidation	Colour
2/7	53	45	2	0.18	53	0	consol.	10 YR 8/2
2/4	53	44	3	0.88	52	1	loose	2.5 Y 8/4
2/1	64	33	3	0.71	39	25	consol.	10 YR 8/6
2/5	54	34	12	1.07	40	14	consol.	2.5 Y 8/4
2/2	48	38	14	1.33	45	3	loose/consol.	2.5 Y 8/4
2/6	67	24	17	1.6	28	39	loose	10 YR 7/8
V5	64	28	18	0.62	33	31	consol.	2.5 Y 8/4
V1	82	7	11	2.2	8	74	consol.	10 YR 7/8
V2	82	4	14	2.2	5	77	consol.	10 YR 7/6
V7	76	12	12	2.37	14	62	consol.	10 YR 8/8
V6	78	12	10	2.93	14	64	consol.	10 YR 7/4
V4	79	8	13	6.98	9	70	consol.	10 YR 6/8

Tab. 1: *Composition of samples.*

by the disintegration of the dolomite.

4 Laboratory analysis of the disintegration of the dolomite

4.1 Chemical analysis

The chemical analysis indicated a decrease of $MgCO_3$ and an increase of $CaCO_3$. In the fresh dolomite, sample 2/7, there is 53% $CaCO_3$ content and 45% $MgCO_3$ content, a normal relationship for dolomite. The sample V4 has been intensively weathered and has a $CaCO_3$ content of 79% and a $MgCO_3$ content of 8%. The $CaCO_3$ content is, therefore, much higher than in normal dolomite. The Fe_2O_3 content in the fresh dolomite sample was 0.19% but 6.98% in the weathered sample.

4.2 Micromorphological analysis

Seven stages can be distinguished from the initiation of disintegration to the total solution of the dolomite:

1. The solution of the dolomite begins in the cement between idiomorphic minerals (fig. 1 and 2). The first weathering pits occur in the spaces between the mineral grains.
2. Orange to brownish gels are sedimented in the weathering pits (fig.3). The size of these deposits is within the clay range.
3. The solution progresses from the initial pits along contacts between mineral grains (fig.4).
4. The iron-rich brownish gel covers the minerals as a thin film (fig.4).
5. The solution of the minerals is now recognizable. Because the mineral lattice is the highest energetic system, chemical solution produces the idiomorphic form of the minerals. The fabric of the dolomite is destroyed due to the large number of solution pits at this stage. Isolated idiomorphic minerals are washed out; they are the main component of the dolomitic sand (fig.5 and fig.6).
6. Passageways for water are created in the rock. Their walls are coated

with iron-rich gel which cements the remaining dolomitic crystals (fig.7).

7. Calcite minerals grow on the iron-rich gel. The gel and the calcite are clearly separated since the deposition of the gel and the production of the calcite crystals does not occur simultaneously (fig.8). At this stage all the dolomite is dissolved and only calcite crystals or calcite cement and iron-rich brownish gel remain.

Stages 1 to 5 can be discerned on the surface of the dolomite.

5 Paleosoils on dolomitic sands

5.1 Red loam on disintegrated dolomite

Reddish-brown clays are developed on disintegrated dolomite on the surface of the etch plains. Thin sections show that this clay contains only some grains of secondary silicon (fig.11). The matrix is of brownish colour under crossed nicols and consists mainly of amorphous material. Kaolinite structure can be seen in some cases. The reddish-brown clays are of the same age as the red loams developed in Cretaceous quartz sands on the same planation surfaces.

In this red clay, the iron oxide content is nearly 40 times the iron oxide concentration in the fresh dolomite: 0.18% in fresh dolomite and up to 6.78% in the brownish loam. Such a concentration would require a layer of dolomite 40 m thick to produce 1 m of iron-rich clay. Since this is not possible, the iron cannot have come directly from the dolomite. The iron-rich gel may be derived from the Cretaceous sands. Zircon, rutile and tourmaline minerals are present in both materials but they are not present in Jurassic dolomite.

5.2 Red loam of the Cretaceous quartz sands

The Cretaceous sands are composed of quartz grains cemented by a skin of orange to brown gel. The dominant mineral is quartz with more than 95%. There are also some feldspars and micas.

The red loam of higher levels (fig.12) has a 60% skeletal portion and contains 15% sand, 67% silt and 18% clay. The skeleton is composed of quartz grains and some very resistant heavy minerals such as zircon, rutile and tourmaline. In contrast to the recent brown loam, the surfaces of the quartz grains are very rough and there are a large number of corrosion pits. The corrosion of the quartz is not as intensive as in other regions of Germany in which intensive corrosion of quartz has been dated pre-Oligocene. In the Franconian Jura there seems to be a mixture of intensively corroded old material and a smaller amount of fresh material.

The matrix has an intense reddish colour and is homogenous. Kaolinite dominates in X-ray analysis. There are also some hematite peaks. The red loam contains 3.7% Fe_2O_3 and 9.8% Al_2O_3.

5.3 Recent brown loam of the valley bottoms

In contrast to the red materials, the brown loam in the valley bottoms (fig.13) has a skeletal portion of 80%. The dominant mineral is quartz. The grain sizes are 60% sand, 20% silt and 20% clay. The largest quartz grains are well-rounded to angular.

Fig.1–10: *Weathering of dolomite*

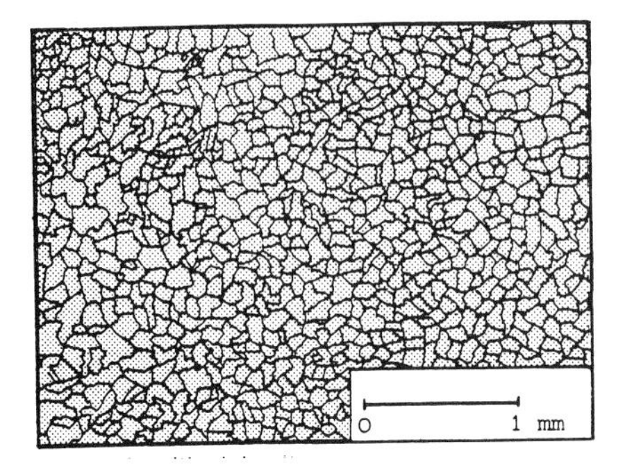

Fig. 1: *Sparitic dolomite.*

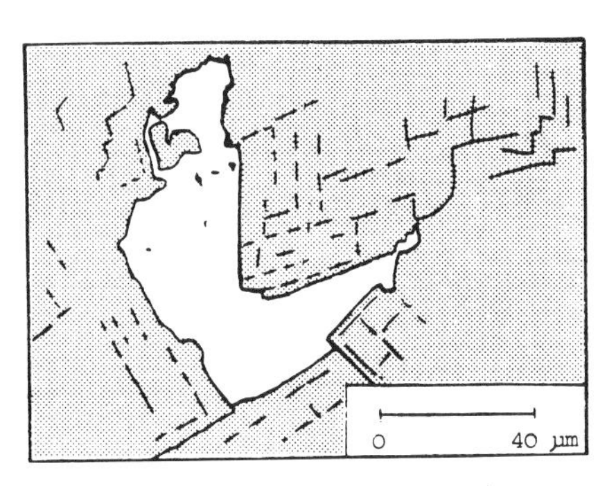

Fig. 2: *Solution in cement between idiomorphic minerals.*

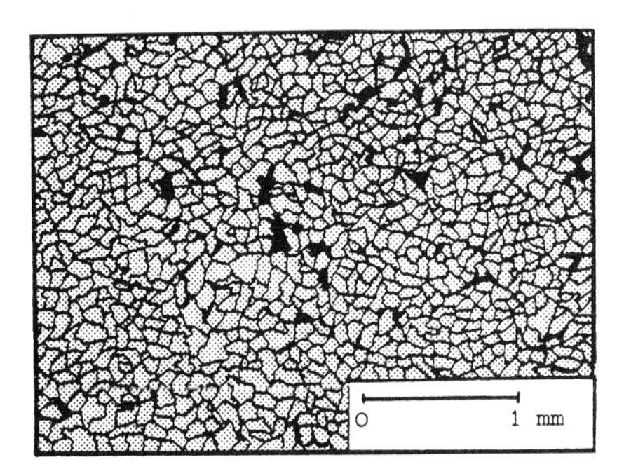

Fig. 3: *Orange to brownish gel in weathering pits.*

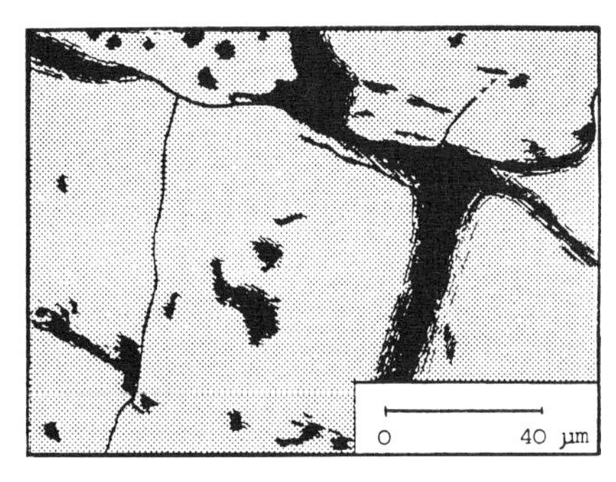

Fig. 4: *Solution progresses along contacts between mineral grains.*

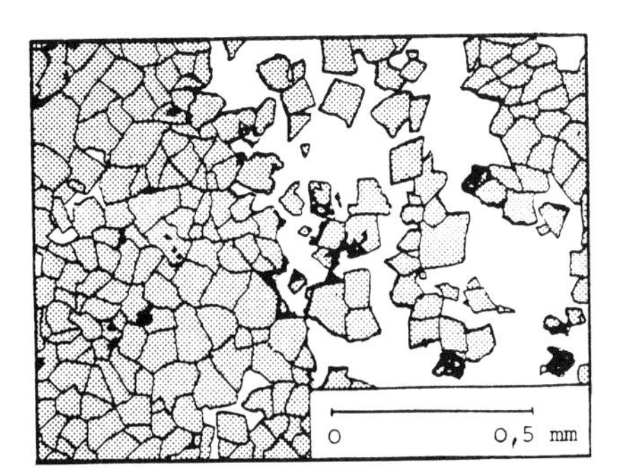

Fig. 5: *Isolated idiomorphic minerals.*

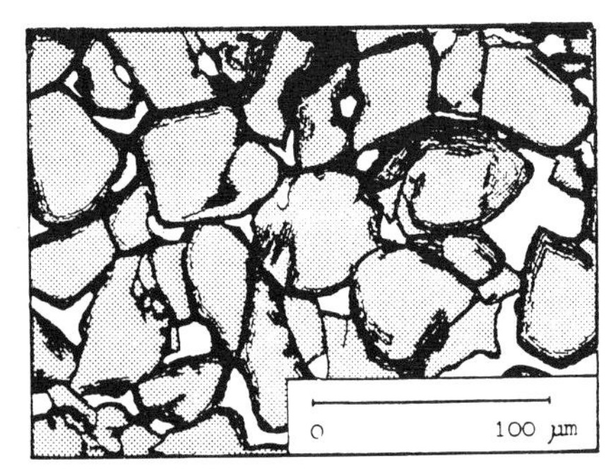

Fig. 6: *Dolomitic sand.*

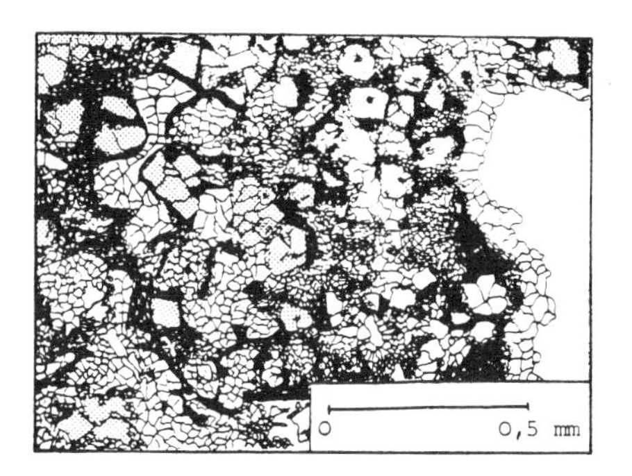

Fig. 7: *Coatings of secondary calcite and iron-rich gel.*

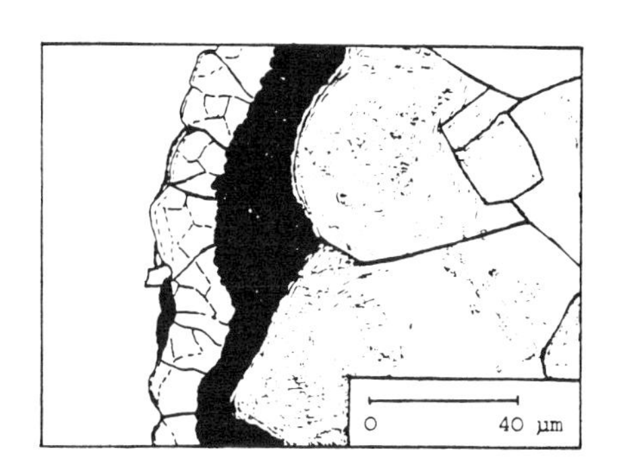

Fig. 8: *Coatings of secondary calcite and iron-rich gel.*

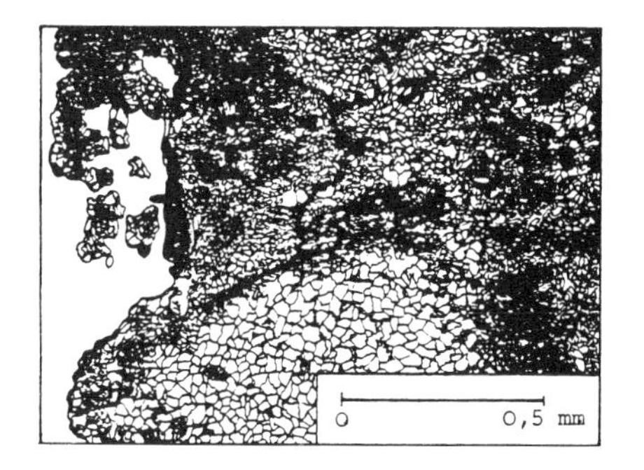

Fig. 9: *Calcite cement and iron-rich gel.*

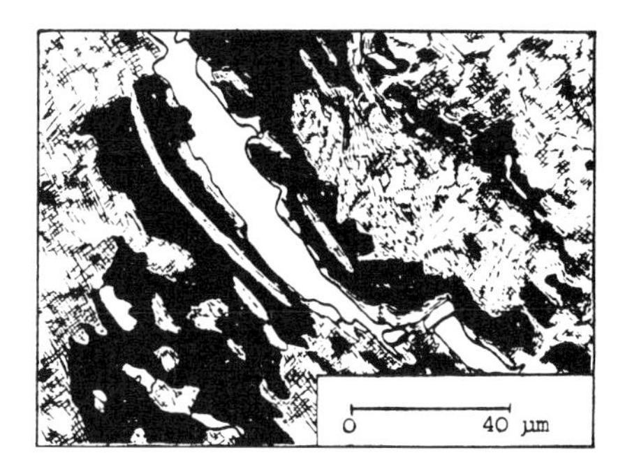

Fig. 10: *Calcite cement and iron-rich gel.*

Dolomitic minerals

Secondary calcite

Iron - rich gel

Fig.1–Fig.10: *Weathering of dolomite*

The matrix is orange and oriented. X-ray analyses show a dominance of chlorite and illite. The content of Fe_2O_3 is 5.2%; the content of Al_2O_3 is 4.6%.

The weathering products from Cretaceous sands can be subdivided into two groups that reflect different intensities of chemical weathering. They differ from one another in the corrosion of primary minerals such as quartz and in the production of clay minerals and iron oxides. Brown loam is, therefore, the weathering product of the Recent and Quaternary periods and red loam was produced during the Tertiary. The red loam could be pre- Oligocene but this has not yet been proven.

Fig.11–13: *Paleosoils*

Fig. 11: *Red loam on disintegrated dolomite.*

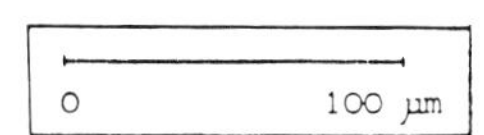

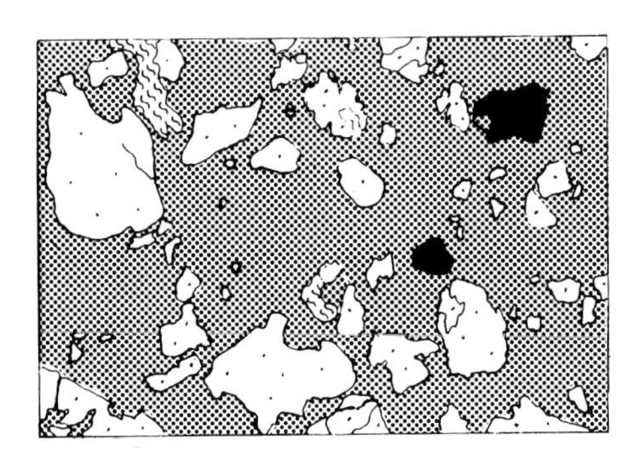

Fig. 12: *Red loam of the Cretaceous quartz sands.*

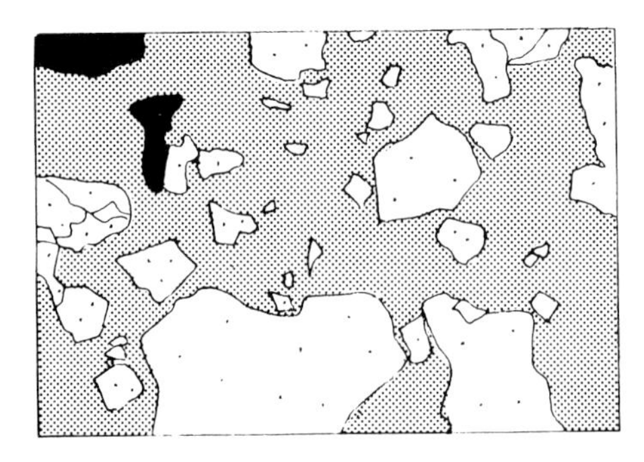

Fig. 13: *Brown loam of the Cretaceous quartz sands.*

References

BAUSCH, W.M. (1965): Dedolomitisierung und Recalcitisierung in fränkischen Malmkalken. N. Jb. Mineral. Mh. 75–82.

FEZER, F. (1969): Tiefenverwitterung circumalpiner Pleistozänschotter. Heidelberger geogr. Arb., H. **24**.

GERSTENHAUER, A. (1963): Beobachtungen über fossile Karsterscheinungen am Spessartrand bei Meerholz. Rhein-Mainische Forschungen **54**.

LEHMANN, H. (1970): Über verzauberte Städte in Carbonatgesteinen Südwesteuropas. Sitzungbericht der Wiss. Gesell. a.d. J.-W. Goethe Universität, Bd **8** Jahrgang 1969, Nr 2.

LEHMANN, O. (1932): Die Hydrographie des Karstes. Leipzig & Wien.

NAGEL, G. (1969): Karst in der Fränkischen Alb bei Pottenstein; Geomorphologische Beschreibung. Landformen in Kartenbild III, **2**, 6–9.

NICOD, J. (1971): Quelques remarques sur la dissolution des dolimies. Bull. Ass. Geogr. Fr.

PFEFFER, K.-H. (1986): Das Karstgebiet der nördlichen Fränkenalb zwischen Pegnitz und Vils. Z.Geomorph. NF, Supp. Bd. **59**, 67–85.

PRIESNITZ, K. (1967): Zur Frage der Lösungsfreudigkeit von Kalkgesteinen in Abhängigkeit von der Lösungsfläche u. ihrem Gehalt an Magnesiumkarbonat. Z. Geomorph. N.F. **11**, 491–498.

SCHNITZER, W.A. (1963): Zum Problem der Dolomitsandbildung auf der südlichen Frankenalb. Mitt. Fränk. Geogr. Ges. Bd. **10**, 292–296.

WAGNER, G. (1954): Der Karst als Musterbeispiel der Verkarstung.

Address of author:
Dr. Dieter Burger
Geographisches Institut der Universität Tübingen
Hölderlinstraße 12
7400 Tübingen
Federal Republic of Germany

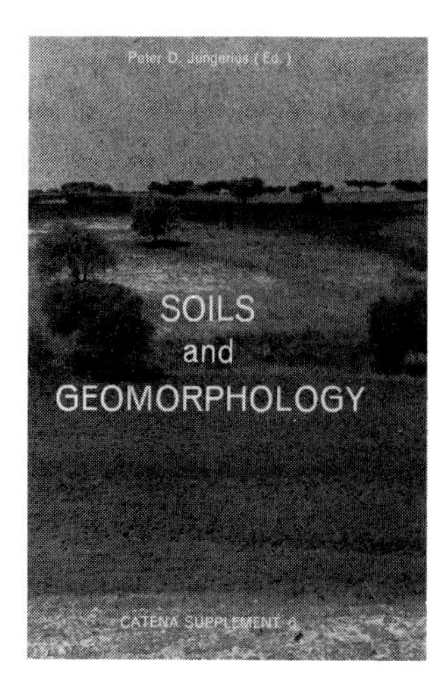

Peter D. Jungerius (Ed.):

Soils and Geomorphology

CATENA SUPPLEMENT 6 (1985)

Price DM 120,-

ISSN 0722–0723 / ISBN 3–923381–05–0

It was 12 years ago that CATENA's first issue was published with its ambitious subtitle "Interdisciplinary Journal of Geomorphology – Hydrology – Pedology". Out of the nearly one hundred papers that have been published in the regular issues since then, one-third have been concerned with subjects of a combined geomorphological and pedological nature. Last year it was decided to devote SUPPLEMENT 6 to the integration of these two disciplines. Apart from assembling a number of papers which are representative of the integrated approach, I have taken the opportunity to evaluate the character of the integration in an introductory paper. I have not attempted to cover the whole bibliography on the subject: an on-line consultation of the Georef files carried out on 29th October, 1984, produced 3627 titles under the combined keywords 'geomorphology' and 'soils'. Rather, I have made use of the ample material published in CATENA to emphasize certain points.

In spite of the fact that land forms as well as soils are largely formed by the same environmental factors, geomorphology and pedology have different roots and have developed along different lines. Papers which truly emanate the two lines of thinking are therefore relatively rare. This is regrettable because grafting the methodology of the one discipline onto research topics of the other often adds a new dimension to the framework in which the research is carried out. It is the aim of this SUPPLEMENT to stimulate the cross-fertilization of the two disciplines.

The papers are grouped into 5 categories: 1) the response of soil to erosion processes, 2) soils and slope development, 3) soils and land forms, 4) the age of soils and land forms, and 5) weathering (including karst).

P.D. Jungerius

CATENA SUPPLEMENT 15 p. 269–278 Cremlingen 1989

SOIL FORMATION IN DISPLACED PLEISTOCENE AEOLIAN SANDS IN THE NÖRDLINGER RIES

Michael **Schieber**, Regensburg

1 Introduction

The Nördlinger Ries, almost 80 km southwest of Nuremberg, is a circular basin more than 20 km in diameter that was produced by meteorite impact in the Tertiary. It is fringed by ejected rock debris and separates the Franconian Jura (Fränkische Alb) from the Swabian Jura (Schwäbische Alb). Although the genesis of the Ries in the Tertiary has been researched, its Quaternary materials and processes have not been investigated in detail. The most important Quaternary deposits are displaced Pleistocene aeolian sands that occur in fairly regular laminar layers up to 10 m or more thick. The geological map of the Nördlinger Ries (1:50,000, 1977) shows the distribution of the Pleistocene and Holocene sediments (GALL et al. 1977) but they are described only briefly in the map explanations. The pedogenesis has also not been investigated apart from a soil interpretation of the Öttingen sheet (1:25,000), one of nine sheets covering the Ries (RÜCKERT 1978). The soil development on the aeolian sands is examined in this paper.

ISSN 0722-0723
ISBN 3-923381-18-2

3-923381-18-4/89/5011851/US$ 2.00 + 0.25

2 Development and distribution of the sediments

The Keuper and Lias sandstones in the upper valley of the Wörnitz river are the source area of the aeolian sands. The river enters the basin from the north and leaves it in the southeast at the Harburger Pforte. The sands were transported mainly by the Wörnitz and its tributaries into the basin from the north and west by fluvial processes. Subsequently they were blown out of the valleys by the westerly winds which prevailed during the Pleistocene. As a result of this aeolian transport the sands occur mainly east of the river, not only within the basin but also beyond the 100 to 120 m high rim, where they cover vast areas of the ejected rock masses (impact breccias). The depth of the sand is 10 m or more in the basin, 1 m and less on the rim and beyond.

The sands from the rim were transported, in part, by streams, slope wash and solifluction back into the basin or to the Wörnitz river which then transported them to the Danube. The distribution area of the sands also includes parts of the Miocene Upper Freshwater Molasse (Obere Süßwassermolasse) and the Monheim High Sands (Monheimer Höhensande, TILLMANN 1983) which are Tertiary sediments of different ages.

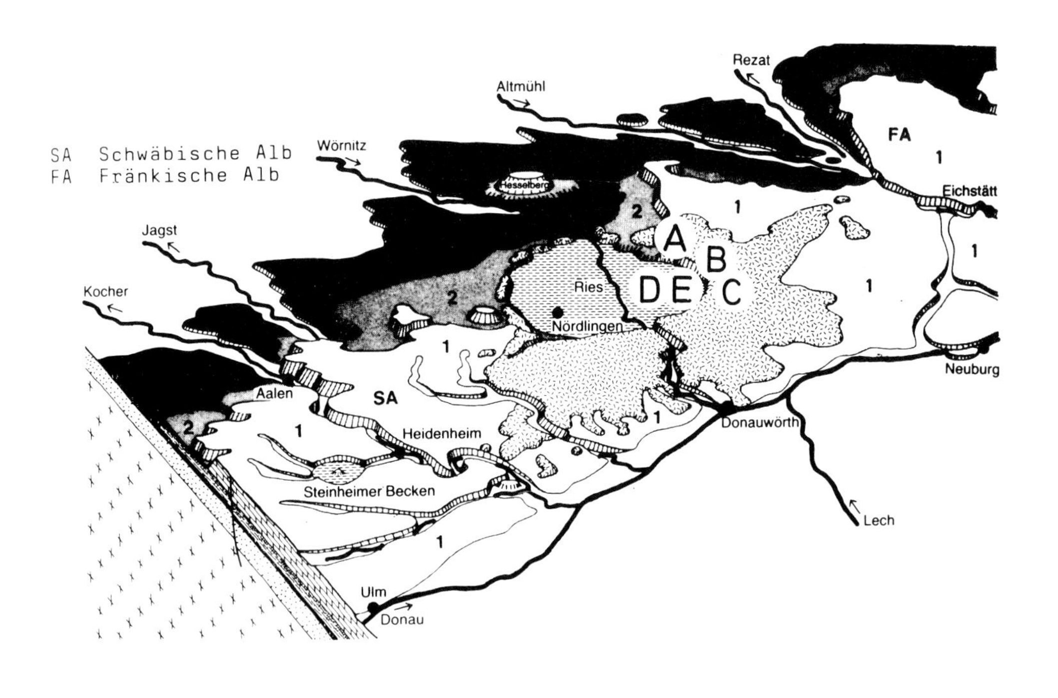

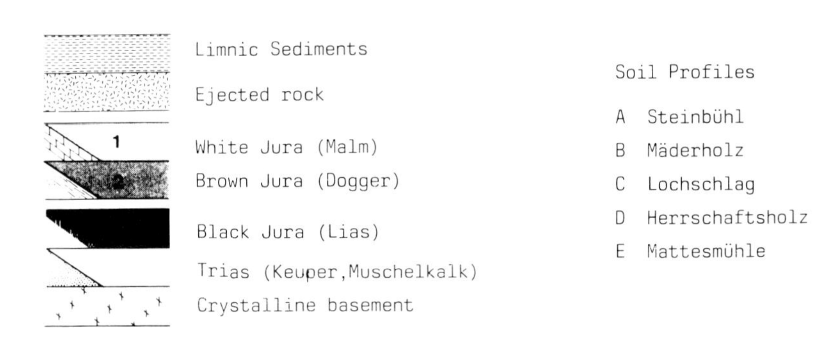

Fig. 1: *Block diagram of the Nördlinger Ries and its environment (taken from LEMCKE 1981, modified).*

A Holocene westward-flowing river system crosses the distribution area so that components from these Tertiary post-Ries sediments may have been incorporated in the displaced aeolian sands.

It should be noted that there is no definitive chronological classification of the sands within the Pleistocene. WEBER (1941) assumes that they date from the Würm and GALL et al. (1977) termed them Riss-Würm sediments. Also, the geological map (1:50,000) shows in situ stratification of sands at the Steinbühl profile (fig.2).

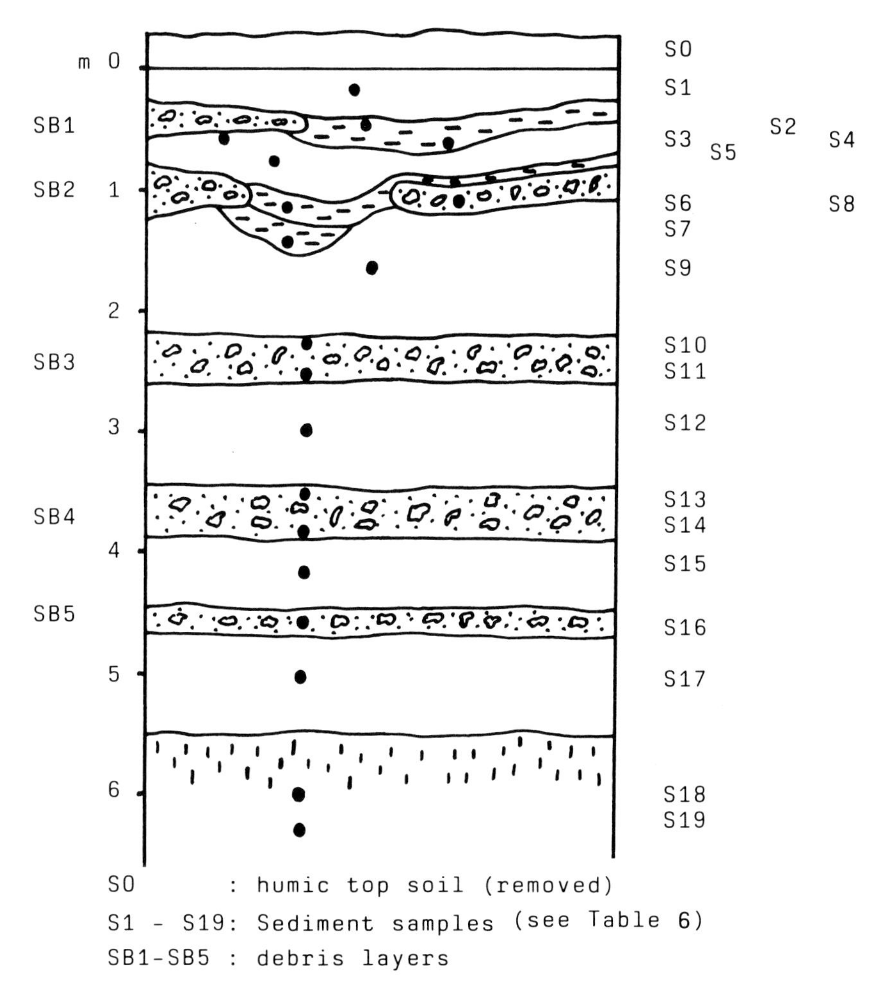

Fig. 2: *The alluvial cone profile at Steinbühl.*

During field work, however, it was found that a large alluvial cone has developed at this site. It must be assumed, therefore, that in situ sands occur only in the form of thin veils (GALL et al. 1977) and that all deeper sand layers consist of retransported material.

3 Soil profiles

Soil profile sites for morphometric and pedogenetic investigation were chosen on the basis of the geological map. They differ with respect to their topography, their sediment depth and their soil development. Present climatic differences within the area are slight: the mean annual temperature ranges from 8°C in the basin to 7°C on the rim; the mean annual rainfall is between 600–650 mm in the basin and 700–750 mm on the rim.

Five profiles were investigated: two in the basin, Mattesmühle and Herrschaftsholz, two on the less dissected rim, Lochschlag and Mäderholz, and one in the slope area connecting rim and basin, Steinbühl (fig.1). All profiles are under forest: coniferous woodland in the basin and mixed woodland on the rim. The depth of the substrate is less on the higher ground than in the basin.

3.1 Profiles in the Ries basin

The Mattesmühle (tab.1) and Herrschaftholz (tab.2) profiles have developed in sand layers that are more than 7 m deep. They are slightly brownish in the upper part of the solum. Podsolization is, however, clearly visible in both profiles.

The substrate samples of both profiles contain between 91% and 99% sand grain sizes and consequently very little silt or clay. Medium sand grain sizes dominate, there are small amounts of heavy minerals including ores, zircon, tourmaline. They also have no lime. More than 95% of the siliceous components of the material consist of quartz. The apparent podzolization is corroborated by the low pH values and the low clay content of the samples. The slight superficial rate of browning reflects the low clay content. The low content of organic matter is characteristic of relatively fast decomposition of humus.

The low content of exchangeable bases indicates the poor ecological quality of the substrates.

3.2 Profiles on the rim of the Ries

The solum in the Lochschlag (tab.3) and Mäderholz (tab.4) profiles on the rim is only about 1 m deep. Soil development has proceeded further than in the basin and there are noticeable differences in the brown earth (Cambisol) sequence. The Lochschlag profile consists of a sandy brown earth underlain by loamy limestone debris. The limestone at the Mäderholz profile is covered by a clayey, water-damming Bt horizon. This has resulted in the development of a brown earth-pseudogley (Cambisol-pseudogley). Neither profile showed any of the primary sediments.

These well-developed soils are characterized by higher silt and clay contents and and a stronger horizon differentiation. Except for the A1 horizon of the Lochschlag profile, the sand content is less than 85%, in many samples below 80%, lower, therefore, than in both basin profiles. There is, consequently, a larger quantity of exchangeable bases. Measureable carbonate contents have been found in the lower parts of the Mäderholz profile, mostly in the form of pieces of limestone. The pH value tends to increase with solum depth. The quantities of heavy minerals in the SdBt horizon of the Mäderholz profile is 1.8% in the medium grained sand fraction, compared to 0.2%–0.4% in the other horizons. This horizon also has a high content (60%) of illuvial clay.

3.3 Profile on the slope of the Ries

The Steinbühl profile (fig.2) has developed as an alternating sequence of sand and limestone debris layers approximately 5.5 m to 6.5 m deep. The two uppermost limestone layers are interrupted, the deeper layers are continuous. The Holocene top soil has been removed so that the deeper sand layers could be used for the production of building stone.

The matrix of the two uppermost layers of limestone debris is the same colour as the sand. Clearly visible browned zones either link the ends of the interrupted debris layers or cover them. The fine-grained matrix of the deeper debris layers has, overall, a more browish colour than the intervening sands.

The location of the profile at the exit

of a small valley on the rim slope and the stratigraphic sequence of layers in the profile indicate that this is a section of an alluvial cone. The cone has been mapped (WEBER, 1941) but was not recorded on the 1:50,000 geological map.

The results of the investigation of the samples from this profile are shown in tab.5. The pH values exceed pH = 8 in the deeper part of the profile and decrease again towards the underlying loamy zone. Such high values are unusual. In the debris layers chert nodules and well-rounded limonite nodules have been found.

4 Interpretation

The interpretation of the profiles and their material properties attempted to answer three sets of questions:

1. Is it possible to determine when and how the alluvial cone at the Steinbühl profile was formed? Are remnants of fossil soils incorporated in the sediment cover?
2. What are the reasons for the differing intensities of pedogenesis in the basin and on the rim? Are there remnants of fossil soil formations?
3. Is it possible to explain the development of the sediment layers using statistical parameters such as sorting and median?

4.1 Dating of the Steinbühl sediments and the question of fossil soils

The Steinbühl profile is located on an alluvial cone at the exit of a small narrow valley on the margin of the Ries. The debris layers on the cone consist predominantly of Jurassic limestone (Malm) and the sand layers of reworked Pleistocene aeolian materials. Both deposits have been transported from the rim. The occurrence of well-rounded limonite boulders (brown iron ore) in the debris layers indicates, however, that there was an additional supply of material from the Miocene Upper Freshwater Molasse. The presence of Monheim High Sands cannot be proved although chert has been found (TILLMANN 1983).

The date of the formation of the alluvial cone cannot be determined conclusively. WEBER (1941) assumes that the deposits are late Würm and GALL et al. (1977) refer more generally to the Riss-Würm period.

Pleistocene mammal fauna that represent both cold climates (mammoth, woolly rhinoceros, reindeer) and warm climates (deer, cattle, horse) have been found in the sands. These fossils cannot be used to date the sediments more precisely because their stratigraphic positions are almost the same, indicating active paleo-morphodynamics in this deposit. The alternation of debris layers and sand layers reflect an alternation of phases of strong runoff (debris layers) and phases of weaker transport (sand layers).

The soil colours and grain sizes in the Steinbühl profile show major variations (tab.5). An obvious weathered layer occurs underneath the alluvial cone sediments (samples S18 and S19) the colours and grain size distribution of which differ from the upper sediment sequences. This substrate is mottled with red and pale, almost white, spots (wet bleaching), the effect of underground and ponded water. The sediment contains no lime and is very loamy (sandy-clayey loam). It also contains strongly weathered pieces of granite. If the alluvial cone dates from

Sample No.	Hor	SD	Size Spectrum >2%	Size Spectrum <2%	Grain Sizes (%) <2 mm Sand				Silt				Clay	Tex	pH	Carb.	OM	EB
			mm	mm	coarse	medium	fine	sum	coarse	medium	fine	sum	T		n/10 KCl	%	%	mval 100 g
MM 2.1	M	3	—	100	15.4	68.8	11.1	95.3	1.5	0.8	0.9	3.2	1.5	S	4.2	—	0.4	—
MM 2.2	A_e	6	—	100	17.1	72.3	8.0	97.4	0.9	0.7	0.8	2.4	0.2	S	3.4	—	0.2	—
MM 2.3	$B_{v(e)}$	18	0.1	99.9	19.0	66.6	10.4	96.0	0.6	0.5	1.5	2.6	1.4	S	4.3	—	0.4	—
MM 2.4	$B_{v(h)}$	35	—	100	22.3	65.3	7.3	94.4	—	—	2.6	2.6	2.5	S	4.5	—	0.6	—
MM 2.5	B_v	55	0.1	99.9	22.7	67.0	6.8	96.5	—	0.7	1.0	1.7	1.8	S	4.6	—	0.3	—
MM 2.6	C_v	70	0.1	99.9	12.9	71.8	14.4	99.1	—	0.1	0.5	0.6	0.3	S	4.9	—	0.1	—

Tab. 1: *Mattesmühle profile.*

Sample No.	Hor	SD	Size Spectrum >2%	Size Spectrum <2%	Grain Sizes (%) <2 mm Sand				Silt				Clay	Tex	pH	Carb.	OM	EB
			mm	mm	coarse	medium	fine	sum	coarse	medium	fine	sum	T		n/10 KCl	%	%	mval 100 g
HH 1	A_{hl}	8	0.9	99.1	9.2	64.9	19.1	93.2	0.4	3.4	0.9	4.7	2.1	S	3.2	—	1.8	—
HH 2	$B_{v(h)}$	18	0.6	99.4	10.5	57.9	23.2	91.6	2.7	1.3	1.2	5.2	3.2	S	3.9	—	1.1	—
HH 3	B_v	35	1.2	98.8	8.3	64.6	22.8	96.0	0.6	1.3	0.8	2.7	1.3	S	4.2	—	1.0	—
HH 4	B_v	50	2.1	97.9	10.8	64.0	21.5	96.3	0.2	1.1	0.4	1.7	2.0	S	4.3	—	0.5	1.0
HH 5	B_v	70	0.9	99.1	13.1	61.8	22.2	97.1	—	1.2	0.8	2.0	0.9	S	4.2	—	0.3	—
HH 6	C_v	90	0.3	99.7	11.8	70.1	15.8	97.7	0.1	2.2	—	2.3	—	S	4.2	—	0.3	—

Tab. 2: *Herrschaftsholz profile.*

Sample No.	Hor	SD	Size Spectrum >2%	Size Spectrum <2%	Grain Sizes (%) <2 mm Sand				Silt				Clay	Tex	pH	Carb.	OM	EB
			mm	mm	coarse	medium	fine	sum	coarse	medium	fine	sum	T		n/10 KCl	%	%	mval 100 g
L 1	A_h	5	—	100	21.5	54.8	10.5	86.8	1.4	8.6	2.1	12.1	1.1	uS	3.4	—	4.8	1.0
L 2	A_l	20	0.4	99.6	9.5	56.7	11.8	78.0	4.8	4.6	10.9	20.3	1.7	uS	4.0	—	1.3	—
L 3	B_v	40	0.2	99.8	8.4	54.5	13.5	76.4	4.9	4.6	0.9	10.4	13.2	tS	5.3	—	0.4	3.1
L 4	B_v	60	0.5	99.5	14.3	55.8	9.2	79.3	3.7	4.5	5.0	13.2	7.5	lS	5.4	—	0.4	4.0
L 5	B_t	80	0.1	99.9	10.0	51.2	9.9	71.1	4.3	4.3	0.7	9.3	19.6	tS	5.7	—	0.6	8.2

Tab. 3: *Lochschlag profile.*

Sample No.	Hor	SD	Size Spectrum >2% mm	<2% mm	Grain Sizes (%) <2 mm Sand coarse	medium	fine	sum	Silt coarse	medium	fine	sum	Clay T	Tex	pH n/10 KCl	Carb. %	OM %	EB mval 100 g
MH 1	A_h	5	0.1	99.9	11.4	59.3	11.7	82.4	4.8	7.2	4.1	16.1	1.5	uS	3.5	—	2.2	—
MH 2	B_{vh}	20	0.2	99.8	20.1	49.0	12.5	81.6	5.6	6.6	3.9	16.1	2.3	uS	3.8	—	1.6	—
MH 3	B_v	50	1.1	98.9	11.2	48.3	21.8	81.3	5.4	6.9	4.6	16.9	1.8	uS	4.2	—	0.3	1.1
MH 4	S_wC_v	70	0.6	99.4	12.9	50.3	11.4	74.6	5.5	6.3	1.7	13.5	11.9	lS	3.9	—	0.3	2.0
MH 5	IIS_dB_t	90	0.9	99.1	1.5	7.3	11.3	20.1	5.9	7.5	6.0	19.4	60.5	lT	6.4	1.3	0.2	75.0
MH 6	C_v	115	75.1	24.9	8.6	12.1	17.7	38.4	13.8	9.5	4.7	28.0	33.6	stL	7.4	13–35	0.5	—

Tab. 4: *Mäderholz profile.*

Sample No.	Hor	SD	Size Spectrum >2% mm	<2% mm	Grain Sizes (%) <2 mm Sand coarse	medium	fine	sum	Silt coarse	medium	fine	sum	Clay T	Tex	pH n/10 KCl	Carb. %	OM %	EB mval 100 g
S 1			0.2	99.8	8.1	75.2	14.6	97.9	0.5	0.1	0.2	0.8	1.3	S	4.4	—	0.5	—
S 2			0.2	99.8	8.8	71.3	12.2	92.3	0.4	0.2	0.6	1.2	6.5	tS	4.8	—	0.3	—
S 3			1.1	98.9	5.4	68.5	22.0	95.9	2.9	—	0.9	3.8	0.3	S	8.0	—	0.3	—
S 4			9.0	91.0	7.7	68.6	16.5	92.8	0.4	0.6	0.5	1.5	5.7	tS	7.6	—	0.2	—
S 5			0.1	99.9	8.6	71.6	16.3	96.5	—	—	—	1.1	2.4	S	7.0	—	0.1	—
S 6			2.1	97.9	10.6	57.4	12.8	80.8	0.2	1.3	12.3	13.8	5.4	lS-uS	6.7	—	0.5	—
S 7			—	100	9.0	68.7	14.8	92.5	2.5	—	—	2.5	5.0	tS-uS	6.4	—	0.3	—
S 8	see		63.7	36.3	11.8	63.3	21.1	96.5	—	—	—	1.4	2.1	S	8.5	2.6	0.3	—
S 9			—	100	5.5	64.7	24.4	94.6	2.3	1.7	—	4.0	1.4	S	6.8	—	—	—
S 10	fig.2		35.4	64.6	8.0	66.8	17.6	92.4	5.7	—	—	5.7	1.9	S	8.0	2.3	0.1	—
S 11			45.6	54.4	8.2	68.3	19.8	96.3	—	2.1	0.7	2.8	0.9	S	8.6	0.6	0.4	—
S 12			0.6	99.4	7.1	69.9	20.4	97.1	1.2	—	1.7	2.9	—	S	8.6	—	0.2	—
S 13			56.5	43.5	7.9	71.2	17.9	97.0	0.7	0.2	0.4	1.3	1.7	S	8.7	2.6	0.5	—
S 14			6.1	93.9	10.6	68.6	17.3	96.5	0.6	0.9	—	1.5	2.0	S	8.6	3.2	0.1	—
S 15			0.2	99.8	6.3	76.0	13.6	95.9	0.2	0.2	0.5	0.9	2.5	S	8.4	—	0.1	—
S 16			3.8	96.2	5.3	67.8	24.3	97.4	0.7	0.1	0.3	1.1	1.5	S	8.7	—	0.4	—
S 17			0.4	99.6	6.3	71.6	19.5	97.4	0.4	0.4	0.1	0.9	1.7	S	8.6	—	0.2	—
S 18			0.5	99.5	5.1	9.5	10.3	24.9	17.3	15.1	15.3	47.7	27.4	stL	6.8	—	—	—
S 19			4.2	95.8	18.2	19.2	12.6	50.0	7.2	6.7	11.0	24.9	25.1	stL-sL	7.1	—	—	—

Tab. 5: *Steinbühl profile.*

the late Würm, the deeper, loamy sediment could represent the erosion remnant of an Eemian interglacial soil, an assumption which would also agree with the suggestion by GALL et al. that the sediment is Riss-Würm.

The colour and structure of the brown zones incorporated in the alluvial cone contrast with the other sediment layers. The brown zones (samples S2, S4, S6, S7 in fig.2) contain a larger proportion of fine grains (less than 0.06 mm diameter) which decreases with depth. The colour is also lighter in the gravel beds (tab.5). It is not known whether these brown zones are the remnants of weathering, a fossil soil in the broadest sense, or whether this is the original colour of the primary sediment.

4.2 Soil development in the basin and on the rim

There is a clear relationship between the location, the depth of the substrate and soil development. In the basin the depth and structure of the sediments have resulted in the development of podzolic soils. On the shallower sands on the rim, pedogenesis has proceeded much further and brown earth-parabrown earth (Cambisols-Luvisols) have developed. There also is a higher degree of loaminess and a larger proportion of silt, both of which indicate a more intensive weathering that is, in turn, related to the site and soil-water budget.

In addition, the Bt (clayey) horizons in the rim profiles lie above loamified ejected rock masses. The high clay content in the Mäderholz profile causes a high bulk density and very low permeability. Seasonal ponding of water has occurred above the Bt horizon and pseudogley has developed in the deeper brown earths (Cambisol). The Lochschlag profile has a lower clay content and pseudogley has not developed.

The Bt horizon in the Mäderholz profile is thought to be the remnant of a fossil soil for the following reasons (tab.4):

1. The transition from the SwCv horizon (Sw = water influenced horizon) to the IISdBt horizon (Sd = water impermeable horizon) contrasts with the more gradual horizon boundaries of the Lochschalg profile. This contrast also applies to the upper horizons of the Mäderholz profile. There is no indication of clay transport from the upper parts of the profile. Moreover, the 60% clay content cannot have been supplied by illuviation from the soil above since it is only 70 cm thick.

2. There are large differences between the clay contents of the Bt horizon in the Mäderholz and Lochschlag profiles.

3. There is a small amount of lime in the Bt horizon of the Mäderholz profile due to the non-uniform distribution of limestone fragments. The Bt horizon of the Mäderholz profile cannot have had the same genesis as the Bt horizon of the Lochschlag profile because the remaining substrate of the Mäderholz profile and the entire Lochschlag profile contain no lime. The Bt horizon in the Lochschlag profile must, therefore, be the result of Holocene soil development and the Bt horizon in the Mäderholz profile is closely related to the loamy Tertiary Ries debris.

4. The presence of heavy minerals in the Bt horizon of the Mäderholz profile also indicates that its origin differs from that of the Lochschlag profile.

Also of significance is the humic horizon, Bv(h) horizon, in the Mattesmühle profile in the basin. The horizon has no sharp boundaries and there are dark vertical streaks, indicating humus transport, in the top Bv(e) horizon. This horizon distribution is apparently the result of Holocene humic podzolization.

4.3 The origin of sediments in the Steinbühl profile

The investigation of the Steinbühl profile has shown that the geological mapping of this area has been incorrect in some places. A statistical analysis of the sediments was made in order that conclusions could be made about their origin. Critical parameters are the sorting index and the median grain size of the sediments (FÜCHTBAUER 1964). The median grain sizes range from 0.25 to 0.40 mm, the sorting index from 1.40 to 4.70. Even if the samples with sorting values higher than 2.5 are excluded, to remove the influence of soil formation, the remaining sorting values are still far too high to indicate a purely aeolian origin. Values of less than 1.23 would be necessary for this. The evidence provided by the few samples investigated is too limited to prove that the sediments have been transported farther than the area indicated on the 1:50,000 map.

5 Summary of results

1. The Pleistocene aeolian sands on the eastern edge of the Nördlinger Ries are less on the rim than on the basin floor. In the slope area, alluvial cones have developed.

2. The sediments of the alluvial cones may contain material derived from the Upper Freshwater Molasse and the Monheim High Sands of post-Ries age.

3. The aeolian sands on the eastern margin of the basin have been almost entirely retransported; there are no clear indications of in situ stratification. The field observations differ from the information on the geological map.

4. The Holocene soils are more intensively developed on the rim than in the basin, as indicated by the larger quantities of silt and clay.

5. Holocene podzolization is more common in the profiles of the basin floor that are under coniferous forests. At similar sites on the rim there are only the initial stages of podzolization in the form of a very thin layer under scattered leaves.

6. Some substrates are fossil in character.

7. It is not possible to recognize a stratigraphic system and a precise date for the fossil substrates. They may be Holocene but could also be older (KOZARSKI 1978).

Acknowledgements

The author thanks Mr. Burtscher of the Kalksteinwerk Wemding (Ries) for providing the results of mineralogical analyses of the substrates of the Mattesmühle and Herrschaftsholz profiles which had been carried out by the Dept. of Mineralogy, Univ. of Giessen, and results of fossil identificatiom which had been carried out by the Bavarian State Collection for Palaeontology and Historical Geology, Munich.

References

FÜCHTBAUER, H. (1964): Sediment-Petrologie, Stuttgart.

GALL, H. et al. (1977): Erläuterungen zur geologischen Karte des Rieses 1:50,000. Geologica Bavarica Nr. **76**, 1976, München.

KOZARSKI, S. (1978): Das Alter der Binnendünen in Mittelwestpolen. In: NAGEL, H. (1978): Beiträge zur Quartär- und Landschaftsforschung. Fink-Festschrift, Wien, 291 pp.

LEMCKE, K. (1981): Das Nördlinger Ries: Spur einer kosmischen Katastrophe. Spektrum der Wissenschaft **1**, 111–121.

RÜCKERT, G. (1978): Bodenkarte von Bayern, Erläuterungen zum Blatt Nr. 7029 Öttingen i. Bayern, München.

TILLMANNS, W. (1983): Zur Verbreitung, Petrographie und Stratigraphie des Monheimer Höhensandes im Östlichen Vorries. Zeitschrift für Geomorphologie, NF. **27**, 3–104.

WEBER, E. (1941): Geologische Untersuchungen in Ries — Das Gebiet des Blattes Wemding. Abh. des Naturkunde- und Tiergartenvereins für Schwaben e.v., Augsburg, Heft III; Geologisch-Paläontologische Reihe, 2. Heft.

Address of author:
Dr. Michael Schieber
Naturkunde Museum Regensburg/Ostbayern
Am Prebrunntor 4
8400 Regensburg
Federal Republic of Germany

CATENA SUPPLEMENT 15 p. 279–287 Cremlingen 1989

LATE PLEISTOCENE AND HOLOCENE DEVELOPMENT OF THE DANUBE VALLEY EAST OF REGENSBURG

Manfred W. **Buch**, Regensburg

1 Introduction

The paleohydrography of the Danube valley in Bavaria between Regensburg and Straubing has been investigated since 1984 (BUCH 1988a and fig.1). These studies indicate that fluvial processes in the late Pleistocene and Holocene are the result of self-control mechanisms (HEY 1979, SCHUMM 1979) that are independent of climatic fluctuations.

2 Geology and morphology of the flood plain and Low Terrace

The Low Terrace of the Danube is composed of Würm deposits and the flood plain of Holocene deposits. There is a scarp in the subsurface so that the base of the earlier Würm gravels lies at a higher elevation than the base of the late Würm(?) and Holocene deposits. The thickness of the Low Terrace Würm gravel varies between 2 m and 11 m. The subsurface is undulating and ranges between 4 m below and 2.5 m above the low water level of the Danube in 1972.

ISSN 0722-0723
ISBN 3-923381-18-2

3-923381-18-4/89/5011851/US$ 2.00 + 0.25

The bedding structures of the Low Terrace gravels vary. In general, the gravels are more or less cross-bedded near the bottom of the sequence and horizontally bedded at the top. The horizontal bedding is typical for braided river channels.

The sequence of the, probably, late Würm and Holocene sediments within the flood plain is known from more than 500 profiles taken from borings. The thickness of the sediments varies between 10 m and 15 m in the relict thalweg of the buried channel. The gravels, which are covered by calcareous silty and sandy flood sediments, can be divided into a lower and upper sequence. The upper sequence is a 7 m thick layer of sediment deposited recently by the Danube which is also present in paleochannels of the Middle Flood Plain Terrace, IIB. The lower sequence also contains boulders and locally a sandy facies dominates.

The Low Terrace, Terrace I in fig.2, can be divided into three levels, IA, IB and IC, the result of differences in the depth of incision of a braided river channel pattern (fig.2).

The highest Low Terrace gravel is on the Cover Level, IA, 8.5 m to 10 m above

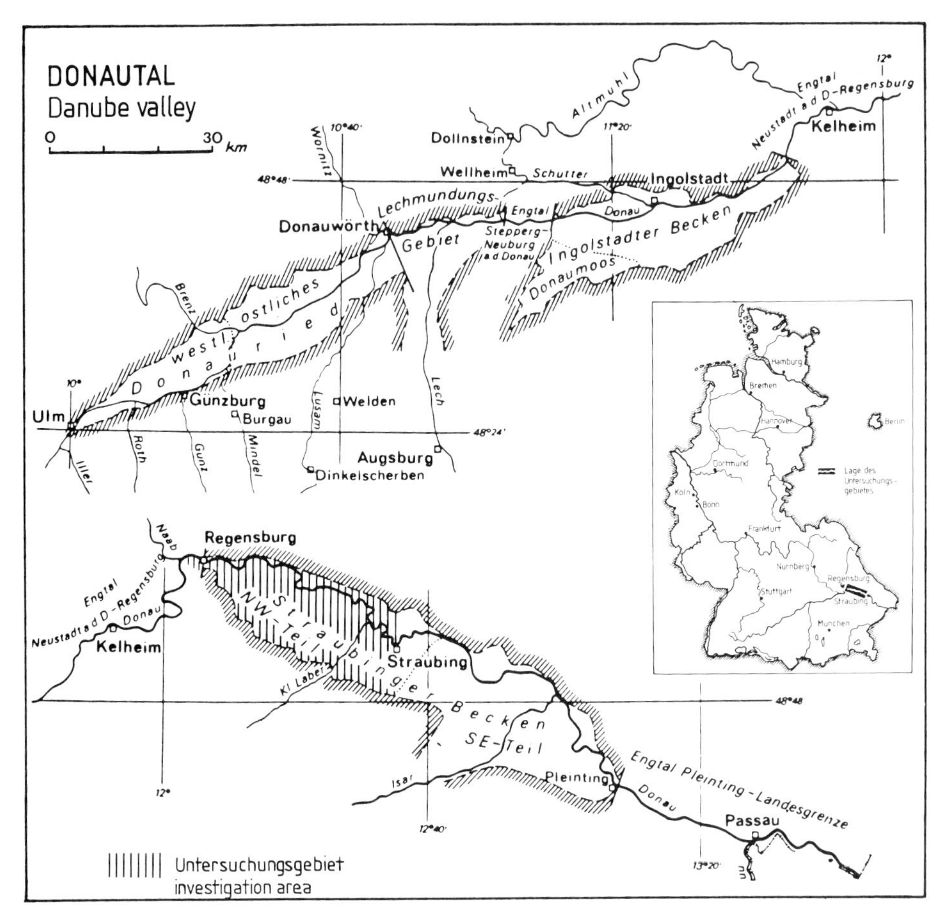

Fig. 1: *Danube River in Bavaria (see HOMILIUS et al. 1983 and author's additions).*

the low water stage of the Danube in 1972. It is covered by 1.5 m to 3.5 m of loess sediments that have a weak brown earth (Cambisol) horizon within the sequence, loess-typical mollusc fauna and parabrown earth (Orthic Luvisol) on top. The surface of the terrace lies between 10 m and 12.5 m above Danube low water. The boundary between the Cover Level and the Main Level of the Low Terrace, IB, is formed by a natural levee on which aeolian loess has been deposited. Parabrown earth has developed on the Main Level of the Low Terrace, IB in fig.2, 7 m to 9 m above Danube low water. Aeolian sands or an aeolian influenced layer cover this terrace. Locally dunes occur on which weak brown earth has developed. However, under the aeolian sands, a well-developed brown earth that shows initial clay migration, has developed on the Würm fluvial gravels and a swamp loess facies. The Disintegration Level of the Low Terrace, IC, 5 m to 8 m above Danube low water, is characterized by a soil catena in which parabrown earth on the gravel ridges and brown earth soils of medium depth on the cal-

FLUVIAL (

terrace dif

Lower Flood
(IIC) since

Middle Flood
(II B)

Upper Flood F

"basal- se
quence" w
floodplain

Disintegrati
of the Low
(IC)

Main Level
Terrace (I

Cover Level
Terrace (I/

Fig. 3:

careous, sandy and silty fill of the paleochannels have developed. Organically rich sediments occur in the bottom of the paleochannels. The deeper incised paleochannels which lie at the level of the highest flood plain terrace, are partially filled by sandy and silty recent flood sediments.

The scarp in the Quaternary subsurface between the Low Terrace and the Flood Plain Terrace, II in fig.2, is the result of a major phase of erosion. There is a difference of 20 m between the top of the Cover Terrace, Low Terrace IA, and the subsurface of the flood plain.

The Flood Plain Terrace II can also be divided into three terrace levels. The Upper Flood Plain Terrace IIA is 4.5 m to 5.5 m above the low water stage of the Danube. It is the highest flood plain terrace and reaches the level of the most deeply incised paleochannels of the Disintegration Level of the Low Terrace. The latter is flooded by the mean annual high water stage of the period 1924/1981. The Upper Flood Plain Terrace is covered by up to 4 m of flood sediments and has only small variations in elevation. The paleochannels under these flood sediments are indicated by the relief on the surface of the gravels, which lie from 1.5 m above to 0.5 m below the Danube low water stage. The paleochannel system was braided. Its shallow channels appear to have been only 1.5 m deep and more than 100 m wide. Organically rich fine-grained sediments occur in the zone of contact with the gravel. The fine-grained fluvial sediments are covered by a well-developed brown earth with calcareous concretions at the base. This brown earth is covered by younger calcareous rich flood deposits on which a weak brown pararendzina is developed.

Since the development of the Middle and the Lower Flood Plain Terrace, levels IIB and IIC, the river has flowed in a pattern comparable to the present one, with a sinuous course between the mouth of the Naab river and the village of Kiefenholz and a meander course upward to Straubing.

The course of the Danube on the Lower Flood Plain Terrace IIC from the beginning of the 18th century can be reconstructed on the basis of historical maps. In the area of the village of Pfatter, three meander generations, which have incised into one another on the Lower Flood Plain Terrace, can be reconstructed for this period. The meanders have migrated a distance of 3 km in only 65 years. Pararendzinas and gleys have developed on the silty and fine sandy flood sediments.

On the flood sediments of the Middle Flood Plain Terrace IIB brown pararendzinas and gleys have developed, indicating that the Middle Flood Plain Terrace is more closely related to the Lower Flood Plain Terrace than to the Upper Flood Plain Terrace on which there are well-developed brown earths. It is not known whether the gravel of the Middle Flood Plain Terrace in the investigation area contains any historical relicts.

3 Fluvial geomorphodynamics of the Danube since the Würm pleniglacial

The Cover Level of the Low Terrace, IA, is covered by a loess sequence that contains snails (*Succinea oblonga elongata* Sndb., *Pupilla muscorum* (L.), *Trichia hispida* (L.) and *Trichia hispida terrena* (Cless.)) which were well-adapted to cold climatic conditions. A weak brown earth

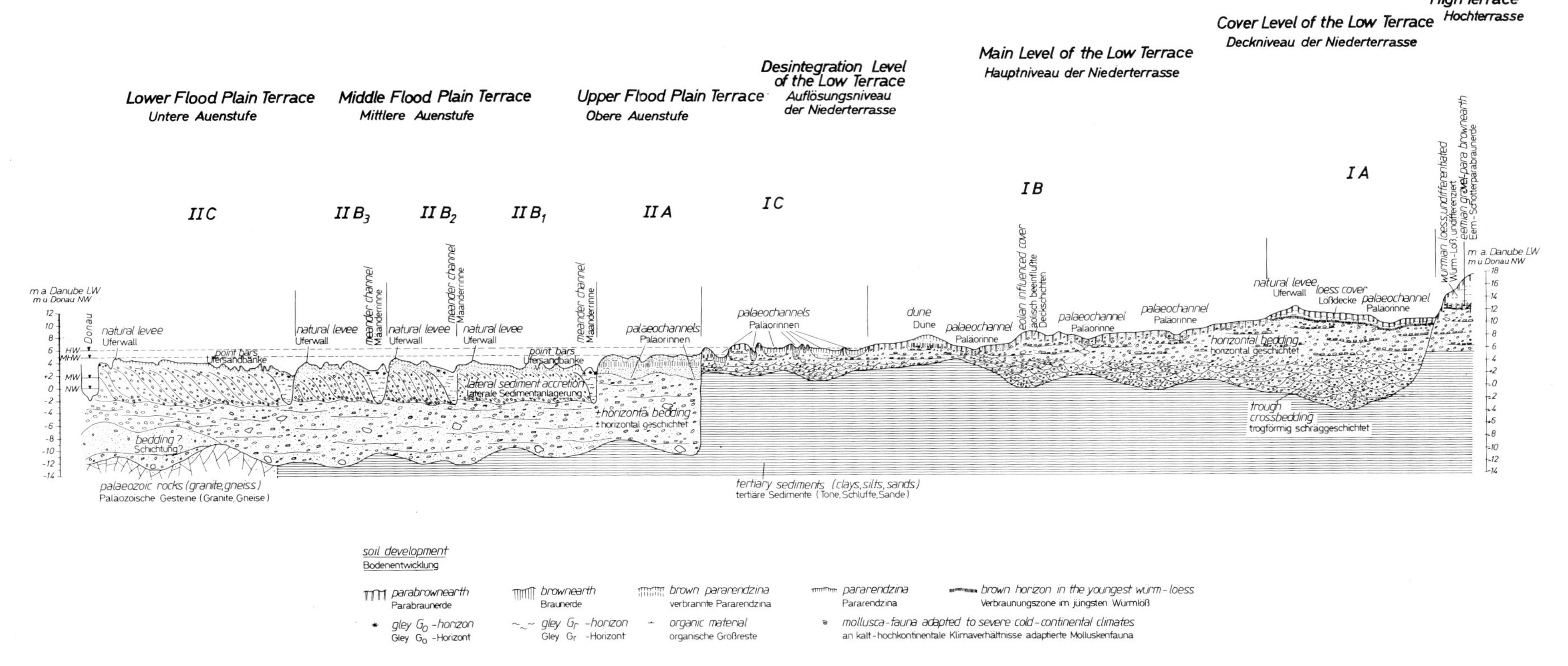

Fig. 2: *Cross-section of the Würm and Holocene system of valley floor terraces of the Danube between Regensburg and Straubing.*

horizon and parabrown earth, which is characteristic for the younger Würm loess sequence in the area of Regensburg, was developed on the surface. It is, therefore, thought that the highest sediments of the Low Terrace were deposited by a braided river system at the beginning of the Würm pleniglacial. During the pleniglacial there was a change from a braided river system to a anastomosing channel pattern in which there was a more or less dominating main channel in a network of braided channels. Comparable river channel patterns are known to have existed in the area of Neustadt on the Danube shortly before the canalization work on the Danube in the 19th century. After the transition from a braided to an anastomosing channel pattern, incision took place. Also, horizontal shifting no longer occurred. The change in the channel pattern is a complex response to the interaction between discharge, sediment load and slope (STARKEL 1983). Complex response was of particular importance during the phase of erosion that resulted in a 20 m difference between the average surface of the Main Level of the Low Terrace and reached more than 9 m below the latter's sub-Quaternary base. The rate of erosion was intensified as soon as the gravels of the Low Terrace had been eroded and incision took place in the cohesive Tertiary clays and silts below. Prehistoric remains, fossil soils and mollusc fauna (*Succinea oblonga elongata* Sndb, *Pupilla muscorum* (L.) and *Pupilla loessica* (Lzk.) in a fine-grained sediment in a paleochannel of the Main Level have been used to date the Main Level of the Low Terrace (IB) and show that some of the paleochannels were abandoned during the pleniglacial. Late Magdalénien artifacts at the base of the dune sands were reported by REISCH (1974) and silex and other sediment structures of the Mesolithic period were found within the dune sands by SCHÖNWEISS & WERNER (1974). Since the swamp loess sediments are dated in the late pleniglacial (± 17,000–10,000 BP) by mollusc fauna and have been covered by a well-developed brown earth with a transition to parabrown earth, which has also been dated to the late pleniglacial because of the dune sand cover, the fluvial development of the Main Level of the Low Terrace must have ceased, at the latest, during the late pleniglacial.

The period of pronounced erosion was followed by a period of continuous gravel aggradation. Results obtained in the area of the Iller river alluvial cone in the upper part of the Danube basin agree with this reconstruction (BECKER 1982). The maximum level of aggradation is indicated by the gravel on top of the Upper Flood Plain Terrace, IIA, which is only 1–3 m below the Disintegration Level of the Low Terrace. The deep paleochannels of that level were occasionally flooded. The gravels were deposited by a braided channel pattern which is still recognizable.

The bouldery basal sequence of the flood plain sediment is part of the new gravel aggradation. A subfossil oak tree within the top layer of the basal sequence is dated 1951±10 BC (BECKER 1982). It is not yet certain when the aggradation of the new gravel began. The oldest subfossil trees in the Danube system are dated 12,000±220 BP (Austrian section of the Danube, FINK 1977) and 9700±120 BP (Upper Danube, Iller river alluvial cone, BECKER 1982). It is assumed that the aggradation began at the transition from the late pleniglacial to the early Holocene.

With the change from a braided to a sinuous river system, the fluvial processes shifted downward from the Upper Flood Plain Terrace and lateral sediment redeposition began within the Middle Flood Plain Terrace, IIB. This change occurred in the first third of the Subboreal period at the same time as the first settlements appeared on the Upper Flood Plain Terrace (Late Neolithic and Early Bronze Age settlements near Öberau, NW of Straubing, BÖHM & SCHMOTZ 1979). Fluvial conditions must have been very stable from the Bronze Age at least until the Roman period, since there is a well-developed brown earth on the Upper Flood Plain Terrace IIA.

The main accretion of the alluvial cone of the Kleine Laaber river west of Straubing had been completed when Middle Bronze Age, Urnfield Age and Roman settlements were founded on the alluvial cone. A Roman road from Regensburg to Straubing also crossed the cone. There was extensive cultivation in the region during this stable period in the Bronze and Hallstatt Ages.

The change in the river channel pattern also meant that the fine-grained sediments became more important in the ecology of the valley floor. They provided improved growth conditions for the oak communities on the flood plain (BECKER 1982).

Calcareous flood sediments on the brown earth soil of the Upper Flood Plain Terrace IIA, indicate a new phase of frequent high floods. Near the village of Kreuzhof, east of Regensburg, a 2nd century AD Roman urnfield was found covered by flood sediments on the edge of the Main Level of the Low Terrace. In the sediments of the alluvial cone of the Naab river, the youngest sediments on well-developed brown earth contain iron slag grains which date from the time of iron industry and extensive forest clearings that began in the 14th century. Historical sources show that, particularly during the "Little Ice Age" from 1560 to 1860, winter floods combined with ice jams were very frequent. The morphological response of the river to the period of extensive high floods from the late Middle Ages until the present varied. In the sinuous reach the channel pattern has shown a tendency to braid with mid-channel bars at the time of the Lower Flood Plain Terrace, in the 18th century. Within the meandering course, meander migration is restricted mainly to the area of Pfatter, just below the sinuous course. From the middle of the 19th century, the Danube has been affected by artificial regulation of the channel (BAUER 1965).

4 The cause of changes in the fluvial environment of the Danube

The fluvial development of the Danube is different from other well-investigated mid-European fluvial systems (fig. 3 and SCHIRMER 1983a, 1983b); the most important differences are:

1. The oldest of three Würm terraces in the Main and Regnitz valley, the Reundorf Terrace (SCHIRMER 1983a), is thought to correspond with the main glacier advance during the Würm pleniglacial. By contrast, the accumulation of the gravels of the Danube Low Terrace, east of Regensburg, was almost completed at the beginning of the pleniglacial, as shown by the existence of the Cover Level of the

Low Terrace IA, which has a differentiated loess cover. A terrace of similar age to a younger Würm loess has been reported as T6-Terrace by SEMMEL (1972, 1974) only in the lower Main valley.

2. During the pleniglacial and late Glacial, the Danube eroded continuously downward, with a maximum phase of deep incision in the transition from the Würm to the Holocene. The differentiation of the three Würm Low Terraces IA, IB, and IC, in the Danube valley was caused by the change from a braided river to an anastomosing river during which the three distinct topographic levels were formed. The Main Level as well as the Disintegration Level of the Low Terrace, IB and IC, are, therefore, erosion terraces that have been cut into the accumulation of the Würm Low Terrace gravel. This contrasts with observations by SCHIRMER (1983a) and SEMMEL (1972, 1974) in the valleys of the Regnitz and Main.

3. The combination of deep incision and lateral erosion of the Würm gravels followed by a new progressive aggradation in the recent flood plain of the Danube is a development that has so far been reported only for the river Gipping near Sproughton in East Anglia (ROSE & BOARDMAN 1983). The new aggradation again took place with a braided channel system.

These findings and the detailed knowledge of Late Pleistocene and Holocene climate and the history of glaciation in in the Alps raise the question whether the commonly assumed causal relationship between climatic fluctuations and fluvial processes (SCHIRMER 1983a, STARKEL 1985) can be accepted for the Danube. The fluvial geomorphodynamic of the Danube during the late Pleistocene and Holocene is thought to be the result of a self-control mechanism that is independent of the climatic rhythm.

Clearly the changes from glacial to interglacial during the Pleistocene created the terrace sequence typical for the valleys in mid-European periglacial areas. The subdivision of the Alpine glaciation is based on this relationship (PENCK & BRÜCKNER 1901/1909). During the late Pleistocene and Holocene period, however, there is evidence that terraces can be formed without external changes. Among recent models of fluvial development, those by HEY (1979) on the dynamic process reponse of river channel development and by SCHUMM (1979) on geomorphic thresholds appear most appropriate to explain the fluvial development of the Danube since the late Pleistocene (BUCH 1988b). The most important ideas of these models are, first, that even rather insignificant changes of the variables of a fluvial system can induce major adjustments in river channel development (SCHUMM 1979) and, second, that upstream and downstream feedback mechanisms operate between sections of the river course in such a way that variables continually change their values both in time and space (HEY 1979).

The self-control mechanism of the Danube is largely determined by the segmentation of the river course into alternating narrow valley stretches and basins, which produces local erosion and accumulation reaches (fig.1). Within these reaches the influence of the tributaries on the discharge regime and the sediment load of the Danube differs considerably. The individuality of the reach investigated between Regensburg and Straubing results from its location below the exit of the narrow valley stretch between Neustadt on the Danube and the mouth of the Naab (fig.1) where the Alpine discharge regime, controlled by the Lech river, is combined with the upland discharge regime of the left bank northern tributaries, the Naab and the Regen. The narrow valley stretch a short distance above the investigation area is also thought to be of major importance because it controls the sediment supply and the available grain size in the reach below.

Every reach of the Danube contains its particular association of internal and external factors and thresholds. Changes of such factors therefore produce different reactions in different reaches. This concept corresponds to findings by FINK (1977) for the Austrian section of the Danube.

Acknowledgements

The investigations were supported by the University of Regensburg from 1984 to 1986 and by the Deutsche Forschungsgemeinschaft (Fluvial Geomorphodynamics in the Holocene, Project He 722/18 1+2, Bu 659/1). Special thanks are due to Prof. Dr. K. Heine (Regensburg), Dr. J. Kovanda (Prague) for mollusc fauna determinations and the Rhein-Main-Donau AG for the register of borings in the Danube valley.

References

BAUER, F. (1965): Der Geschiebehaushalt der bayerischen Donau im Wandel wasserbaulicher Maßnahmen. Die Wasserwirtschaft **55 (4/5)**, 1–16.

BECKER, B. (1982): Dendrochronologische und Paläoökologie subfossiler Baumstämme aus Flußablagerungen. Ein Beitrag zur nacheiszeitlichen Auenentwicklung im südlichen Mitteleuropa. Mitt. d. Komm. f. Quartärforschung d. Österr. Akad. d. Wiss. **5**, Wien, 120 pp.

BÖHM, K. & SCHMOTZ, K. (1979): Die vorgeschichtliche Besiedlung des Donautales nordwestlich von Straubing und ihre geologischen Voraussetzungen. Jahrbuch Histor. Verein Straubing **81**, 39–88.

BUCH, M.W. (1988a): Spätpleistozäne und holozäne fluviale Geomorphodynamik im Donautal zwischen Regensburg und Straubing. Regensburger Geogr. Schrft. **21**, 240 pp. and Supp.

BUCH, M.W. (1988b): Zur Frage einer kausalen Verknüpfung fluvialer Prozesse und Klimaschwankungen im Spätpleistozän und Holozän — Versuch einer geomorphodynamischen Deutung von Befunden von Donau und Main. Z. Geomorph. N.F., Suppl. Bd. **70**, 131–162.

FINK, J. (1977): Jüngste Schotterakkumulation im Österreichischen Donauabschnitt. Erdwissenschaftliche Forschungen XIII, (Steiner) Wiesbaden, 190–211.

HEY, R.D. (1979): Dynamic process-response model of river channel development. Earth Sur. Proc. and Landf. **4**, 59–72.

HOMILIUS, J., WEINIG, H., BROST, E. & BADER, K. (1983): Geologische und geophysikalische Untersuchungen im Donauquartär zwischen Ulm und Passau. Geologisches Jahrbuch E 25, 3–73.

PENCK, A. & BRÜCKNER, E. (1901/1909): Die Alpen im Eiszeitalter. Tauchnitz, Leipzig, 3 vols.

REISCH, L. (1974): Eine spätjungpaläolithische Freilandstation im Donautal bei Barbing, Ldkr. Regensburg. Quartär. **25**, 53–71.

ROSE, J. & BOARDMAN, J. (1983): River activity in relation to short-term climatic deterioration. Quaternary Studies in Poland **4**, 189–198.

SCHIRMER, W. (1983a): Die Talentwicklung am Main und Regnitz seit dem Hochwürm. Geologisches Jahrbuch **A 71**, 11–43.

SCHIRMER, W. (1983b): Holozäne Talentwicklung — Methoden und Ergebnisse. (Ed.), Geologisches Jahrbuch **A 71**, 370 pp.

SCHÖNWEISS, W. & WERNER, H.J. (1974): Mesolithische Wohnanlagen von Sarching,

Ldkr. Regensburg. Bayerische Vorgeschichtsblätter **39**, 1–29.

SCHUMM, S.A. (1979): Geomorphic thresholds: concepts and its applications. Inst. of Brit. Geogr., Trans, NF **44**, 485–515.

SEMMEL, A. (1972): Untersuchungen zur jungpleistozänen Talentwicklung in deutschen Mittelgebirgen. Z. Geomorph. NF, Supp. Bd. **14**, 105–112.

SEMMEL, A. (1974): Der Stand der Eiszeitforschung im Rhein-Main Gebiet. Rhein-Mainische Forschungen **78**, 9–56.

STARKEL, L. (1983): The reflection of hydrologic changes in the fluvial environment of the temperate zone during the last 15,000 years. In: GREGORY, K.J., Background to Palaeohydrology, Wiley & Sons, Chichester, 214–235.

STARKEL, L. (1985): Lateglacial and postglacial history of river valleys in Europe as a reflection of climatic changes. Zeitschr. für Gletscherkunde u. Glazialgeologie **21**, 159–164.

Address of author:
Dr. Manfred W. Buch
Geographisches Institut der Universität Regensburg
Postfach 397
8400 Regensburg
Federal Republic of Germany

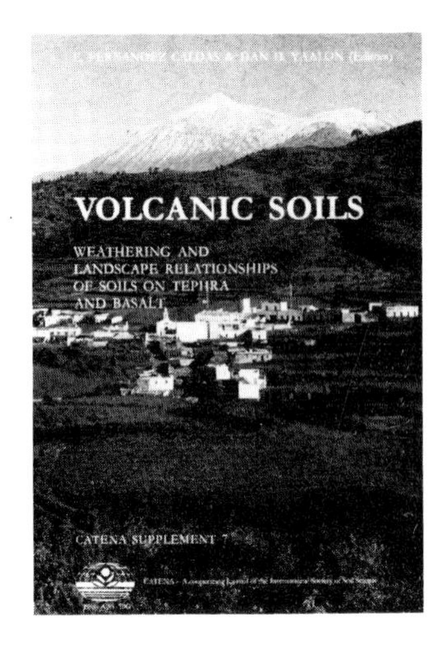

E. Fernandez Caldas & Dan H. Yaalon (Editors):

VOLCANIC SOILS

Weathering and Landscape
Relationships of Soils on Tephra and Basalt

CATENA SUPPLEMENT 7, 1985

Price DM 128,—

ISSN 0722–0723 / ISBN 3–923381–06–9

PREFACE

This CATENA SUPPLEMENT contains selected papers presented at the International Meeting on Volcanic Soils held in Tenerife, July 1984. The meeting brought together over 80 scientists from 21 countries, with interest in the origin, nature and properties of soils on tephra and basaltic parent materials and their management. Some 51 invited and contributed papers and 8 posters were presented on a wide range of subjects related to volcanic soils, many of them dealing with weathering and landscape relationships. Classification was also discussed extensively during a six day excursion of the islands of La Palma, Gomera and Lanzarote, which enabled the participants to see the most representative volcanic soils of the Canary Archipelago under a considerable range of climatic regimes and parent material ages.

Because volcanic soils are not a common occurrence in regions where pedology developed and progressed during its early stages, recognition of their specific properties made an impact only in the late forties. The name **Ando** soils, now recognized as a special Great Group in all comprehensive soil classification systems, was coined in 1947 during reconnaissance soil surveys in Japan made by American soil scientists. Subsequently a Meeting on the Classification and Correlation of Soils from Volcanic Ash, sponsored by FAO and UNESCO, was held in Tokyo, Japan, in 1964, in preparation for the Soil Map of the World. This was followed by meetings of a Panel on Volcanic Ash Soils in Latin America, Turrialba, Costa Rica, in 1969 and a second meeting in Pasto, Colombia, in 1972. At the International Conference on Soils with Variable Charge, Palmerston, New Zealand, 1981, the subject of Andosols was discussed intensively. Most recently the definitions of Andepts, as presented in the 1975 U.S. Soil Taxonomy, prompted the establishment of an International Committee on the Classification of Andisols (ICOMAND), chaired by M. Leamy from C.S.I.R., New Zealand, which held a number of international classification workshops, the latest in Chile and Ecuador, in January 1984. The continuous efforts to improve and revise the new classification of these soils is also reflected in some of the papers in this volume.

While Andosols or Andisols formed on tephra (volcanic ash), essentially characterize by low bulk density (less than 0.9 g/cm^3) and a surface complex dominated by active A cover worldwide an area of about 100 million hectares (0.8% of the total land area), the va basaltic plateaus and their associated soils cover worldwide an even greater area, frequent with complex age and landscape relationships. While these soils do not generally belong t the ando group, their pedogenetic pathways are also strongly influenced by the nature an physical properties of the basalt rock. The papers in this volume cannot cover the wide varie of properties of the soils in all these areas, some of which have been reviewed at previou meetings. In this volume there is a certain emphasis on some of the less frequently studie environments and on methods of study and characterization as a means to advance th recognition and classification of these soils.

The Tenerife meeting was sponsored by a number of national and international organ zations, including the Autonomous Government of the Canary Islands, the Institute of Iber American Cooperation in Madrid, the Directorate on Scientific Policy of the Ministry of Edu cation and Science, Madrid, the International Soil Science Society, ORSTOM of France, an ICOMAND. Members and staff of the Department of Soil Science of the University of L Laguna had the actual task of organizing the meeting and the field trips. In editing the boo we benefitted from the manuscript reviews by many of our colleagues all over the world, an the capable handling and sponsorship of the CATENA VERLAG. To all those who hav extended their help we wish to express warm thanks.

La Laguna and Jerusalem,
Summer 1984

E. Fernandez Caldas
D.H. Yaalon
Editors

CONTENTS

CATENA SUPPLEMENT 15 p. 289–295 Cremlingen 1989

ASPECTS OF THE QUATERNARY IN THE TERTIARY HILLS OF BAVARIA

Horst **Strunk**, Regensburg

1 Introduction

A well-differentiated loess profile containing eight interglacial soils is being excavated in a brickearth pit in Hagelstadt, near Regensburg. The Höhenhofer gravels, the oldest known Danube gravels in this area, lie at the base of the profile. A podzol was developed on the gravels. Several paleosoils from this area can be related to accumulation terraces of the Danube so that information provided by the loess profile in Hagelstadt will be of importance in any reconstruction of the relief development during the Quaternary in the upper Danube basin.

The Hagelstadt pit lies about 15 km south of Regensburg in the area of Tertiary hills. To the northeast of Regensburg are the Jurassic limestones of the Franconian Jura and to the northwest the crystalline rocks of the Bohemian Massif. The Danube flows from west to east approximately along the boundary of the Jurassic and the Tertiary deposits. Near Regensburg, at the northernmost point of its course, it is joined by two major tributaries from the north, the Naab and the Regen. East of Regensburg, the Danube follows the Danube marginal fault which, by a throw of several hundred meters (BAUBERGER et al. 1969), has separated the sunken southern block of Mesozoic rocks that form the foothills of the Alps from the crystalline rocks to the north.

At Hagelstadt, the Jurassic and upper Cretaceous sediments dip east and southeast below younger, mainly upper Tertiary and Quaternary strata. The dip at the boundary of the Cretaceous and Tertiary deposits does not exceed 1.5 percent in this area. Nevertheless the overlying Tertiary and Quaternary strata have an average thickness of 10–15 m and 25–35 m, respectively, at the northern margin of the Tertiary hills between Regensburg and Straubing. In some areas the Pleistocene loess deposits attain 30–35 m, although few loess deposits have been examined so far. There are several exposures in brickearth pits in the area of Regensburg of which the pit at Hagelstadt is the most important that has been investigated so far. The profile comprises a sequence of seven paleosoils from interglacials that were developed by weathering on loesses covered by deciduous forest. The oldest paleosoil is of lower Pleistocene age.

2 Development of relief during the Quaternary

The oldest gravel in this area, the Höhenhofer gravels, were gravel fans of

ISSN 0722-0723
ISBN 3-923381-18-2

3-923381-18-4/89/5011851/US$ 2.00 + 0.25

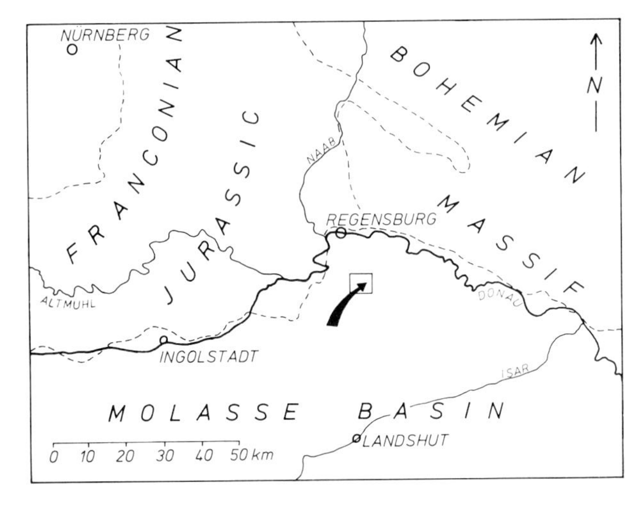

Fig. 1: *Location of study area.*

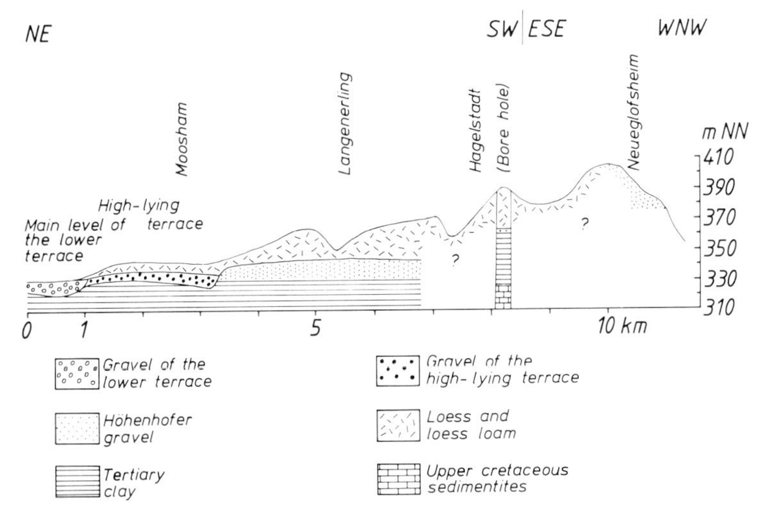

Fig. 2: *Cross-section of the Tertiary hills and Danube terraces southeast of Regensburg.*

the Danube (fig.2). They have a high quartz content and an average thickness of 15-20 m. The varying elevations of their subsurfaces indicate that the gravels were deposited as fills in an area of irregular relief. In addition to a dip of about one percent to the east due to tectonic influence at the base of the gravel, the elevations of the gravel bases vary between 330 m and 380 m above sea level. Using indicator components, the Höhenhofer gravels can be subdivided into deposits by the Danube from the west in which there are radiolarites from the Alps, and deposits of the Naab river in which there are black and light brown lydites from the Variscan Saxothuringicum in the north. These deposits are interbedded in the Hagelstadt profile.

The age of the Höhenhofer gravels

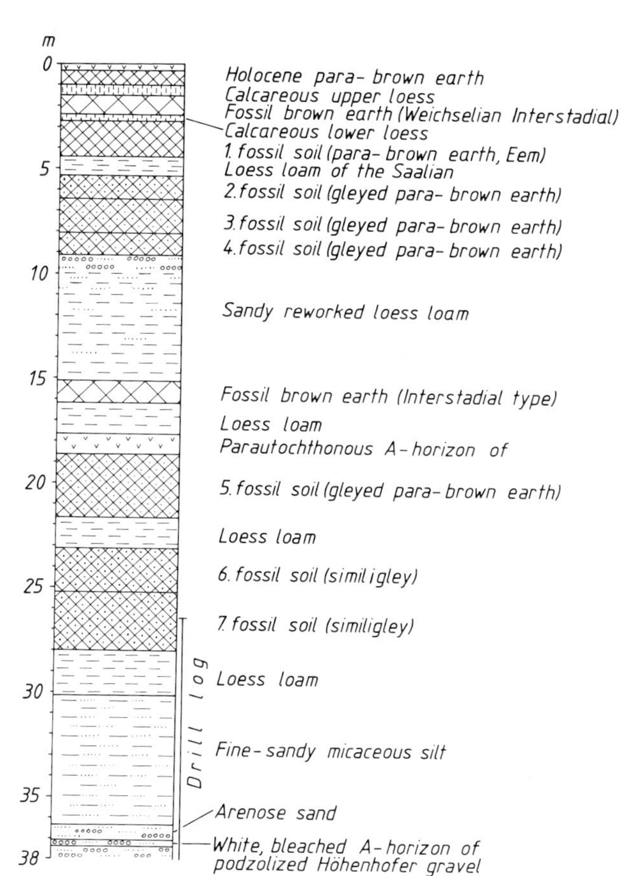

Fig. 3: *Generalized section of the average thickness of the Quaternary layers in the brickearth pit at Hagelstadt.*

is uncertain. OSCHMANN (1958) believes them to be Pliocene (Pontian). BAUBERGER et al. (1969) have divided the gravels into several deposits and suggested that they are lower Pleistocene. TILLMANNS (1977), however, divides them into a middle Pliocene and a middle to upper Pliocene deposit. At Hagelstadt, the gravels were found in a borehole at the base of the Quaternary overburden, covered by eight interglacial soils (fig.3).Consequently, the minimum age of the gravels could be more than 800,000 years BP. At Hagelstadt and also nearby at Strobel, they are overlain by two fossilized pseudogleys (gleysols). Residuals of a fossilized pseudogley that cover several deposits of Höhenhofer gravel in the area of Regensburg have been described by STÜCKL (1971, 1976). Moreover, at Strobel, which has a profile comparable to Hagelstadt, BRUNNACKER et al. (1976) found that the lowest paleosoils, 5 and 6 in that profile, were fossilized pseudogleys from the Matuyama epoch. Since the Brunhes/Matuyama boundary of the Strobel profile was below the third paleosoil, the 5th and 6th paleosoils must be older than 700,000 BP. Alternating and reverse magnetism was found in the basal layers of that profile. These oscillations represent the Jaramillo event so that both fossilized pseudogleys are from 0.9–1.2 million years old.

The oldest recognizable morphologic aggradation terrace of the Danube is the High Terrace. The top of the gravel deposit on the terrace is nearly 15 m above the present Danube (fig.2). The age of the gravel is uncertain but in some locations this High Terrace is overlain by more than 10 m of loessic cover, in which several fossilized soils have been preserved. BRUNNACKER (1957) describes three parabrown earths, formed during interglacials, that were found in a brickearth pit at Köfering southeast of Regensburg and that cover gravels at the same level and in the same stratigraphic range as the High Terrace. The gravels of the High Terrace appear to be identical with those at Köfering, in which case the terrace would date from the third-youngest (Mindel?) glaciation or earlier.

The youngest Danube terrace, the lower terrace, was accumulated during the last glacial period. Incision into the High Terrace was limited and the gravel deposits were covered by loess only on the margins. During the Pleistocene, however, the wide gravel fans of the Danube, beyond the outlet of its gorge near Regensburg, formed the blow-out area of the loesses which accumulated to thicknesses of up to 35 m and buried the older sediments. These loesses were interstratified by interstadial and interglacial paleosoils.

3 The Quaternary profile at Hagelstadt

The Hagelstadt pit is on the eastern scarp of a nearly flat plateau at 390 m to 395 m above sea level. Based on seismic measurements and a well boring, Quaternary loesses and loess loams in the area appear to be up to 26 m thick underlain by fissured Tertiary clay that is interbedded by Höhenhofer gravel. Only 16 m of this sequence have been exposed at Hagelstadt but because there is a middle Pleistocene channel fill in the eastern part of the exposure and the basal layers have been uplifted by up to 8 m on the western face of the exposure, a total average thickness of up to 38 m is visible (fig.3).

There is a Holocene parabrown earth at the top of the profile developed on Upper Weichselian loess under mixed oak forest. The unweathered Upper Weichselian loess is preserved underneath and has a calcium carbonate content of 25 percent. It is separated from the Lower Weichselian loess by a interstadial soil, a brown earth developed during the Middle Weichselian interstadial that is comparable to the Lohner soil of the Kassel Basin in Hessen (SCHÖNHALS et al.1964, ROHDENBURG & MEYER 1966), the Stillfried-B soil in the neighbouring Austrian area of the Danube (FINK 1979), and soils in Czechoslovakia (KLIMA et al. 1961). This soil has been termed the Brown Weathering Horizon in Bavaria (BRUNNACKER 1956). At the present time it is found in areas that have an average annual precipitation of less than 650 mm. If the precipitation is higher, tundra gleys or pseudogleys develop (BRUNNACKER 1957).

The loesses of the Lower Weichselian overlie the parabrown earth of the last glacial period, the A-horizon of which has been completely eroded. The two humic Lower Weichselian soils that are widespread in the Rhine-Main basin and in Franconia (BRUNNACKER 1970; SEMMEL 1968; SEMMEL & STÄBLEIN 1971) and that were developed during the interstadials of Amersfoort and Brörup, appear to be absent in the Regensburg area. The parabrown earth is separated from the loesses by a solifluction horizon of reworked loess which contains lumps of the parabrown earth that is below, indicating an erosional disconformity in the Lower Weichselian. On the other hand, the parabrown earth of the last interglacial is well preserved and has developed on top of a Saalian loess which has changed to loess loam at the base of this soil.

Beneath the loess loam of the Saalian glaciation is a complex of three fossilized gleyed parabrown earths developed during interglacials (fig.3, nos 2, 3 and 4). All are strongly eroded and are separated by clear discordances. For this reason the A-horizon of none of these soils has been preserved. This soil complex has not yet been integrated into the Pleistocene stratigraphy but TL dating and palaeomagnetic investigation is being undertaken. The Brunhes/Matuyama boundary (700,000 BP) in the Strobel profile, which can be compared to the Hagelstadt profile, is below the third fossilized soil. However, in this area the High Terrace of the Danube is older than the second paleosoil at Hagelstadt since nearby at Köfering it is overlain by three interglacial soils (fig.3 nos 2, 3 and 4) which in turn overlie a channel fill, up to 8 m thick, of reworked sandy loess loam with interbedded gravel layers and sand lenses. In the lower part of this loam, a fossilized interstadial brown earth has developed on a light yellow, loess-like silt which is preserved below the soil. The loess loam is younger than the 5th paleosoil, a bright reddish brown gleyed parabrown earth up to 4 m thick, developed from a loess loam which is preserved below.

The paleosoils 6 and 7 are pseudogleys in an unstratified sandy loam. In contrast to the overlying gleyed parabrown earths these pseudogleys show no lessivation but iron and manganese concretions and greyish reduction patches are present. Both soils developed from sandy loess loams.

The 7th paleosoil is the basal loess facies in the Hagelstadt profile. Below it a micaceous, fine-sandy greenish grey silt and a horizon of reworked arenose sand have been deposited on the Höhenhofer gravel. The Höhenhofer gravel is podzolized and has an intensively bleached white A-horizon about 20 cm thick at the top. Neither the age of the Höhenhofer gravel nor of the podzolic soil is known.

4 Profile at Neueglofsheim

A sand pit was open in 1986 near Neueglofsheim, 2 km west of Hagelstadt, in which Höhenhofer gravel is exposed in the upper zone of the excavation. The bedding of this pit is shown in fig.4. Because of tectonically induced dipping of the Tertiary layers, the top of the gravel is at 382 m above sea level, 10 m higher than in the Hagelstadt pit. It has also been podzolized to a depth of 50 cm from its surface in the Neueglofsheim sand pit. The gravel was deposited as a channel fill above a discordance that cut the bleached layer. It is overlain by alternating beds of silt and clay, possibly high flood sediments. Above is a grey clay which changes gradually to alternating laminated beds of grey, humic and brown clay horizons. The sequence seems to represent a quiescent area facies that was followed by alluviation. The upper 1.6 m of the profile was disturbed by solifluction during the Pleistocene and has alternate beds of clay and loess loam.

The profile of Neueglofsheim is impor-

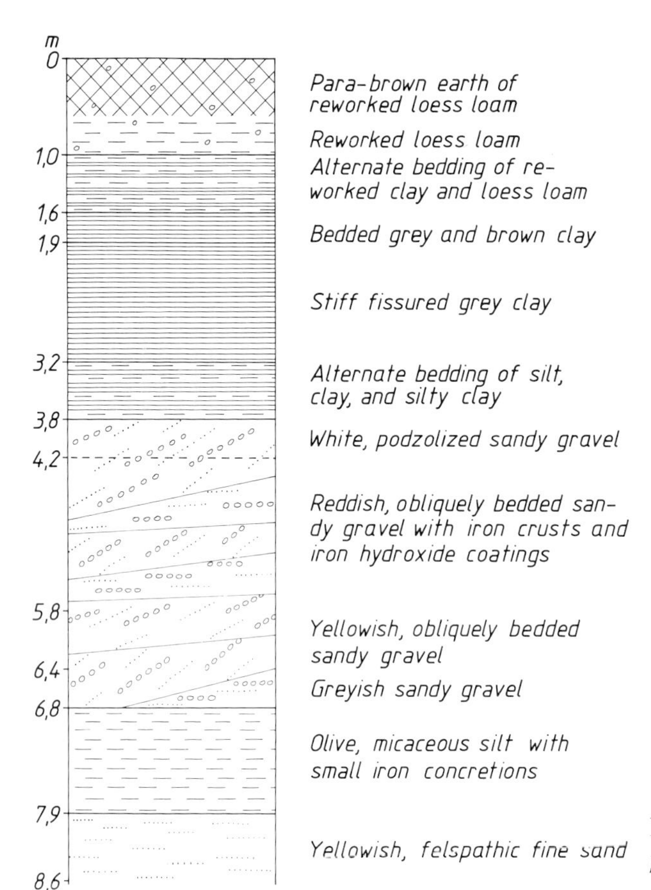

Fig. 4: *Section of the sand pit at Neueglofsheim.*

tant because of the possibility to determine the age of the podzolization on the Höhenhofer gravels by pollen analysis of its clayey overburden and thereby to fit the Hagelstadt profile into the Pliocene and Pleistocene stratigraphy of the German Danube area.

5 Conclusions

Of the estimated 17 interglacials in the Quaternary in Central Europe (FINK & KUKLA 1977), 8 interglacials are represented in the Hagelstadt pit by parabrown earths or gleyed parabrown earths. This soil sequence overlies the Höhenhofer gravels, the oldest gravel of the upper Danube area which must, therefore, be older than its overburden. The position of the paleosoils and the Höhenhofer gravels within the Quaternary is uncertain. If it is assumed that the length of the average glacial and interglacial period, including the formation of an interglacial soil, is about 100,000 years, the age of the Höhenhofer gravel at the base of this profile would be more than 800,000 years. However, the ages of the paleosoils cannot be determined because they are separated by erosional disconformities. Interglacial soils may have been destroyed by removal during each of these erosional events. Also, soils at the surface at the top of a sequence with no younger overburden may have been superimposed by the pedogenetic processes of several glacial-interglacial cycles and several interglacial periods may be represented by one interglacial soil.

Dating of the Quaternary sequence in the area of Regensburg will enable comparisons to be made with adjacent areas in Czechoslovakia (KUKLA 1975) and the Austrian Danube area (FINK 1979).

References

BAUBERGER, W., CRAMER, P. & TILLMANN, H. (1969): Erläuterungen zur Geologischen Karte von Bayern 1:25,000, Blatt Nr. 6938, Regensburg, 414 pp.

BRUNNACKER, K. (1956): Regionale Bodendifferenzierungen während der Würmeiszeit. Eiszeitalter und Gegenwart **7**, 43-48.

BRUNNACKER, K. (1957): Die Geschichte der Böden im jüngeren Pleistozän in Bayern. Geologica Bavarica **34**, 1-94.

BRUNNACKER, K. (1970): Zwei Lößprofile extremer Klimabereiche Bayerns. Geologica Bavarica **63**, 195–206.

BRUNNACKER, K. (1982): Äolische Deckschichten und deren fossile Böden im Periglazialbereich Bayerns. Geologisches Jahrbuch F 14, 15–25.

BRUNNACKER, K., BOENIGK, W., KOCI, A. & TILLMANNS, W. (1976): Die Matuyama/Brunhes-Grenze am Rhein und an der Donau. Neues Jahrbuch für Geologie und Paläontologie, Abh. 151, 358–378.

FINK, J. (1979): Stand und Aufgaben der österreichischen Quartärforschung. Innsbrucker Geographische Studien **5**, (Leidlmair-Festschrift), 79–104.

FINK, J. & KUKLA, G.J. (1977): Pleistocene climates in Central Europe: at least 17 interglacials after the Olduvai event. Quaternary Research **7**, 363–371.

KLIMA, B., KUKLA, J., LOZEK, V. & de VRIES, H.(1962): Stratigraphie des Pleistozäns und Alter des paläolithischen Rastplatzes in der Ziegelei von Dolni Vestonice (Unter-Wisternitz), Anthropozoikum **11**, 93–145.

KUKLA, G.J. (1975): Loess stratigraphy of Central Europe. In: BUTZER, K.W. & ISAAC, G.L. (eds): After the Australopithecines, Mouton, Paris, 99–188.

OSCHMANN F. (1958): Erläuterungen zur Geologischen Karte von Bayern 1:25,000, Blatt Nr. 7038 Bad Abbach, 184 pp.

ROHDENBURG, H & MEYER, B. (1966): Zur Feinstratigraphie und Paläopedologie des Jungpleistozäns nach Untersuchungen an südniedersächsischen und nordhessischen Lössprofilen. Mitteilungen der Deutschen Bodenkundlichen Gesellschaft **5**, 1–135.

SCHÖNHALS, E., ROHDENBURG, H. & SEMMEL, A. (1964): Ergebnisse neuerer Untersuchungen zur Würmlöss-Gliederung in Hessen. Eiszeitalter und Gegenwart **15**, 199–206.

SEMMEL, A. (1968): Studien über den Verlauf jungpleistozäner Formung in Hessen. Frankfurter Geographische Hefte **45**, 1–133.

SEMMEL, A. & STÄBLEIN, G. (1971): Zur Entwicklung quartärer Hohlformen in Franken. Eiszeitalter und Gegenwart **22**, 23–24.

STÜCKL, E. (1971): Marmorierter Pseudogley als fossile Bodenbildung im Süden Regensburgs. Acta Albertina Ratisbonensia **31**, 151–164.

STÜCKL, E. (1976): Relikte der Pseudogley-Landoberfläche bei Regensburg. Geologische Blätter für Nordost-Bayern **26**, 105–116.

TILLMANNS, W. (1977): Zur Geschichte von Urmain und Urdonau zwischen Bamberg, Neuburg/Donau und Regensburg. Sonderveröffentlichungen des Geologischen Instituts der Universität zu Köln **30**, 1–198.

Address of author:
Dr. Horst Strunk
Institut für Geographie der Universität Regensburg
Universitätsstraße 31
D-8400 Regensburg
Federal Republic of Germany

M. Pécsi (Editor)

LOESS AND ENVIRONMENT

SPECIAL ISSUE ON THE OCCASION OF THE XII th International Congress of the INTERNATIONAL UNION OF QUATERNARY RESEARCH (INQUA) Ottawa 1987

CATENA SUPPLEMENT 9
160 pages / hardcover / price DM 128,— / US $ 75.—

Date of publication: July 15, 1987 ORDER NO. 499/00 108
ISSN 0722-0723/ISBN 3-923381-08-5

CONTENTS

CATENA SUPPLEMENT 15 p. 297–309 Cremlingen 1989

SPATIAL DIFFERENCES OF SOLUTE LOAD OUTPUT IN "MIDDLE BAVARIA"

Robert **Lang**, Lappersdorf

1 Introduction

The analysis of solute load, suspended load and bed load in different catchments is a field of research common to hydrology, geomorphology and geoecology. The solute load output is of geomorphological interest as an indicator of weathering, linear erosion and regional denudation. Runoff data (in m^3/s) and the concentration of solute and suspended load (in g/m^3) measured at stream gauging stations are often converted to $l/s{\cdot}km^2$ or $g/s{\cdot}km^2$ and $t/km^2{\cdot}a$, by dividing them by the area of the catchment. Such values are meaningful only for homogeneous small basins. In heterogeneous basins the spatial variation has to be taken into account by using a more differentiated approach. An important question is "where should the measurements be made and at how many sites, and when in terms of the timing and frequency of data collection" (WALLING 1987). In this paper, a method of assessing the spatial variation of the solute loads of different subcatchments is presented, using a catchment in the Tertiary Hills of Bavaria as an example.

ISSN 0722-0723
ISBN 3-923381-18-2

D–3302 Cremlingen-Destedt, W. Germany
3-923381-18-4/89/5011851/US$ 2.00 + 0.25

2 Theoretical considerations

Geographical statements about an area must be related to that area's size. The methodological and practical problems of the size of areal units are of central importance in geoecology, particularly in the study of the turnover of elements and energy. LESER (1984) and LANG (1984) and, with special reference to hydrology, LEIBUNDGUT (1984) discuss the relationship between homogeneous and heterogeneous areas, the problems concerning the transferability of site-related data into the chorological dimension and also the disaggregation of chorological data to subregions.

Runoff and solute transport of a river basin are examples of a process output which is expressed as spatial information. These data, related to the basin area (output/area), are not average values but total values (integrals) for the whole area.

Only in a relatively homogeneous catchment is the runoff or load value significant as an average value; it has been used for this purpose in many publications. This is, however, only an approximation. Most catchments are heterogeneous, composed of subcatchments with differing relief, soils, rocks, vegetation and land uses. From such areas, varied solute, suspended and bed load supplies are to be expected. The recording of the

Fig. 1: *Water quality measuring stations in "Middle Bavaria".*

runoff and the sampling of the stream load has, therefore, to be based on an areally differentiated approach which allows conclusions to be drawn from the data of the chorological unit that can be applied to the subareas. Decisive is the number and location of measuring stations within the catchment (see also WALLING 1987).

The stream network can be used as the basis for the subdivisions of drainage basins in which the relationships between the amount of solute and the runoff can change slowly or rapidly. Runoff and load can be measured wherever the stream network branches or where inflows occur. A hierarchical ranking of the subcatchments is possible for investigations of large areas but for investigations of small areas it is "impossible from an operational standpoint" (WALLING 1987). In the analysis of a large area every drainage basin can be subdivided into subcatchments which then serve as the basic areal units.

Even the smallest subcatchments are not homogeneous geoecological units but consist of sub-units termed physiotopes or ecotopes which have specific turnovers of elements, water and energy and for which solute output should be evaluated. This involves the measurement of vertical and lateral movements of water (surface runoff, interflow and,

Spring	conductivity μS (25°C)	pH	Ca mg/l	K mg/l	Na mg/l	Mg mg/l	Cl mg/l	CO_3 mg/l	NO_3 mg/l	SO_4 mg/l
Quaternary										
Irler Augraben	933.4	7.8	136.1	1.3	10.8	26.8	64.1	354.0	37.0	75.0
Seegraben	1027.7	7.6	153.8	4.0	18.9	30.3	85.9	396.6	5.0	90.0
Barbinger Augraben	1010.4	7.2	134.8	2.4	44.2	35.6	97.7	368.3	31.0	70.0
Tertiary										
Höllquelle (Pfatter)	721.6	7.5	118.0	1.0	6.4	31.4	56.0	262.4	22.9	35.0
Aumühlquelle 1 (Pfa.)	680.1	7.8	93.6	0.9	8.2	28.3	49.1	251.9	37.0	10.2
Schanzquelle (Pfa.)	619.6	8.1	83.4	0.5	6.0	28.3	41.3	274.6	8.0	19.0
Teichquelle Paring	787.6	7.6	109.1	2.0	4.9	32.0	45.3	282.2	27.0	26.0
Tenacker (Aub.)	961.2	7.4	112.5	2.6	24.0	32.5	136.3	286.3	14.9	40.0
Augrabenquelle (Aub.)	1105.2	7.7	170.3	1.2	31.5	34.8	180.7	345.1	1.9	56.0
Schwefelquelle 1 (Aub.)	710.7	7.4	70.4	6.0	25.3	33.7	26.9	368.1	0.4	5.0
Mesozoic										
Irlbrunnerbachl	313.4	7.8	40.3	0.8	3.5	8.2	8.3	102.4	1.9	17.0
Quelle Essing	539.2	7.8	79.0	1.2	4.1	4.6	14.1	240.2	23.0	3.2
Hauptquelle Oberf.	527.1	7.5	63.1	3.1	16.8	19.5	17.4	262.4	2.0	19.8
Bachmühlbach	595.1	7.7	80.6	0.9	3.3	17.9	12.8	304.8	13.1	1.5
Oberfecking 2	506.4	7.6	59.2	2.9	14.7	12.5	14.4	228.8	2.5	18.4
Paleozoic										
Steinseige (Otterb.)	154.7	6.8	7.5	1.7	6.9	2.4	5.3	18.8	12.3	10.2
Steinbrünnl (Otterb.)	147.6	6.9	4.8	1.0	9.6	3.7	5.7	18.0	8.0	18.2
Diebsgraben (Otterb.)	147.3	7.0	7.2	1.8	6.4	3.6	5.2	13.1	12.2	23.5
Birnmahdgraben (Ellb.)	172.7	7.1	11.9	1.8	6.8	4.7	5.6	24.4	9.5	40.5
Gänsgraben (Ellb.)	240.0	7.0	12.3	2.7	12.3	4.2	10.3	21.4	32.0	22.8

Tab. 1: *Cation and anion concentrations of representative springs in "Middle" Bavaria (see fig.1).*

if possible, groundwater flow) and the related transport of elements at instrumented test plots.

This important type of research has been carried out by the physiography and geoecology research group in Basel (SEILER 1983, MOSIMANN 1985) and in Braunschweig (ROHDENBURG et al. 1983). The author was only able to work at the level of the smallest catchments that could be distinguished on the basis of the stream network.

3 The study area and the measuring programme

In this study the runoff and solute load of several small catchments and subcatchments in the hills on Tertiary sediments (Molasse) south of Regensburg are investigated (fig.1). Additional water samples were taken at selected springs and gauging stations in the area of the Quaternary Danube terraces, on the Mesozoic rocks of the Franconian Jura and in the upland of Paleozoic rocks northeast of Regensburg, for a comparison of water quality in the areas.

The results of this comparison are

shown in tab.1. The individual regions have a wide range of cation and anion values. The highest SO_4 values are in samples from springs in the Quaternary deposits; these also have high conductivity and total hardness values.

The springs show a much greater variation within the Tertiary deposits than in other areas. However, the mean values of conductivity and of the solute concentration are exceeded in the Quaternary materials. The springs in the area of Mesozoic rocks have lower values and those in the Paleozoic rocks of the Bavarian Forest the lowest.

The Aubach catchment in the Tertiary Hills was selected for an analysis of the spatial differences of solute load in subcatchments because its springs showed the greatest variation (tab.1).

The measuring stations (fig.4) were placed in such a way that hierarchical sampling could be carried out, taking all stream orders, all springs and all manmade groundwater exits into account.

The investigations were carried out during the hydrological years 1984, 1985 and 1986. In addition to the available official limnograph, discontinuous flow gaugings were made at springs and in streams. Water samples were taken weekly at the main measuring stations, fortnightly at the subordinate stations and monthly at the springs.

4 Classification of the subcatchments

Any analysis of runoff and solute load in a large catchment must distinguish between subcatchments in which the runoff is essentially natural in origin and the solute load is derived entirely from chemical weathering and denudation, and subcatchments where some direct inflows and part of the solute load are the result of man-made inputs into the system. In particular, it is necessary "to isolate the nondenudational component" (PETTS & FOSTER 1985).

Tab.2 lists 25 springs lying within the 21.5 km^2 Aubach catchment which is situated mainly in Tertiary deposits. Substantial variation in the value for cations and anions and, therefore, also for the sum of the solute contents occur, indicating that the water quality varies. The springs are grouped into classes (SYMADER 1980). A principal component analysis of the 25 spring waters was made in order to reduce the number of variables. The four factors obtained helped considerably to explain the total variation of 88%. The first factor, with a variance explanation of 39.5%, had high positive loadings of calcium (0.90), chloride (0.85) and sulphate (0.89), the second, with a variance explanation of 21%, had positive loadings of nitrate (0.79) and H^+-ion concentration, the third (14.4%) of natrium (0.79) and potassium (0.80) and the fourth (13.2%) only of magnesium. The calcium, nitrate, natrium and magnesium variables were standardized and then used for cluster analysis which permitted the classification of springs and made it possible to deduce natural and perhaps also human influence. The resulting dendrogram is shown in fig.2.

At a threshold value of 40% of the error-square increase as a percentage of the greatest increase, five different groups of springs are distinguished.

Fig.3 illustrates the classification of the springs in relation to the geology of the catchment. One group consists of the four sulphur springs (S1, S2, S3, S4), the water of which has a very long residence

	Spring	conductivity μS (25°C)	pH	Ca mg/l	K mg/l	Na mg/l	Mg mg/l	Cl mg/l	HCO_3 mg/l	NO_3 mg/l	SO_4 mg/l
Ma	(Marienholz)	402.0	7.7	53.0	0.4	4.6	12.2	17.5	164.7	6.2	20.0
Ui1	(Unterisling 1)	725.3	7.4	100.2	1.0	4.2	22.8	42.5	268.3	37.5	33.0
Ui3	(Unterisling 3)	541.9	7.4	78.1	0.9	4.4	20.2	50.2	230.9	31.0	32.0
Au	(In der Au)	652.0	7.1	99.8	1.6	4.5	23.0	42.0	274.6	24.8	28.0
Oi3	(Oberisling 3)	906.0	7.0	126.2	2.7	4.7	37.3	92.5	335.2	37.0	38.0
Ui2	(Unterisling 2)	749.8	7.4	106.6	0.9	4.7	24.7	56.0	277.6	35.0	27.9
Oi2	(Oberisling 2)	619.4	7.3	89.3	0.7	4.7	19.0	33.2	274.6	17.0	30.0
Oi5	(Oberisling 5)	678.5	7.2	102.8	0.5	4.8	15.8	43.1	271.6	31.0	26.0
Ja	(Jagdhaus)	750.0	7.2	129.0	0.8	5.1	9.0	49.0	292.3	20.0	37.0
Po	(Posthof)	530.0	6.8	65.0	0.9	6.0	19.4	35.5	202.0	43.0	14.0
Neu	(Neudorf)	705.0	6.8	90.0	0.8	6.5	29.0	78.0	244.0	42.0	18.0
Ohi	(Oberhinkofen)	630.0	7.2	108.0	0.9	7.3	19.0	52.0	277.0	24.8	32.0
Sch2	(Scharmassing 2)	480.0	6.5	52.1	0.7	10.0	17.0	46.0	152.0	37.0	5.0
Sch1	(Scharmassing 1)	615.0	7.3	68.0	3.3	12.0	20.0	67.4	201.4	37.0	5.0
Wie	(Wiesgraben)	948.0	7.7	122.0	1.1	15.0	35.0	80.0	364.0	49.0	24.0
Oi4	(Oberisling 4)	798.4	7.5	109.5	13.3	12.8	23.2	58.1	329.5	32.0	25.0
Oi1	(Oberisling 1)	836.6	7.1	114.8	1.7	25.6	24.0	53.0	353.6	31.0	38.0
Ten	(Tenacker)	961.2	7.4	112.5	2.6	24.0	32.5	136.3	286.3	14.9	40.0
Hog	(Hohengebrach)	1105.2	7.7	170.3	1.2	31.5	34.8	180.7	345.1	1.9	56.0
Hö	(Hölkering)	945.0	7.3	149.0	2.0	24.0	16.0	106.0	347.0	18.0	49.0
Ui4	(Unterisling 4)	698.5	7.3	102.6	1.4	5.2	28.8	39.7	359.5	6.1	28.0
S1	(Schwefelquelle 1)	710.7	7.4	70.4	6.0	25.3	33.7	26.9	368.1	0.4	5.0
S2	(Schwefelquelle 2)	837.7	7.5	85.3	5.4	41.6	30.0	63.3	396.6	0.4	4.0
S3	(Schwefelquelle 3)	907.6	7.3	102.2	4.5	43.4	33.9	70.7	422.0	0.4	4.7
S4	(schwefelquelle 4)	838.0	7.1	93.0	5.5	41.0	32.0	60.0	407.0	1.8	2.2

Tab. 2: *Cation and anion concentrations of the springs in the Aubach catchment.*

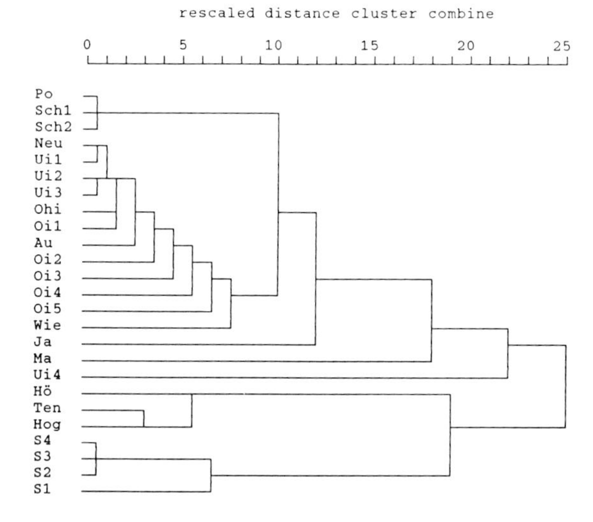

Fig. 2: *Dendrogram of the hydrochemical cluster analysis for the Aubach catchment.*

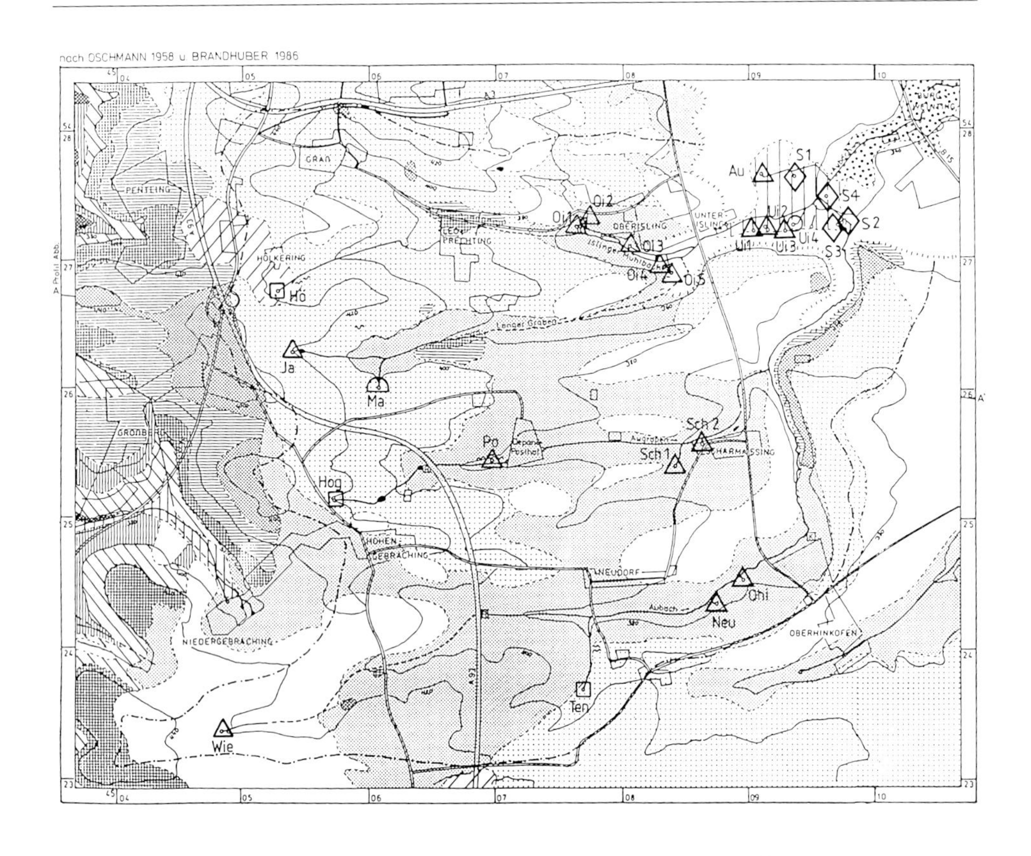

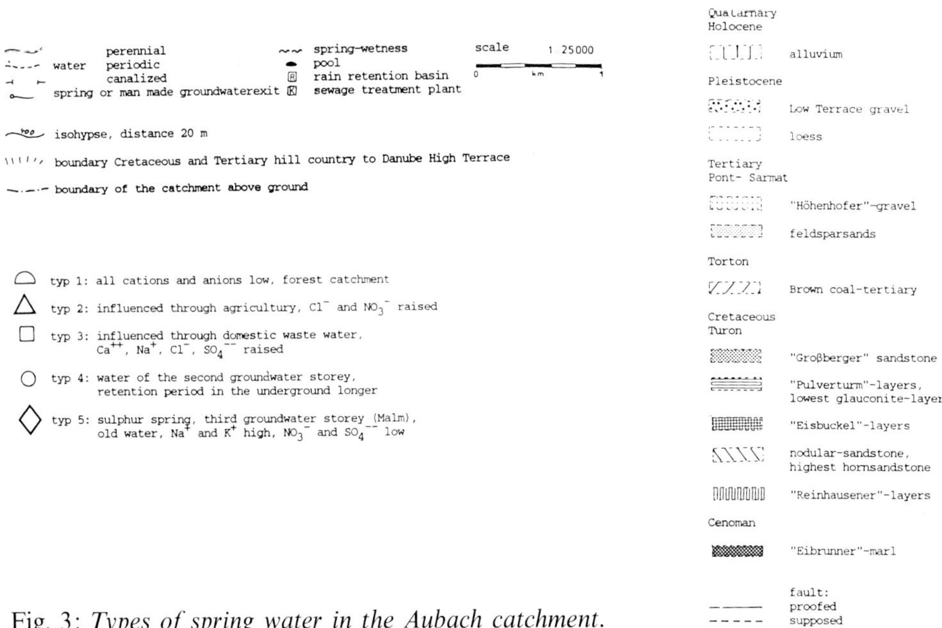

Fig. 3: *Types of spring water in the Aubach catchment.*

time in the aquifer. Their discharges and, therefore, also their solute contents are ignored when making water and element balances because only the internal water is used.

The Marienholz spring forms a "group" on its own in the cluster analysis because it has the lowest values for all ions. Its catchment area is entirely covered by woodland and its values can be compared to those of springs influenced by human activity. The Unterisling 4 spring (Ui4) also forms a group on its own because of its very low nitrate content and high magnesium content. The water of this spring comes from a second aquifer (permeable layers over Eibrunner marl) and its catchment extends beyond the limits of the Aubach catchment. As a result, its flow is very large: 2.2–2.6 l/s during groundwater discharge.

All other springs are strongly affected by human activity. Increased chloride and nitrate values indicate agricultural influence. Two springs, Hohengebraching and Hölkering, are grouped together because they are influenced by waste water and show very high calcium, natrium, chloride and sulphate concentrations. Similar analyses have to be made for the streams (fig.4) in order to assess other influences on the water system.

It was possible to localize the influence of human activity fairly precisely and to define it qualitatively and quantitatively. Substantial inflows resulted from the salt input during the winter months following the salting of roads, from the waste water of a potato distillery which affects part of the Aubach during the autumn distilling and from diffuse agricultural effluents and a sewage treatment plant which cleans only mechanically (LANG & BRANDHUBER 1988).

Human activity influenced the total water system of Aubach very strongly and the lower part of the Augraben strongly. The Islinger Mühlbach was influenced only moderately and the Langer Graben hardly at all. In late September 1985, during a period of low flow investigated for spatial differentiation, only the influence of the distillery's waste water was observed.

5 Spatial variation of runoff, concentration and load in the Aubach

In late September 1985, the flow, concentration and load of all springs and streams were measured, using BRANDHUBER's (1986) flow values. The hydrograph is separated so that the runoff contains only groundwater flow and from the measured flow value a daily average value is calculated. Tab.3 lists the flow, solute concentration and the load data of all springs and other measuring stations (fig.4).

Considerable differences in discharge exist within the catchment. The Islinger Mühlbach contributes 30% to the total runoff, the sulphur ditch and the Upper Aubach 20% each. The remaining 30% is contributed by the large number of springs in the transition zone between the Quaternary and Tertiary areas and the groundwater inflows in the Quaternary. In order to obtain the area water balance in the catchment, the discharge of all sulphur springs and the Unterisling 4 (Uis 4) spring have to be subtracted from the catchment runoff because they are fed from outside the river basin. Also, two measuring stations, Langer Graben embouchure and Islinger Mühlbach above Oberisling, which are at the exits of larger subcatchments,

gauging station / *spring*	area km^2	runoff m/s	solute concentration mg/l	load g/s	load per area g/s*km^2
Aubach (Burgweinting)	21.50	39.00	447.0	17.433	0.811
Schwefelgraben	1.00	8.31	402.3	3.343	3.343
Au	0.75	4.09	397.8	1.627	2.169
Schwefelquelle 1	—	3.78	375.5	1.419	—
Islinger Mühlbach	6.51	11.58	460.4	5.331	0.819
Oberisling 1+2	0.30	0.23	478.3	0.110	0.367
Oberisling 3	0.20	2.42	576.6	1.395	6.975
Islinger Mühlb. v. Ois	3.08	—	—	—	—
Hölkeringer Einzugsg.	0.73	1.91	648.1	1.238	1.696
Hölkering	0.22	0.20	737.5	0.147	0.668
Langer Graben Mündung	2.42	—	—	—	—
Oberisling 4+5	?2.42	2.89	469.4	1.357	4.112
Langer Graben früher	1.07	1.79	330.8	0.592	0.553
Marienholz	0.23	1.29	236.9	0.306	1.330
Marienholz	0.12	1.09	223.5	0.244	2.033
Jagdhaus	0.21	0.31	384.4	0.119	0.567
Unterislinger Quellen	0.22	0.51	—	—	—
Unterisling 1	0.07	0.12	505.1	0.061	0.871
Unterisling 2	0.07	0.20	518.5	0.104	1.486
Unterisling 3	0.08	0.20	406.8	0.081	1.012
Unterisling 4	—	2.54	420.2	1.067	—
Schwefelquelle 2	—	0.20	388.9	0.078	—
Schwefelquelle 3	—	0.04	531.9	0.021	—
Schwefelquelle 4	—	0.08	438.1	0.035	—
Aubach Weintinger Hölzl	12.62	8.35	540.9	4.517	0.358
Aubach Scharmassing	6.61	5.58	639.2	3.567	0.540
Aubach Oberhinkofen	5.39	3.04	1332.1	4.050	0.751
Oberhinkofen	0.08	0.04	460.4	0.018	0.225
Neudorf	0.06	0.51	442.5	0.226	3.767
Zulauf Tenacker	0.90	0.86	460.4	0.396	0.440
Tenacker	0.21	0.55	605.9	0.334	1.590
Aubach Tenacker	3.13	0.62	4318.0	2.677	0.855
Wiesgraben	0.20	0.04	630.3	0.025	0.125
Augraben Scharmassing	4.49	2.42	415.7	1.006	0.224
Scharmassing 2	0.13	0.08	304.0	0.024	0.185
Scharmassing 1	1.02	0.08	371.0	0.030	0.029
Augraben Posthof	1.79	2.11	335.2	0.707	0.395
Posthof	0.16	0.62	339.7	0.211	1.319
Hohengebraching	0.09	0.31	818.0	0.254	2.822
Aubach Burgweinting without extraneous water	21.50	32.40	456.0	14.774	0.687
Aubach Burgweinting without e.w. and reduced	21.50	32.40	442.0	14.308	0.665

Tab. 3: *Runoff, solute concentration, solute load and areal-related solute load of all gauging stations and springs in the Aubach catchment (daily average for 25.9.1985).*

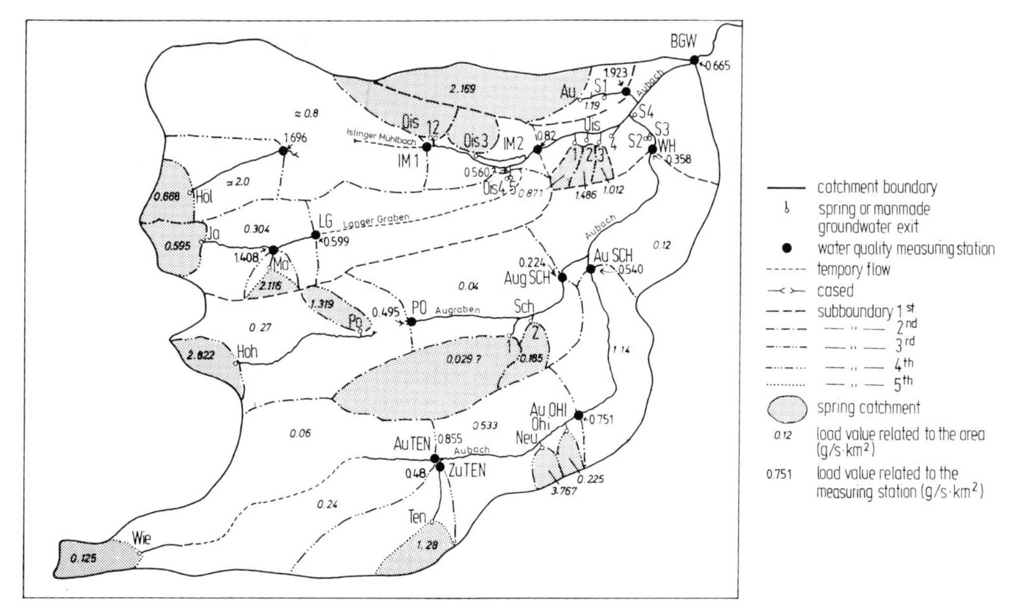

Fig. 4: *Solute load values of the Aubach catchment related to the area and to the measuring station (daily average value September 1985).*

are without water in the autumn during the period of groundwater flow because the water which flows in their upper course infiltrates in the middle course before reaching the stations. In addition, the water which infiltrates in the subcatchment Langer Graben probably reemerges in the man-made groundwater exits (Oberisling 4 and 5) near the embouchure of the Langer Graben.

The solute concentrations of the streams are very high in the Aubach subcatchment at the Tenacker, Oberhinkofen and Scharmassing measuring stations at the beginning of September because of the industrial waste water input from the potato distillery. The solute concentrations measured at all other periods at these stations at low water flow is much lower.

The load values (g/s) assigned to the individual springs and streams and the load values (g/s·km^2) related to the areas vary considerably. A value of 0.811 g/s·km^2 has been recorded at the Burgweinting station which is valid as the average for the whole catchment. If the springs fed externally are omitted the value is 0.687 g/s·km^2, and if the waste water from the distillery is also omitted, the value is 0.665 g/s·km^2. The two extreme values in the Aubach catchment are due to exceptional circumstances. The basin of the Scharmassing spring 2 which has the lowest value is very small and the large delivery at the Oberisling 3 spring may be the result of garden watering in a suburban area or, perhaps, of a larger subterranean catchment.

Fig.4 shows the solute load output of the Aubach catchment for a day with low water flow. These values are related both to the springs and streams and to the subareas. There are clear differences

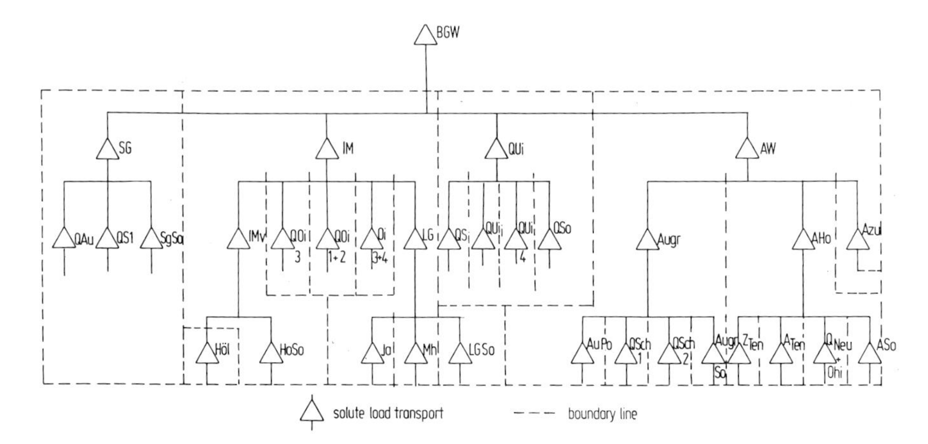

Fig. 5: *Flow diagram of the solute load in the Aubach catchment.*

between the subareas which would be obscured in the sum value at the main measuring station of the catchment. This demonstrates the doubtful significance of the "average" values that are often used for entire catchments.

In the other river basins in the Tertiary and on the Paleozoic basement complex northeast of Regensburg monitored (fig.1), the conditions were similar although differences between catchments were generally smaller in the basement complex.

6 Annual solute load output

The evaluation of annual load output related to the drainage areas is more difficult because the discharge could only sometimes be measured at all stations. There is only one limnograph in the Aubach catchment so that the total annual output of solute elements had to be determined first. The chorological data was then disaggregated. There is no clear connection between runoff and solute concentration; instead there are annual and storm-period hystereses. The complexity, causes and controls of hysteresis and its geomorphological significance have been shown in detail by WEBB, DAVIS & KELLER (1987). Because an automatic sampler was not available to measure at regular intervals and in all storm periods, a weekly measuring interval was chosen. The runoff process was divided into 5 runoff phases. The measured solute concentrations were related to the corresponding runoff phases: the rising limb, the peak flow, the steeply descending limb, the weakly inclined descending limb and the terminal limb.

The relationship between concentration and runoff was significant only for the peak flows and these only for November, December, January, May, June and July. The data obtained were used to calculate the annual sum of solute output. Average values of the concentration were used for the other months to that the annual value is approximate (RAUCH 1982, SCHRIMPF 1983). As a control, the discharge was measured three times during the year with a current meter at all stations. The solute concentration was evaluated weekly at the receiving streams, fortnightly at the springs that were easy to reach and monthly at the re-

gauging station / *spring*	area km^2 %	runoff m^3 %	solute concentration mg/l	load t %	load per area t/km^2
Aubach (Burgweinting)	21.50	2624832	475	1247.51	58.02
Schwefelgraben	4.7	14.00	82.	11.49	247.00
Au	3.5	6.04	82.	4.94	142.0
Schwefelquelle 1	—	5.04	79.	4.00	—
Islinger Mühlbach	30.3	30.00	104.	31.25	103.0
Oberisling 1+2	1.4	0.43	100.	0.43	31.0
Oberisling 3	0.9	4.66	118.	5.49	590.0
Islinger Mühlb. v. Ois	14.3	7.94	133.	10.58	74.0
Hölkeringer Einzugsg.	3.4	4.76	130.	6.19	182.0
Hölkering	1.0	0.04	150.	0.53	52.0
Langer Graben Mündung	11.3	2.03	73.	1.47	
Oberisling	+	4.79	99.	4.73	55.0
Langer Graben früher	5.0	3.75	65.	2.44	49.0
Marienholz	1.1	2.22	52.	1.16	109.0
Marienholz	0.6	1.46	48.	0.70	126.0
Jagdhaus	1.0	0.49	87.	0.42	44.0
Aubach Weintinger Hölzl	58.7	36.08	127.	46.79	80.0
Aubach Scharmassing	30.7	23.05	111.	25.45	83.0
Aubach Oberhinkofen	25.1	17.98	119.	21.40	85.0
Oberhinkofen	0.4	0.16	94.	0.15	42.0
Neudorf	0.3	0.88	91.	0.80	285.0
Zulauf Tenacker	4.2	5.88	101.	5.97	142.0
Tenacker	1.0	1.08	113.	1.22	125.0
Aubach Tenacker	14.6	6.20	144.	8.90	61.0
Wiesgraben	0.9	0.17	131.	0.22	23.0
Augraben Scharmassing	20.9	11.03	164.	18.85	86.0
Scharmassing 2	0.6	0.33	65.	0.21	35.0
Scharmassing 1	4.7	0.54	77.	0.42	8.9
Augraben Posthof	8.3	6.87	156.	10.72	128.0
Posthof	0.7	1.12	71.	0.79	107.0
Hohengebraching	0.4	0.57	171.	1 0.99	237.0

Tab. 4: *Annual sum of runoff, solute concentration, solute load and areal-related solute load of gauging stations and springs in the Aubach catchment in relation to the base station (hydrological year 1985).*

mote springs. All the values are relative percentages of the quantities measured at the gauging station.

With the help of a multi-stage mixing model based on the input-output flow diagram in fig.5, the discharge was calculated on the basis of the known concentration for individual measuring stations. The load values of a catchment of nth order are equal to the sum of the load values of the subcatchments of (n+1)th order, the total basin being regarded as an area of zero order. The calculation uses an inhomogeneous linear equation system with n unknown discharge values q1.....qn of the subcatchments and the known concentrations c11.....cmn of the respective measuring stations and as the discharge q and the concentrations C1...Cm of the superior measuring station.

$$c_{11}q_1 + \ldots + c_{1n}q_n = C_1q$$
$$\vdots \qquad \vdots$$
$$c_{m1}q_1 + \ldots + c_{mn}q_n = C_mq$$

A definite solution is obtained if the number of linear independent runoff concentration equations equals the number of unknown discharge values. The independent variables were obtained through principal component analysis, which was made in connection with the cluster analysis of the springs and receiving streams. In the calculations, n was at no time greater than 5.

After the calculation of the discharge values the load values of all stations were determined with the help of the known concentration data for all measuring dates. The evaluations were completed as relative values in respect of the value of the total river basin. Subsequently, the annual solute load was determined, also relative to the base station at Burgweinting. The results for the hydrological year 1985 are shown in tab.4 as relative values to the base station. They have not been adjusted to allow for anthropogenic inputs and for wet and dry atmospheric fallout. Only after such adjustment would it be possible to show a balance of the input and output of the solute elements within the catchment. The wide range of values in tab.4 shows the heterogeneity of the river basin with regard to the different flow and concentration values of the subcatchments.

Very similar values of solute output per areal unit were obtained for the catchments in the Paleozoic basement complex because the discharges per unit area are about twice as high and the concentration values about half as high as in the Aubach catchment.

7 Suggestions for the aggregation and disaggregation of solute data of landscape units

The example presented deals with the problem of obtaining accurate areal-related data for the subareas in place of undifferentiated total value for a whole catchment.

It has been shown with reference to the solute output value per unit area that both the disaggregation of the total catchment value and the aggregation of subareal values are possible if precise values for discharge and solute concentration exist for the subareas.

A multi-stage mixing model was applied in which, with the help of inhomogeneous linear equation systems, the station runoffs were calculated that were unknown at the respective level and in the monitored catchment.

The application of algorithms for the error of estimation could improve the method further or perhaps serve as an alternative; this has been done by STREIT (1984) for areal and subareal precipitation. On the other hand, the application of the principles of geoecological description could be helpful in this case.

Mapping of the geoecological elements for the whole area and the subsequent delimitation of homogeneous areal units such as pedotopes and hydrotopes as well as integrated physiotopes can provide the basis for the evaluation of the spatial element turnover within the subareas. Digital charts and data processing would clearly be helpful for all these applications.

References

BRANDHUBER, R. (1986): Untersuchungen

zur geogenen und anthropogenen Stoffbelastung kleiner Fließgewässer. Diploma Dissertation, University of Regensburg, unpublished.

LANG, R. (1984): Probleme bei der zeitlichen und räumlichen Aggregierung topologischer Daten. Geomethodica **9**, 67–114.

LANG, R. & BRANDHUBER, R. (1988): Der Einfluß von Landwirtschaft, Verkehr, Brennereien und Kanalisation auf die Gewässergüte eines Einzugsgebietes — der Aubach südlich von Regensburg. Regensburger geogr. Schriften, in preparation.

LEIBUNDGUT, CH. (1984): Zur Erfassung hydrologischer Meßwerte und deren Übertragung auf Einzugsgebiete verschiedener Dimensionen. Geomethodica **9**, 141–170.

LESER, H. (1984): Das neunte "Basler Geomethodische Colloquium": Umsatzmessungs- und Bilanzierungsproblem bei topologischen Ökosystemforschungen. Geomethodica **9**, 5–29.

MOSIMANN, T. (1985): Untersuchungen zur Funktion subarktischer und alpiner Geoökosysteme (Finnmark (Norwegen) und Schweizer Alpen). Physiogeographica **7**, Basel.

PETTS, G. & FOSTER, I. (1985): Rivers and Landscape. Arnold, London, 274 pp.

RAUSCH, R. (1982): Wasserhaushalt, Feststoff- und Lösungsaustrag im Einzugsgebiet der Aich. Dissertation, University of Tübingen.

ROHDENBURG, H. (1983): Forschergruppe "Wasser- und Stoffhaushalt landwirtschaftlich genutzter Einzugsgebiete unter besonderer Berücksichtigung von Substrataufbau, Relief und Nutzungsform". Tischvorlagenband für das wissenschaftliche Kolloquium am 17. und 18.11.1983 in Braunschweig. Braunschweig.

SCHRIMPF, E. (1983): Kreisläufe und Bilanzen von ausgewählten Umweltgiften in Niederschlagsgebieten Nordostbayerns. Forschungsbericht für das Umweltbundesamt, UFOPLAN-Nr 83-106 07 024, Bayreuth.

SEILER, W. (1983): Bodenwasser- und Nährstoffhaushalt unter Einfluß der rezenten Bodenerosion am Beispiel zweier kleiner Einzugsgebiete im Basler Tafeljura bei Rothenfluß und Anwil. Physiogeographica **5**, Basel.

STREIT, U. (1984): Angleich-Verfahren für geschätzte Teilgebiets-Niederschläge bei bekanntem Gesamtgebietsniederschlag. Deutsche Gewässerkundliche Mitteilungen **28**, 11–16.

SYMADER, W. (1980): Zur Problematik landschaftsökologischer Raumgliederungen. Landschaft + Stadt **12**, 81–89.

WALLING, D.E. (1987): Physical hydrology. Progress in Physical Geography **11, 3**, 112–120.

WEBB, B.W., DAVIS, J.S. & KELLER, H.M. (1987): Hysteresis in stream solute behaviour. International Geomorphology 1986 Part 1, 767–782.

Address of author:
Dr. Robert Lang
Gotenstraße 2
8417 Lappersdorf
Federal Republic of Germany

Jan de Ploey (Ed.)

RAINFALL SIMULATION, RUNOFF and SOIL EROSION

CATENA SUPPLEMENT 4, 1983

Price: DM 120,-

ISSN 0722–0723 ISBN 3–923381–03–4

This CATENA-Supplement may be an illustration of present-day efforts made by geomorphologists to promote soil erosion studies by refined methods and new conceptual approaches. On one side it is clear that we still need much more information about erosion systems which are characteristic for specific geographical areas and ecological units. With respect to this objective the reader will find in this volume an important contribution to the knowledge of active soil erosion, especially in typical sites in the Mediterranean belt, where soil degradation is very acute. On the other hand a set of papers is presented which enlighten the important role of laboratory research in the fundamental parametric investigation of processes, i.e. erosion by rain. This is in line with the progressing integration of field and laboratory studies, which is stimulated by more frequent feed-back operations. Finally we want to draw attention to the work of a restricted number of authors who are engaged in the difficult elaboration of pure theoretical models which may pollinate empirical research, by providing new concepts to be tested. Therefore, the fairly extensive publication of two papers by CULLING on soil creep mechanisms, whereby the basic force-resistance problem of erosion is discussed at the level of the individual particles.

All the other contributions are focused mainly on the processes of erosion by rain. The use of rainfall simulators is very common nowadays. But investigators are not always able to produce full fall velocity of waterdrops. EPEMA & RIEZEBOS give complementary information on the erosivity of simulators with restricted fall heights. MOEYERSONS discusses splash erosion under oblique rain, produced with his newly-built S.T.O.R.M-1 simulator. This important contribution may stimulate further investigations on the nearly unknown effects of oblique rain. BRYAN & DE PLOEY examined the comparability of erodibility measurements in two laboratories with different experimental set-ups. They obtained a similar gross ranking of Canadian and Belgian topsoils.

Both saturation overland flow and subsurface flow are important runoff sources und the rainforests of northeastern Queensland. Interesting, there, is the correlation between s colour and hydraulic conductivity observed by BONELL, GILMOUR & CASSELLS. Runc generation was also a main topic of IMESON's research in northern Morocco, stressing t mechanisms of surface crusting on clayish topsoils.

For southeastern Spain THORNES & GILMAN discuss the applicability of erosi models based on fairly simple equations of the "Musgrave-type". After Richter (German and Vogt (France) it is TROPEANO who completes the image of erosion hazards in Europe vineyards. He shows that denudation is at the minimum in old vineyards, cultivated wi manual tools only. Also in Italy VAN ASCH collected important data about splash erosion a rainwash on Calabrian soils. He points out a fundamental distinction between transpo limited and detachment-limited erosion rates on cultivated fields and fallow land. For representative first order catchment in Central-Java VAN DER LINDEN comments co trasting denudation rates derived from erosion plot data and river load measurements. He too, on some slopes, detachment-limited erosion seems to occur.

The effects of oblique rain, time-dependent phenomena such as crusting and runc generation, detachment-limited and transport-limited erosion including colluvial depositic are all aspects of single rainstorms and short rainy periods for which particular, predicti models have to be built. Moreover, it is argued that flume experiments may be an econom way to establish gross erodibility classifications. The present volume may give an impetus further investigations and to the evaluation of the proposed conclusions and suggestions.

Jan de Ploey

CATENA SUPPLEMENT 15 p. 311–328 Cremlingen 1989

THE PLEISTOCENE ILLER GLACIERS AND THEIR OUTWASH FIELDS

Karl Albert **Habbe**, Erlangen & Konrad **Rögner**, Trier

1 Introduction (H. & R.)

The meltwater deposits of the Pleistocene Iller glaciers in the region between the rivers Iller and Lech, the "Iller-Lech Platte", have been a challenge to south German Quaternary geomorphologists since A. PENCK demonstrated using the example of the gravel aggradations of the Memmingen area (PENCK & BRÜCKNER 1901/09; fig.1) that the ice age in the Alps and their foreland can be subdivided into four glacial periods, Günz, Mindel, Riss, Würm. Subsequently relics of older glaciations, the multiple Donau glaciation (EBERL 1930) and the still older Biber glaciation (SCHAEFER 1956, 1957, 1968) were also detected in this region. After ZAGWIJN (1957) proved that early Pleistocene cold periods had also occurred in Northern Europe, WOLDSTEDT (1958) tentatively correlated the six north European cold phases known at that time with the six cold phases of the Iller-Lech Platte.

The Iller-Lech Platte is characterized by deposits of the cold phases of the entire Pleistocene lying at or near the surface. In the area of the Iller-Lech Platte deposits of the cold phases of the entire Pleistocene lie at or near the surface. Warm period relics have been identified with reasonable certainty only from young interglacials. The deposits of different ages are generally not laid down in a vertical sequence but lie adjacent to each other and contain erosional disconformities. The relative age of the deposits can be determined from their altitudes. Absolute dating is possible only rarely and then mainly of the younger Pleistocene deposits.

In the last thirty years investigations in the area have generally followed one of two approaches. One approach has been to identify the glaciofluvial deposits of the Iller-Lech Platte more precisely on the basis of morphology and sedimentological analysis (GRAUL 1949 et seq., SCHAEFER 1950 et seq., SINN 1972, LÖSCHER 1976 et seq. JERZ et al. 1975, JERZ & WAGNER 1978, RÖGNER 1979 et seq.). The other approach has been based on detailed geomorphological mapping of the topographic sheet Grönenbach 1:25,000 (HABBE 1985a et seq.) which can be considered as the type region for the pleniglacial of the Würm glaciation in the area of the Iller glacier but which also contains deposits from older glaciations. These deposits are mainly glacial or near-glacier sediments, the stratigraphy of which had to be reexamined. Both approaches were used in this paper.

ISSN 0722-0723
ISBN 3-923381-18-2

3-923381-18-4/89/5011851/US$ 2.00 + 0.25

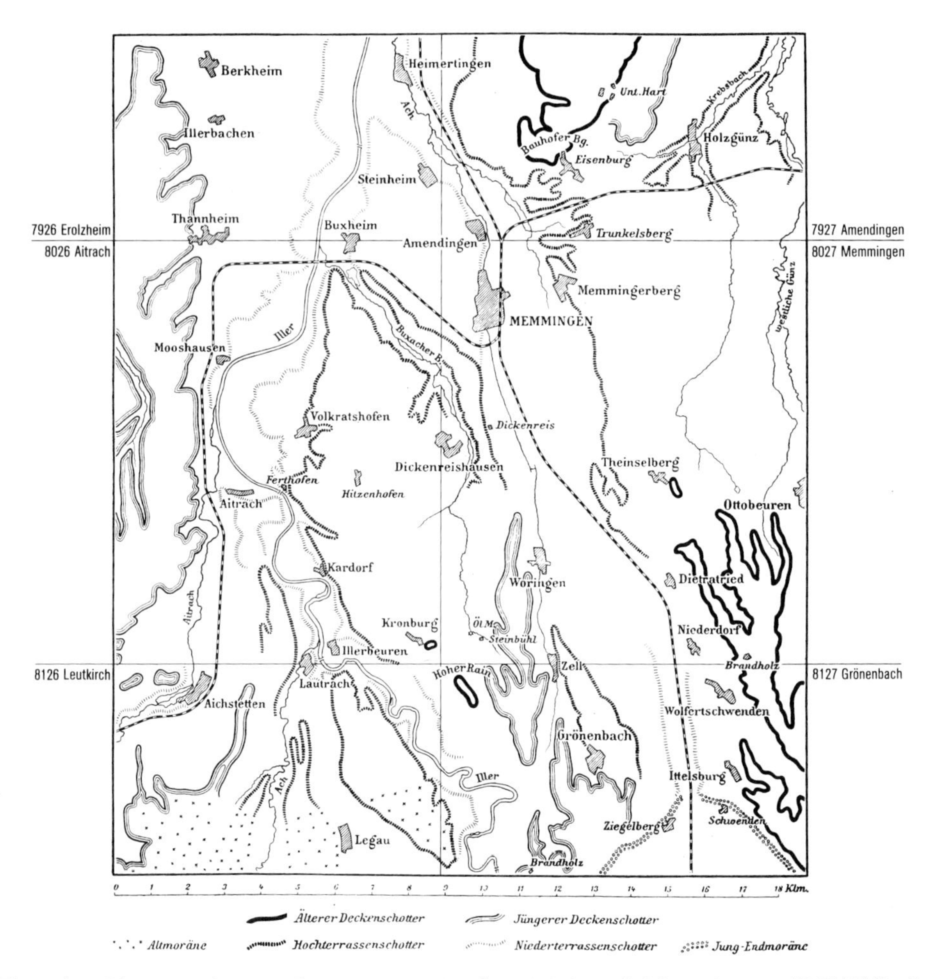

Fig. 1: *The gravel aggradation areas in the vicinity of Memmingen (PENCK & BRÜCKNER 1901/1909, from HABBE 1986a).*

2 Paleogeography and stratigraphy of the oldest Pleistocene in the northern and eastern parts of the Iller-Lech Platte (R.)

The investigation is based on the reconstruction of the Pleistocene drainage systems and of the "glacial series" (A. PENCK 1901/1909), that is, the spatial and functional association of the glacier lobe basin, its terminal moraine and its fluvial outwash deposits. The uplift of the Alpine Foreland after the deposition of the Miocene Upper Freshwater Molasse (Obere Süßwassermolasse, OSM) caused the rivers to incise their beds in several distinct phases that were separated by phases of outwash accumulation. The outwash gravels lie, therefore, on a series of terraces, the highest of which are the oldest. The youngest sediments fill the valley bottoms.

PENCK termed this step-like occurrence of the Pleistocene terraces the Swabian type of glaciofluvial outwash plains and used the terraces together with the moraines, as evidence of the Würm, Riss, Mindel and Günz glaciations. EBERL showed that each glaciation consisted of two or three stages and that there was, in addition, an older group of glaciations, the Donau glaciations, which differed in terms of the lithologic composition of their sediments, their degree of weathering and the gradients of their drainage systems. The Donau gravels lie at elevations that are above the local baselevels of the Würm, Riss, Mindel and Günz deposits near Memmingen and must, therefore, be older (EBERL 1930).

SCHAEFER (1957) mapped a group of deposits on the northern edge of the Staudenplatte in the Staufenberg area that belonged to neither the Günz glaciation nor the Donau gravels. They were termed the Biber deposits.

SINN (1972), LÖSCHER (1976) and RÖGNER (1979) have mapped the middle, old and oldest Pleistocene sediments in the northern and eastern areas of the Iller-Lech Platte and reconstructed the paleogeographic changes that took place during the Quaternary (fig.2). An exact dating of the chronostratigraphic boundaries between the Biber and Donau and between the Donau and Günz deposits has not yet been accomplished. The Pleistocene sediments described in sections 2.1–2.5 were deposited by glaciofluvial rivers during the Biber and Donau glaciations. These rivers were the paleo-Iller, and perhaps the paleo-Danube, in the north and west, the paleo-Iller in the southwest and the united drainage system of the paleo-Iller and paleo-Lech in the northeast, the latter mainly in the Aindlinger terrace series.

Of these sediments, the Staufenberg series, the Staudenplatte and Zusamplatte deposits and the glaciofluvial gravels of the younger Donau glaciation, together with its moraines near Bickenried and Königsried, were accumulated during the reverse Matuyama period, earlier than 690,000 years BP. The intermediate terrace gravel (Zwischenterrassenschotter) in the northwest of the Iller-Lech Platte is probably of a similar age.

2.1 The Staufenberg terrace series (Biber group of deposits)

The oldest Pleistocene deposits on the Iller-Lech Platte are the gravels at the top of the Staufenberg terrace series west of Augsburg (SCHAEFER 1957, SCHEUENPFLUG 1974, LÖSCHER 1979). The High Gravel (Hochschotter) of the Aindlinger terrace series is probably of the same age (TILLMANNS, BRUNNACKER & LÖSCHER 1983). The Staufenberg series dates from the Biber glaciation. Its four gravel deposits, from oldest to youngest, Staufenberg, Achselberg, Reitenberg and Batzenghau, are separated by phases of intensive erosion (LÖSCHER 1979).

The Biber glaciation consisted of several cold stages or ice advances. The Staufenberg terrace series resulted from a lateral southeastward shift of the paleo-Iller. The High Gravel of the Aindlinger terrace series was deposited by the combined paleo- Iller/paleo-Lech system that shifted northwestward during the following phases of accumulation and erosion. During the Biber glaciation the paleo-Iller flowed from the southwest and the paleo-Lech maintained an approximately south to north direction of flow. For this

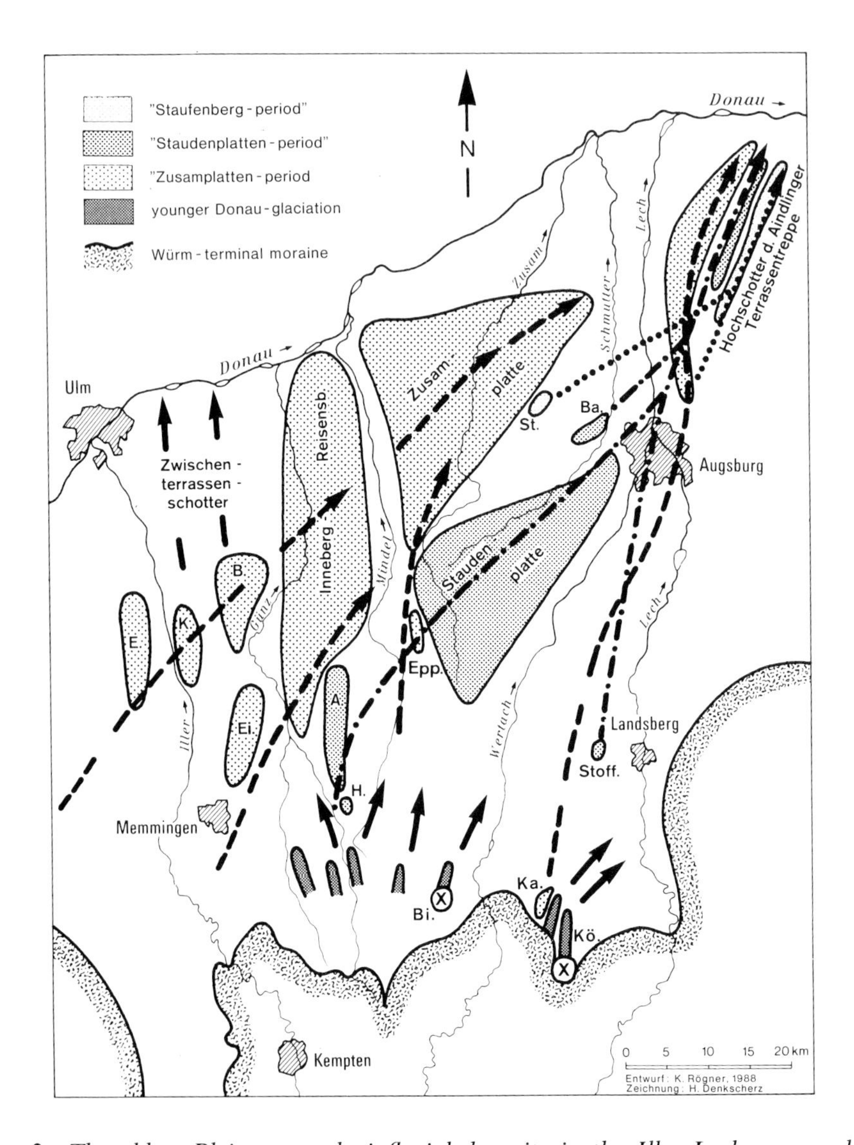

Fig. 2: *The oldest Pleistocene glaciofluvial deposits in the Iller-Lech area and the paleo-drainage directions of the glaciofluvial meltwaters (SINN 1972, LÖSCHER 1976, 1979, RÖGNER 1979, 1980, 1986b, RÖGNER, LÖSCHER & ZÖLLER 1988, TILLMANNS, BRUNNACKER & LÖSCHER 1983, LÖSCHER & RÖGNER in prep.).*

Abbreviations for gravel deposits: E = Erolzheimer, K = Kellmünzer, Ka = Kanzel, B = Bucher, Ei = Eisenburger, A = Arlesrieder, H = Hochfirst, Epp = Eppishauser, St = Staufenberg, Ba = Batzenghau, Stoff = Stoffersberg, X in a circle: west = glacial series of Birkenried-Irsee (Bi), east = glacial series of Königsried-Stocken (Kö).

reason there are no widespread Biber sediments in the area between Mindel/Wertach and Lech in front of the Lech glacier. By contrast, the westward shift of the paleo-Iller caused the oldest Pleistocene Iller deposits to be preserved.

No glaciofluvial sediments or moraines of Biber age have yet been found near the margins of the Pleistocene Iller and Lech glaciers, south of a line from Memmingen to Landsberg, probably because of later glaciofluvial, periglacial and glacial erosion. The early stages of the Donau glaciation are also missing for the same reason.

2.2 The Staudenplatte and the youngest group of Biber deposits

The Staudenplatte and other smaller remnants of the same age, the Arlesrieder gravels and the Hochfirst gravels (A and H in fig.2) lie southwest of Augsburg. They are separated from the Zusamplatte by the Dinkelscherben Old Divide (Dinkelscherbener Altwasserscheide, GRAUL 1949) and probably continue in the Mittlere Deckschotter (Intermediate cover gravels; TILLMANNS, BRUNNACKER & LÖSCHER 1983) of the Aindlinger terrace series. The Staudenplatte converges on the northeast with the lowest of the Staufenberg series, the Batzenghau gravel (Ba) and can, therefore, be correlated with the last period of the Biber group (LÖSCHER 1979). [1]

The Hochfirst and Arlesrieder gravels and the Staudenplatte indicate that the paleo-Iller flowed first from the south and then from the southwest and united with the paleo-Lech near Augsburg. (LÖSCHER & RÖGNER in prep.). The only Biber sediment of the paleo-Lech occurs in the Stoffersberg (Stoff) gravel (RÖGNER 1979). The Arlesrieder gravels, the Hochfirst gravels and the Stoffersberg gravels all have a lithologic composition that is identical with that of the younger glaciofluvial sediments, indicating that the Biber gravel accumulations are also of glaciofluvial origin (RÖGNER 1986b).

SINN (1972) and LÖSCHER (1976) suggested that the paleo-Iller, which at that time flowed from the region between the present Mindel and Günz valleys, had additional inflows from further west that deposited sediments almost devoid of dolomite pebbles and can, therefore, be distinguished from glaciofluvial sediments. They were derived mainly from the Miocene Upper Freshwater Molasse conglomerates in the Adelegg area west of Kempten. They occur underneath the glaciofluvial gravel and have been termed subjacent facies (Liegendfazies, GRAUL 1953, SINN 1972).

2.3 The Zusamplatte, the Jüngere Deckschotter (Younger cover gravels) and the older period of the Donau glaciation

North of a line from Memmingen to the Staufenberg there are extensive gravel deposits that have been termed the Jüngere Deckschotter (Younger cover gravels, LÖSCHER 1976). The Zusamplatte is the largest of these areas. Others are the Eppishauser (Epp) gravels, the Kanzel (Ka) gravels and the Untere Deckschotter (Lower cover gravels;

[1] Editor's note: German terminology distinguishes between "Deckenschotter" and "Deckschotter", the former were deposited during the Günz and Mindel, the latter during the Donau glaciation. The English translation in either case is "cover gravels". To avoid confusion, the German terms Deckenschotter and Deckschotter have been retained.

TILLMANNS et al. 1983) in the Aindlinger terrace series. The Erolzheimer (E) gravels, west of the present Iller, and the Eppishauser gravels, between the Arlesrieder and Staudenplatte, indicated that there were other drainage systems in addition to the paleo-Iller, perhaps a paleo-Riss from the Rhine glacier area and a paleo-Mindel or paleo-Wertach. The direction of the Lech has hardly changed since the Staufenberg period.

LÖSCHER (1976) and SCHEUENPFLUG (1970, 1971) suggested that the oldest course of the Pleistocene Danube was located farther to the south than the present course and that this explains the frequent occurrence of Malm limestone pebbles (white Jurassic facies) from the Swabian Jura at the base of the Zusamplatte gravels (LÖSCHER & SCHEUENPFLUG 1981). The former more southerly position of the Danube has been disputed (SCHAEFER 1980).

The Zusamplatte gravels and other gravels of the same age cover large areas in the northern part of the Iller-Lech Platte. LÖSCHER (1976) suggested that the preservation of these gravels was the result of a northward lateral shift of the Danube during the earliest Pleistocene which left a large slightly inclined plain. Another explanation is that the paleo-Iller and other paleo-rivers overflowed the Dinkelscherben Old Divide.

The Zusamplatte is of significance because the fluvial floodwater loams that overlie the gravels have been dated paleomagnetically, by pollen and by fossils. This has been possible at only a few locations in the German Alpine Foreland, the most important of which was Uhlenberg, west of Augsburg (SCHEUENPFLUG 1979). Paleomagnetic measurements prove that the Zusamplatte gravels date at least from the Jaramillo event of the Matuyama epoch, about 930,000 years BP (TILLMANNS, KOCI & BRUNNACKER 1986).

Pollen analysis of a fossil peat layer above the floodwater loam (SCHEDLER 1979) showed that the peat was older than the Holstein or Cromer interglacials. It cannot be placed with any certainty in the Waal interglacial either and may be of even greater age. The fauna (DEHM 1979) indicates that the peat was formed, at the latest in the Günz/Donau interglacial. The Zusamplatte gravels were probably, therefore, deposited before the Jaramillo event, during an early stage of the Donau glaciation.

The Zusam valley is incised into the Zusamplatte and the Staudenplatte and four continuous periglacial-fluvial levels have been mapped on the slopes of this valley below the baselevel of the Zusamplatte (ESSIG 1979). This is a further indication that the Zusamplatte originated before the Günz glaciation.

It seems probable that at the beginning of the Zusamplatte period, there was an inclined plain north of the Biber gravels that was partially covered by gravels of white Jurassic limestone. The Biber sediments, which had been deposited by the paleo-Iller during the Staudenplatte age, extended from southwest to northeast in the direction of drainage. They were then dissected and removed by a south to north flowing drainage system which incised a valley between the Arlesrieder gravels and the Staudenplatte. This valley later became the Mindel valley of the present. Evidence of the south to north drainage system is provided by the Eppishauser gravels (Epp in fig. 2) which were deposited in a south-north direction subsequent to the phase of incision. The change in the course of the paleo-Lech and paleo-Iller in the Aindlinger Platte can be similarly interpreted.

2.4 The Zwischenterrassenschotter (Intermediate Terrace gravels) of the later Donau group

These deposits occur mainly in the northwestern part of the Iller-Lech Platte near Ulm (fig.2). They differ in lithologic composition from the glaciofluvial gravels (LÖSCHER 1976), lie entirely to the north of the gravels of Zusamplatte type and were deposited by a river system that flowed from south to north. The Intermediate Terrace gravels are the earliest indication of a change from a southwest to northeast drainage direction to one oriented from south to north in the northwest of the Iller-Lech Platte. This was similar to a reorientation that occurred in the area of the Eppishauser gravels. The present valley directions were also roughly established. There are in addition several small levels of Intermediate Terrace gravels on both sides of the lower course of the Lech (LÖSCHER 1976, TILLMANNS, BRUNNACKER & LÖSCHER 1983). The Intermediate Terrace gravels can be dated only approximately into the later part of the Donau glaciation. They are not glaciofluvial gravels and cannot, therefore, be dated more precisely. They have not yet been correlated with the moraines and glaciofluvial gravels of the younger Donau glaciation.

2.5 The glaciofluvial gravels and the moraines of the younger Donau glaciation

The youngest deposits of the Donau group are the glaciofluvial gravels that lie south of a line from Memmingen to Landsberg. They are the highest Pleistocene deposits in the area (RÖGNER 1979, 1980, 1986a, SINN 1972). They were also deposited in the valleys of the north Iller-Lech Platte but were subsequently eroded completely. These sediments are important because the glaciofluvial gravels are interbedded with moraines at two locations, Königsried (Rögner 1979) and Bickenried (Rögner 1980), providing thereby the first direct proof of a fifth glaciation in the Alpine Foreland. The relationship of these gravels and moraines to the Intermediate Terrace gravels has not yet been explained. The drainage direction of the younger Donau gravels is similar to the direction of the present rivers which follow, in general, the Low Terraces (Niederterrassen) of the last glaciation. Apart from changes of the glaciofluvial meltwater flow from one valley system to another, the valley systems show the south to north direction that was developed in the early and younger Donau glaciations.

3 The front of the Iller glaciers and the root zone of the Iller-Lech Platte (H.)

The glaciofluvial sediments of the Iller-Lech Platte were deposited by the meltwater streams that emerged from the Pleistocene Iller glaciers. These glaciers came from the Allgäu High Alps where they had reached considerable thicknesses, for example, 750 m in the Oberstdorf basin during the last glaciation (WEINHARDT 1973). They remained, however, relatively small since their tributary area was limited to the Limestone Alps. Their terminus did not reach beyond the pre-Alpine Molasse mountains (fig.3). The exact position of the glacier

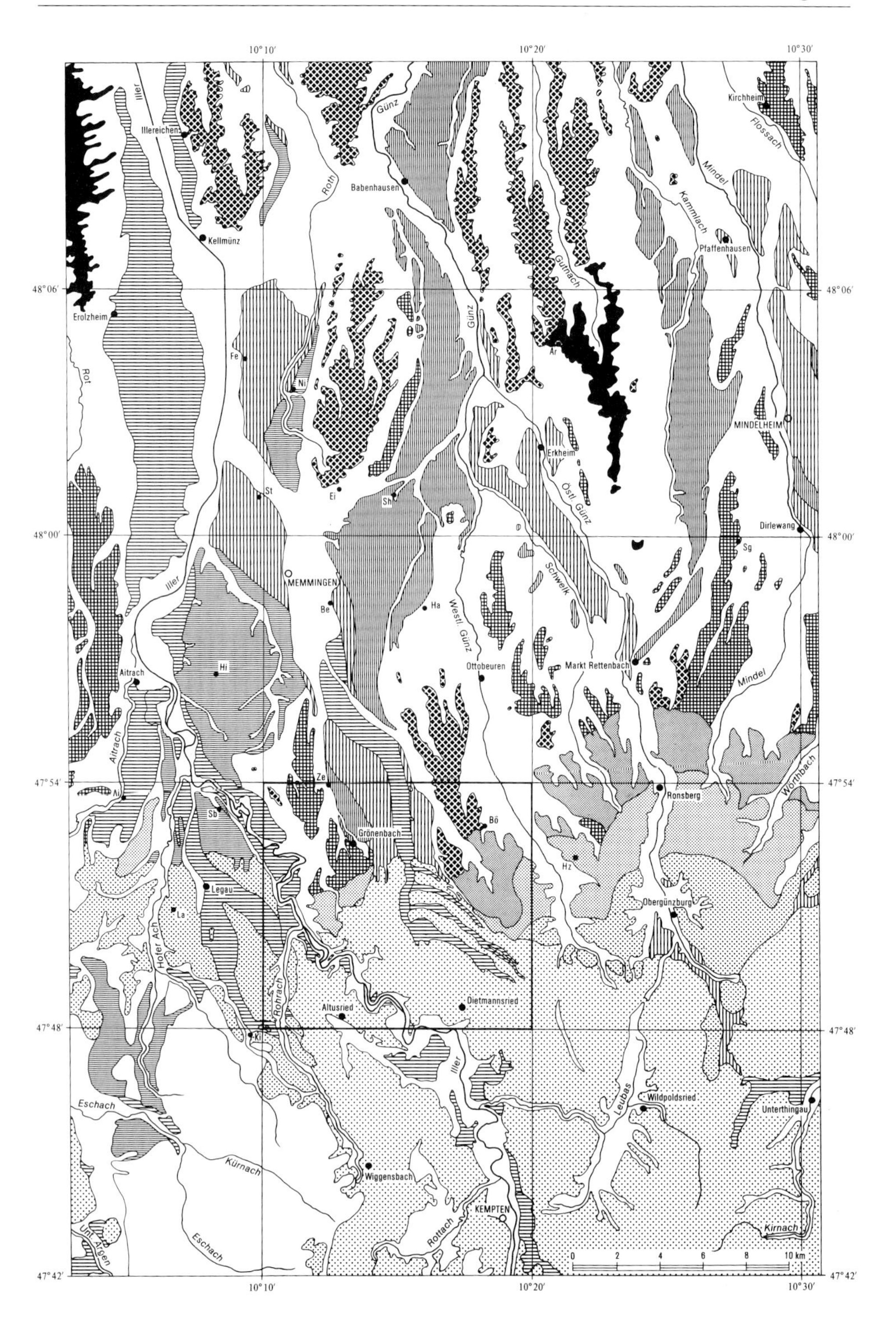
10°10'
10°20'
10°30'
48°06'
48°00'
47°54'
47°48'
47°42'
Iller
Illereichen
Kellmünz
Erolzheim
Rot
Roth
Günz
Babenhausen
Kirchheim
Flossach
Mindel
Kammlach
Pfaffenhausen
Gutnach
Ar
Fe
Ni
St
Ei
Sh
MINDELHEIM
Erkheim
Östl. Günz
Dirlewang
Sg
Schwelk
MEMMINGEN
Be
Ha
Westl. Günz
Ottobeuren
Markt Rettenbach
Hi
Aitrach
Aitrach
Ai
Ze
Ronsberg
Wörthbach
Sb
Grönenbach
Bö
Hz
Legau
La
Obergünzburg
Hofer Ach
Rohrach
Altusried
Dietmannsried
Ki
Iller
Leubas
Wildpoldsried
Unterthingau
Eschach
Kürnach
Wiggensbach
KEMPTEN
Rottach
Kirnach
Eschach
Unt. Argen
0 2 4 6 8 10 km

Fig.3

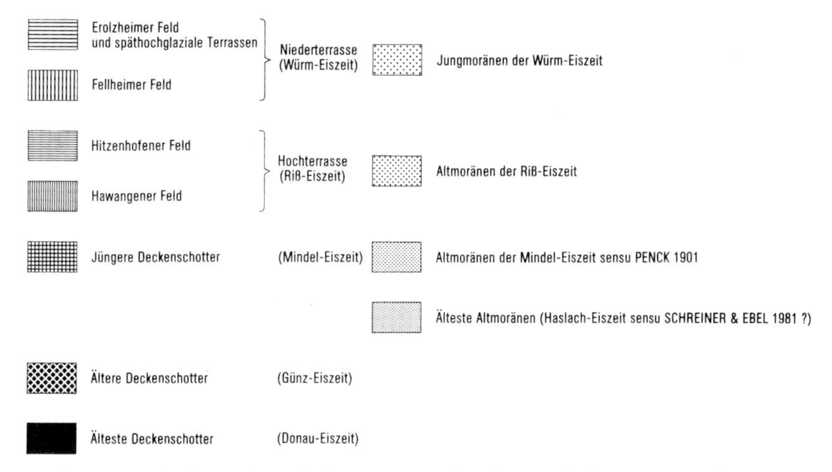

Fig. 3: *Glacial and glaciofluvial deposits in the front of the region Pleistocene Iller glaciers (SINN 1972, 1974, JERZ et al. 1975, SCHOLZ & ZACHER 1983, from HABBE 1986a). The degree sections show the sheet lines of the 1:25,000 topographical map.*
Abbreviations: Ai = Aichstetten, Ar = Arlesried, Be = Benningen, Bö = Böhen, Ei = Eisenburg, Fe = Fellheim, Ha = Hawangen, Hi = Hitzenhofen, Hz = Hinterschmalholz, Ki = Kimratshofen, La = Lausers, Ni = Niederrieden, Sb = Maria Steinbach, Sg = Saulengrain, Sh = Schwaighausen, St = Steinheim, Ze = Zell.

front was determined by the location of the Iller valley and by the elevations of the divides.

During the glacier maxima, meltwaters spilled over from the high sides of the glacier through wind gaps in the divides into adjacent autochthonous valleys that had been formed by periglacial erosion and denudation. In some cases the glaciers also flowed over the divides so that the direction of the glacial advances shifted. The glaciers moved first to the north-northeast, via the Wildpoldsried branch basin to the headwaters of the Günz and the Mindel (fig.3), then to the north, via the Dietmannsried branch basin in the direction of the large Memmingen dry valley and finally to the north-northwest, via the Altusried branch basin to the catchment area of the Aitrach river.

Some of the lateral meltwater outlets were so intensively cut down by headward erosion that the runoff from the central glacier lobe basin in the Kempten area during the following interglacial flowed through these newly developed gaps rather than through those used in the preceding warm period. The glaciofluvial sediments of the following glacial were also deposited along the new main outlet until new meltwater spillovers developed during the maximum stage of the glacier and deposited glaciofluvial sediments in adjacent valleys.

The alternation between interglacial runoff in one drainage way and pleniglacial runoff in several drainage ways, which caused the very large number of younger Pleistocene gravel trains in the west of the Iller -Lech Platte, had two reasons:

1. The Iller glaciers of the Pleistocene

cold periods did not extend into the low relief Alpine Foreland but ended in the marginal mountain zone. Consequently, they were more compressed and thicker and could flow easily across low points on their lateral divides but were, however, not thick enough to overflow an entire divide.

2. Headward erosion was more effective in the autochthonous Molasse valleys during early and late glacial climatic conditions than in the allochthonous main valley where the meltwater rivers were loaded to capacity with glacigenous gravels (SCHAEFER 1950). The divides of the main valleys were, therefore, lowered by headward erosion from neighbouring periglacial-fluvial valleys and the resulting cols or wind gaps could be crossed during the maximum stage of the glacier, either by meltwater flows or by the glacier itself.

3.1 The Iller glacier and its outwash fields during the Würm glaciation

During the early Würm the meltwaters of the Iller glacier drained into the Memmingen dry valley which had also been the main outlet of the Kempten basin in the Mindel and Riss glaciations (fig.3). The related gravel track can be traced by way of the Steinheim-Fellheimer Feld (Feld means aggradation area in the sense used by PENCK) northward to the valley of the Roth. During the advance of the glacier two more meltwater outlets developed: in the west via the Altusried branch basin and Legau-Aitrach-Erolzheimer Feld to the present valley of the lower Iller and in the east via the Wildpoldsried branch basin to the valley of the Eastern Günz (fig.3). These three independent meltwater channels, which flowed parallel to one another towards the Danube and accumulated separate gravel deposits, existed already before the glacial maximum. At the maximum, additional outlets developed: northeastward to the valley of the Western Günz and westward to the eastern Rhine Glacier, via the Kürnach, Eschach and Argen rivers. Because of their limited supply of water, high gradient and, generally, small gravel load, these meltwater flows of the maximum stage, for the most part, only eroded. The Rhine glacier, about 10 km to the west, advanced parallel to the Iller glacier and forced the Argen and Eschach rivers to flow northward to the Aitrach valley where they combined with runoff from the Iller glacier that flowed to the Danube. These meltwaters took part in the aggradation of the Erolzheimer Feld from the beginning.

Because of its periglacial-fluvial origin, the Iller-Aitrach valley floor was lower than the Steinheim-Fellheimer Feld adjacent to the east. Lateral erosion cut through the interfluve in the Memmingen area and a spillway was opened from the higher, eastern, to the lower, western, level as an equalizing channel ("Ausgleichsrinne") which became the channel of the present Memminger Ach. The northern segment of the original main meltwater channel of the Iller glacier along the Steinheim-Fellheimer Feld and the Roth valley became dry (GRAUL & SCHAEFER 1953, HABBE 1986a). V-shaped erosion channels were found recently at the base of the Erolzheimer Feld and below the deposits of the Aitrach valley ("Aitrach deep channel"). These old channels which can be traced to the terminal moraines of the Rhine glacier and to the Molasse mountains of the Adelegg, have not yet been fully explained (KUPSCH 1982, ELLWANGER 1988). A branch of these channels lies beneath the aggraded floor of the Memminger Ach. Although the details are uncertain, this means that the opening of the spillway from the Steinheim-Fellheimer Feld into the valley of the present Iller took place before the aggradation of the Erolzheimer Feld and that the opening formed before the Iller glacier reached its maximum extent.

The history of the glacier maximum and of the following late pleniglacial is shown by the terminal moraines of the triple maximum stage (1a–c), the type locality of which lies near Ziegelberg at the southern rim of the Memmingen dry valley, by six internal stages (2–7) and by related meltwater deposits. This subdivision of the Würm pleniglacial is identical with that of the western Rhine glacier which has been investigated in great detail (SCHREINER 1970) and is, therefore, of more than local significance (HABBE 1985a, 1985b). The following observations are notable:

1. Distinct rampart-shaped terminal moraines exist only at the outermost margin of the former glacier. Even these are not continuous. The location of the glacier margin can, however, be clearly identified by, for example, kame terraces. The moraines consist only in part of typical till. Meltwater deposits or meltwater-influenced material are more frequent.

2. The glacier of the pleniglacial advanced over its permanently frozen glaciofluvial deposits which remained in place beneath the overlying moraines and were not, or only partly, eroded by the advancing glacier. Because of the permafrost conditions, glaciofluvial accumulation took place despite gradients of 0.015 and higher on the outwash field. The influence of permafrost ceased during the internal phase 6 of the glacier and, as a result, drumlin development took place (HABBE 1988).

3. When the glacier retreated from its maximum stage, dissection began at the inner margins of the glaciofluvial deposits because of the high relief relative to the glacier lobe basin, which was gradually becoming uncovered, and because of the concentration of runoff along confined discharge channels that were fixed by the position of the gaps in the terminal moraine ridge. As a result, the large trumpet valley (TROLL 1926) at the southern end of the Memmingen dry valley near Ziegelberg developed. However, the runoff of the meltwaters in the late pleniglacial flowed in the same three channels that had been used in the early pleniglacial. This pattern changed when a spillway that had been formed in the northwest during the glacier maximum, began to influence the discharge of the entire glacier. Because of its higher gradient, this spillway was incised more rapidly than those leading from the Dietmannsried and Wildpoldsried branch basins. Since internal phase 6, at the latest, the entire discharge of the Kempten basin has drained through this drainage channel, the Iller canyon of the present-day.

3.2 The Iller glacier and its outwash fields during the Riss glaciation

During the Riss the Iller glacier advanced along the interglacial and early glacial valley system to the Memmingen dry valley where its meltwaters deposited the glaciofluvial gravels of the Zell high terrace spur, and, farther north, the Hawangener Feld (fig.3) and the High Terrace of the Günz valley. There are, however, no Riss terminal moraines in this central area. They were overridden by the Würm glacier. Probably the valley along which the Riss glacier advanced into the Alpine Foreland at that time was so narrow, similar to the present Iller canyon, that the movement was more impeded than during the Würm glaciation.

Other conditions existed at the glacier front beyond the two branch basins in the east and in the west. Near Obergünzburg north of the Wildpoldsried basin, and well beyond the outermost terminal moraines of the Würm glaciation, the glacier was dammed by older gravels and moraines and accumulated a distinct double rampart terminal moraine, similar to the Riss terminal moraines at the type locality near Biberach in the area of the eastern Rhine glacier. The High Terrace in front of this terminal moraine has, however, been removed by erosion in the Eastern Günz valley. Related fluvioglacial deposits occur first ten kilometers down valley on the eastern slopes near Markt Rettenbach (Fig.3). They continue with increasing width, along the valley of the Auerbach brook, the rivers Kammlach and Mindel and finally to the High Terrace of the Danube. Since the main runoff of the advancing glacier flowed to the Memmingen dry valley, this lateral outlet of the Riss meltwaters must have been formed by a spillway across the divide to the valley of an autochthonous river during the Riss maximum. The valley of the present Eastern Günz, which runs north-northwest, did not develop before the last phase of the Riss glaciation.

The ice on the western flank of the glacier advanced as far as the vicinity of Legau, also far beyond the Würm terminal moraines, and occupied a Molasse area of relatively low relief that had previously only been affected by periglacial processes. As a result, the terminal moraines are not prominent. However, the related Steinbach High Terrace (fig.3) is well-developed and continues north of the Iller in the Hitzenhofener Feld and in the valley of the Roth.

At its maximum during the Riss, the Iller glacier also had three independent meltwater outflows with related outwash gravel fields of which the two western ones developed in a manner similar to those of the Würm glacier (SINN 1972). The narrow divide between the Hitzenhofener Feld and the Hawangener Feld was breached by lateral erosion from the lower Hitzenhofener Feld near Memmingen, a spillway was opened and an equalizing channel was formed. The two western meltwater flows of the glacier then both drained via the valley of the Roth. The Hawangener Feld and its northern continuation, the Günztal High Terrace, became dry. In addition, during the late phases of the Riss glaciation, the equalizing channel in the Memmingen dry valley was eroded headward so rapidly and incised so deeply that it became the only outlet for the Kempten basin during the following interglacial and the early Würm glacial.

3.3 The Iller Glacier and its outwash fields during the Mindel glaciation; the Haslach glaciation

PENCK's type region for the Mindel glaciation is the Brandholz-Manneberg terminal moraines and the Grönenbacher Feld (fig.3). The "Jüngere Deckenschotter" (younger cover gravels, PENCK & BRÜCKNER 1901/1909) of the Grönenbacher Feld continue on the western side of the Günz valley north of Memmingen as the Schwaighausen gravel (fig.3). EBERL (1930) and SINN (1972) have confirmed PENCK's interpretation, but SCHAEFER (1973) ascribed the Grönenbacher Feld to a series of 10 different cold periods, EICHLER & SINN (1975) and LÖSCHER (1976) assigned it to a fourth last glaciation, (PENCK's Günz), and SINN (1972), JERZ et al. (1975) and ROPPELT (1988) associated the related terminal moraines east of the Memmingen dry valley with the Riss glaciation. The controversy (HABBE 1986a, 1986b) has been focused on the area between the eastern and western Günz rivers and especially on a fossil soil detected by SINN (1972) near Hinterschmalholz (Hz in fig.3). This soil developed in a thin till overlying a cemented gravel, is overlain by a decalcified fine-grained sediment of, at least, Mindel age (RÖGNER & LÖSCHER 1987, RÖGNER, LÖSCHER & ZÖLLER 1988) and the Mindel terminal moraine. The subjacent till extends as a moraine cover outward beyond the terminal moraines. The boundary of the Mindel glaciation can now be fixed approximately in the position assumed by EBERL (1930). What, however, is the age of the moraine beyond the Mindel terminal moraines?

The Iller glacier of the Mindel glaciation must have advanced, unlike its successors, mainly to the north-northeast and followed the direction of the oldest drainage paths in the Kempten basin in a straight-line continuation of the Alpine Iller valley between Immenstadt and Kempten. However, the high elevations of the bases of all older Pleistocene deposits in the headwater area of the Günz and Mindel rivers indicate that the main meltwater outlet of the Kempten basin was no longer in that area at that time (SINN 1972). It was farther to the west in the area of the present Memmingen dry valley where the meltwaters flowed northward towards the Schwaighauser gravel train. During the Mindel maximum that outlet was filled and overlain by the gravels of the Grönenbacher Feld. At the same time, the meltwaters from the front of the glacier in the north-northeast drained toward the Mindel valley in which the Kirchheim-Burgauer gravel track was accumulated.

Sections made from west to east across the glaciofluvial deposits of the Iller-Mindel area approximately in the latitude of Mindelheim (EICHLER & SINN 1975, JERZ et al. 1975) show that the Schwaighauser gravel lies about 25 m lower than the Kirchheim-Burgauer gravel near Mindelheim. The elevation difference is significant because the distance from the receiving stream, the Danube, is nearly 50 km in both cases. This suggests that the Schwaighauser gravels might be younger than the Kirchheim-Burgauer gravel. In any case the Schwaighauser gravel cannot be assigned to a glaciation older than that during which the Kirchheim-Burgauer gravel was deposited, as EICHLER & SINN (1975) and LÖSCHER (1976) had assumed. On the other hand, the

Kirchheim-Burgauer gravel at its northern end near Offingen was determined as being of Mindel age by means of its cover sediments (RÖGNER, LÖSCHER & ZÖLLER 1988).

EICHLER & SINN (1975) indicated and SCHREINER & EBEL (1981) later proved that there are deposits of two glaciations between Riss moraines and Günz gravels in the area of the eastern Rhine glacier. Consequently they divided the intervening period into the younger Mindel glaciation and an older glaciation which they named the Haslach glaciation. The fossil soils on the Haslach deposits have characteristics and a thickness that are similar to the Hinterschmalholz fossil soil. The thin till underlying the Hinterschmalholz soil and the moraine cover beyond the Mindel terminal moraines at that site can, therefore, be correlated with the moraines of the Haslach glaciation in the eastern Rhine glacier area. At the type locality the gravels of the Haslach glaciation are overlain down valley by the Tannheimer gravels of the Mindel glaciation (HAAG 1982) and cannot, therefore, be traced more than 15 km from their area of origin. This crossing of the terraces, which seems to be a common phenomenon, may be a reason for the Haslach glaciation not having been identified before. A crossing of the terraces might also explain the conspicuously high position of the Kirchheim-Burgauer gravel which can possibly be divided into an older gravel of Haslach age at the base and a younger Mindel gravel on top. Investigations on gravel track topography in the area are in progress (RÖGNER 1986a, RÖGNER, LÖSCHER & ZÖLLER 1988).

The following conclusions can be made for the headwater area of the two Günz rivers:

1. Beneath and in front of the terminal moraines of the Mindel glaciation moraines exist of a glaciation that is older than the Mindel glaciation but younger than the Günz glaciation and that can be correlated with the Haslach glaciation of SCHREINER & EBEL (1981). In this area, which lies adjacent to PENCK's type region for the fourfold subdivision of the Alpine ice age, the Günz glaciation was the fifth, not the fourth, glaciation from the present.

2. Two different "Jüngere Deckenschotter" must exist in this area between the deposits of the Riss glaciation and those of the Günz glaciation. They seem to be represented by the Schwaighauser and the Kirchheim-Burgauer gravels: the Schwaighauser, as the normal Mindel gravel and the Kirchheim-Burgauer, because of a crossing of the terraces, with a base gravel of Haslach age.

3. The terrain conditions during the Haslach glaciation were different from those that prevailed later. Haslach deposits are absent along the Memmingen dry valley. The drainage and glacier advance must, therefore, have followed the original path to the north-northeast. As a result, Haslach moraines occupy a large area in the headwater region of the Günz rivers and extend far beyond the terminal moraines of the Mindel glaciation. They have been preserved because the main meltwater outlet of the Kempten basin and the direction of the glacier advances shifted to the west.

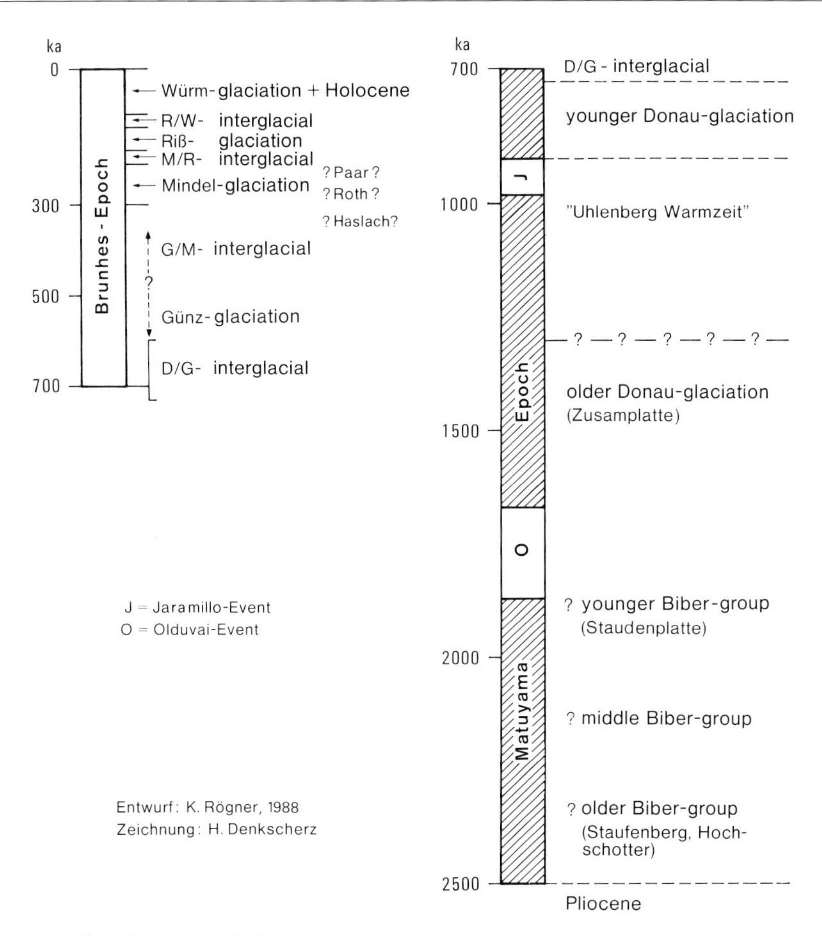

Fig. 4: *Possible division of the Quaternary in the Iller-Lech Platte (BRUNNACKER et al. 1976, TILLMANNS et al. 1986, RÖGNER, LÖSCHER & ZÖLLER 1988).*

3.4 The older Pleistocene glaciations

The Würm, the Riss, the Mindel and, with restrictions, the Haslach glaciations can be traced morphologically by their glacial and fluvioglacial deposits, classified morpho-stratigraphically and dated absolutely by means of their covering strata (RÖGNER, LÖSCHER & ZÖLLER 1988). The lower limit of the Würm glaciation fits to the limit of stage 5d/5e of the oxygen isotope curves of SHACKLETON & OPDYKE (1976), the Riss glaciation to stage 6, the Mindel glaciation to stage 8, and the Haslach glaciation presumably to stage 10. Thus the geomorphological history of the Iller-Lech Platte during the past 350,000 years can be reliably fitted into the standard subdivision of the Pleistocene. However, the sequence of events in the older Pleistocene, before the Haslach glaciation, is uncertain.

Absolute dating of older Pleistocene deposits is possible at only a few locations. Two paleomagnetic dates have been obtained by BRUNNACKER et al. (1976) and TILLMANNS et al. (1986). Glacial deposits are absent at the surface so that a spatial correlation of moraines and gravels is difficult or impossible. The older Pleistocene gravels can, therefore, only be dated relatively. On the other hand, they represent the entire period (BRUNNACKER & TILLMANNS 1982). Thus the subdivision of the older Pleistocene gravels into deposits of the Günz (PENCK & BRÜCKNER 1901/1909), Donau (EBERL 1930) and Biber (SCHAEFER 1956, 1957) glaciations has to be regarded as not yet complete (see RÖGNER in this paper and fig.4).

References

BRUNNACKER, K., BOENIGK, W., KOCI, A. & TILLMANNS, W. (1976): Die Matuyama/Brunhes-Grenze am Rhein und an der Donau. Neues Jahrb. Geol. Paläontol. Abh. **151**, 358–378.

BRUNNACKER, K. & TILLMANNS, W. (1982): New results on Quaternary stratigraphy in the lower Rhine area and in the Northern foreland of the Alps. Quaternary glaciations in the Northern Hemisphere, Report No. **7**, 33–35.

DEHM, R. (1979): Artenliste der altpleistozänen Molluskenfaunen vom Uhlenberg bei Dinkelscherben. Geologica Bavarica 80, 123–125.

EBERL, B. (1930): Die Eiszeitenfolge im nördlichen Alpenvorlande — Ihr Ablauf, ihre Chronologie auf Grund der Aufnahmen im Bereich des Lech- und Illergletschers. Augsburg.

EICHLER, H. & SINN, P. (1975): Zur Definition des Begriffs "Mindel" im schwäbischen Alpenvorland. Neues Jahrb. Geol. Paläontol., Monatsh., 705–718.

ELLWANGER, D. (1988): Würmeiszeitliche Rinnen und Schotter bei Leutkirch/Memmingen. Jahresh. Geol. Landesamt Baden-Württemberg **30**, 207–229.

ESSIG, W. (1979): Die periglazial-fluviatilen Schotterablagerungen des Zusamtales. Heidelberger Geogr. Arb., **49**, 139–163.

GRAUL, H. (1949): Zur Gliederung des Altdiluviums zwischen Wertach-Lech und Flossach-Mindel. Ber. Naturforsch. Gesellsch. Augsburg **2**, 3–31.

GRAUL, H. (1953): Über die quartären Geröllfazien im deutschen Alpenvorlande. Geologica Bavarica **19**, 266–280.

GRAUL, H. & SCHAEFER, I. (1953): Zur Gliederung der Würmeiszeit im Illergebiet. Geologica Bavarica **18**.

HAAG, T. (1982): Das Mindelglazial des nordöstlichen Rheingletschergebietes zwischen Riss und Iller. Jahresber. Mitteil. Oberrhein. Geolog. Verein **64**, 225–266.

HABBE, K. A. (1985a): Das Späthochglazial der Würm-Eiszeit im Illergletscher-Gebiet — Ergebnisse einer geomorphologischen Kartierung. Quartär **35/36**, 55–68.

HABBE, K. A. (1985b): Erläuterungen zur Geomorphologischen Karte 1:25,000 der Bundesrepublik Deutschland — GMK 25, Blatt 18, 8127 Grönenbach, Berlin.

HABBE, K. A. (1986a): Zur geomorphologischen Kartierung von Blatt Grönenbach (I) - Probleme, Beobachtungen, Schlußfolgerungen. Erlanger Geogr. Arb. **47** (= Mitt. Fränk. Geogr. Gesellsch. **31/32**, 1984/85, 365–479).

HABBE, K. A. (1988): Zur Genese der Drumlins in süddeutschen Alpenvorland — Bildungsräume, Bildungszeiten, Bildungsbedingungen. Z. Geomorph., NF, Supp. Bd. **70**, 33–50.

JERZ, H., STEPHAN, W., STREIT, R. & WEINIG, H. (1975): Zur Geologie des Iller-Mindel-Gebietes. Geologica Bavarica **74**, 99–130.

JERZ, H. & WAGNER, R. (1978): Geologische Karte von Bayern 1:25,000, Blatt Nr 7927 Amendingen mit Erläuterungen, München.

KUPSCH, F. (1982): Geologie. In: KUPSCH, F. & WILLIBALD, D.: Hydrogeologische Karte von Baden-Württemberg — Oberschwaben: Erolzheimer Feld/Illertal, Erläuterungen, Freiburg/Karlsruhe, 12–23.

LÖSCHER, M. (1976): Die präwürmzeitlichen Schotterablagerungen in der nördlichen Iller-Lech-Platte. Heidelberger Geogr. Arb. **45**.

LÖSCHER, M. (1979): Abschlußbericht des DFG Projektes LÖ 247/1, unpublished.

LÖSCHER, M. & RÖGNER, K. (in prep.) Zur Paläogeographie und Stratigraphie von Staudenplatte, Arlesrieder und Eppishauser Schotter.

LÖSCHER M. & SCHEUENPFLUG, L. (1981): Der altpleistozäne Donaulauf und der untere Deckschotter in der nördlichen Iller-Lech-Platte. Jahresber. Mitt. Oberrhein. Geolog. Verein **63**, 335–343.

PENCK, A. & BRÜCKNER, E. (1901/1909): Die Alpen im Eiszeitalter. Leipzig, 3 vols.

RÖGNER, K. (1979): Die glaziale und fluvioglaziale Dynamik im östlichen Lechgletschervorland — Ein Beitrag zur präwürmzeitlichen Pleistozänstratigraphie. In: GRAUL, H. & LÖSCHER, M. (eds): Sammlung quartärmorphologischer Studien II, Heidelberger Geogr. Arb. **49**, 67–138.

RÖGNER, K. (1980): Die pleistozänen Schotter und Moränen zwischen Mindel- und Wertachtal (Bayerisch-Schwaben). Eiszeitalter u. Gegenwart **30**, 125–144.

RÖGNER, K. (1986a): Die quartären Ablagerungen beiderseits des östlichen Günztals zwischen den Marktorten Rettenbach und Ronsberg (Bayerisch-Schwaben). Jahresber. Mitt. Oberrhein. Geol. Verein **68**, 177–188.

RÖGNER, K. (1986b): Genese und Stratigraphie der ältesten Schotter der südlichen Iller-Lech-Platte (Bayerisch-Schwaben). Eiszeitalter u. Gegenwart **36**, 111–119.

RÖGNER, K. & LÖSCHER, M. (1987): Quartäre Sedimentations- und Verwitterungsphasen bei Hinterschmalholz (Regierungsbezirk Schwaben). Mitt. Geogr. Gesellsch. München **72**, 161–170.

RÖGNER, K., LÖSCHER, M. & ZÖLLER, L. (1988): Stratigraphie, Paläogeographie und erste Thermolumineszenzdatierungen aus der westlichen Iller-Lech-Platte (Nördliches Alpenvorland, Deutschland). Z. Geomorph. NF, Supp. Bd. **70**, 51–73.

ROPPELT, T. (1988): Die Geologie der Umgebung von Obergünzburg im Allgäu mit sedimentpetrographischen Untersuchungen der glazialen Ablagerungen. Diss. rer. nat., TU München.

SCHAEFER, I. (1950): Die diluviale Erosion und Akkumulation — Erkenntnisse aus Untersuchungen über Talbildung im Alpenvorlande. Forsch. z. dt. Landeskunde **49**.

SCHAEFER, I. (1956): Sur la division du Quaternaire dans l'avant-pays des Alpes en Allemagne. Actes IV Congres INQUA, Rome/Pise 1953, vol. **2**, 910–914.

SCHAEFER, I. (1957): Erläuterungen zur Geologischen Karte von Augsburg und Umgebung 1:50,000. München.

SCHAEFER, I. (1968): The succession of fluvioglacial deposits in the Northern Alpine Foreland. Proc. of the VIII Congress INQUA, Boulder-Denver/Colorado, vol. 14, 9–14.

SCHAEFER, I. (1973): Das Grönenbacher Feld — Ein Beispiel für Wandel und Fortschritt der Eiszeitforschung seit Albrecht Penck. Eiszeitalter u. Gegenwart **23/24**, 168–200.

SCHAEFER, I: (1980): Der angeblich "altpleistozäne Donaulauf" im schwäbischen Alpenvorland. Jahresber. Mitt. Oberrhein. Geol. Verein **62**, 167–198.

SCHEDLER, J. (1979): Neue pollenanalytische Untersuchungen am Schieferkohlevorkommen des Uhlenberges bei Dinkelscherben (Schwaben). Geologica Bavarica **80**, 159–164.

SCHEUENPFLUG, L. (1970): Weißjurablöcke und -gerölle der Alb in pleistozänen Schottern der Zusamplatte (Bayerisch- Schwaben). Geologica Bavarica **63**, 177–194.

SCHEUENPFLUG, L. (1971): Ein alteiszeitlicher Donaulauf in der Zusamplatte. Ber. Naturforsch. Gesellsch. Augsburg **27**, 3–10.

SCHEUENPFLUG, L. (1974): Zur Stratigraphie Altpleistozäner Schotter südwestlich bis nordwestlich Augsburg (Östliche Iller-Lech-Platte). Heidelberger Geogr. Arb. **40**, 87–94.

SCHEUENPFLUG, L. (1979): Der Uhlenberg in der östlichen Iller-Lech-Platte (Bayerisch-Schwaben. Geologica Bavarica **80**, 159–164.

SCHEUENPFLUG, L. (1987): Die quartäre Eintiefung des Gewässernetzes und Ausräumung im Augsburger Umland. Ber. Naturwissensch. Verein Schwaben. **91**, 82–86.

SCHOLZ, H. & ZACHER, W. (1983): Geologische Übersichskarte 1:200,000, Blatt CC 8726 Kempten (Allgäu). Hannover.

SCHREINER, A. (1970): Erläuterungen zur Geologischen Karte des Landkreises Konstanz mit Umgebung 1:50,000. (2. berichtigte Aufl. 1974). Stuttgart.

SCHREINER, A. & EBEL, R. (1981): Quartärgeologische Untersuchungen in der Umgebung von Interglazialvorkommen im östlichen Rheingletschergebiet (Baden-Württemberg). Geol. Jahrb. **A 59**, 3–64.

SHACKLETON, N. J. & OPDYKE, N. D. (1973): Oxygen isotope and palaeomagnetic stratigraphy of equatorial Pacific core V28-238: oxygen isotope temperatures and ice volumes on 10^5 and 10^6 scale. Quaternary Research **3**, 39–55.

SHACKLETON, N. J. & OPDYKE, N. D. (1976): Oxygen-Isotope and Paleomagnetic Stratigraphy of Pacific Core V28-229 - Late Pliocene to Latest Pleistocene. Geol. Soc. Am. Memoir **145**, 449–464.

SINN, P. (1972): Zur Stratigraphie und Paläogeographie des Präwürm im mittleren und südlichen Illergletscher-Vorland. Heidelberger Geogr. Arb. **37**.

SINN, P. (1974): Glazigene, fluvioglaziale und periglazialfluviatile Dynamik in ihrem Zusammenwirken an der präwürmzeitlichen Talgeschichte der Eschach zwischen Rhein- und Illergletscher. In: Hans-Graul-Festschrift, edited by H. EICHLER & H. MUSALL, Heidelberger Geogr. Arb. **40**, 95–120.

TILLMANNS, W., BRUNNACKER, K. & LÖSCHER, M. (1983): Erläuterungen zur Geologischen Karte der Aindlinger Terrassentreppe zwischen Lech und Donau 1:50,000. Geologica Bavarica **85**, 3–31.

TILLMANNS, W., KOCI, A. & BRUNNACKER, K. (1986): Die Brunhes/Matuyama-Grenze in Roßhaupten (Bayerisch-Schwaben). Jahresber. Mitt. Oberrhein. Geol. Verein, NF **68**, 241–247.

TROLL, C. (1926): Die jungglazialen Schotterfluren im Umkreis der deutschen Alpen - Ihre Oberflächengestalt, ihre Vegetation und ihr Landschaftscharakter. Forsch. z. dt. Landes- u. Volkskunde **24**, 161–256.

WEINHARDT, R. (1973): Rekonstruktion des Eisstromnetzes der Ostalpennordseite zur Zeit des Würmmaximums mit einer Berechnung seiner Flächen und Volumina. In: GRAUL, H. & EICHLER, H. (eds): Sammlung quartärmorphologischer Studien I, Heidelberger Geogr. Arb. **38**, 158–178.

WOLDSTEDT, P. (1958): Das Eiszeitalter — Grundlinien einer Geologie des Quartärs. Zweiter Band: Europa, Vorderasien und Nordafrika im Eiszeitalter. Stuttgart.

ZAGWIJN, W. H. (1957): Vegetation, climate and time correlations in the Early Pleistocene of Europe. Geol. en Mijnbouw. **19**, 233–244.

Addresses of authors:
Prof. Dr. Karl Albert Habbe
Institut für Geographie der Universität Erlangen-Nürnberg
Kochstr. 4
D-8520 Erlangen
Prof. Dr. Konrad Rögner
Fachbereich III Geographie/Geowissenschaften der Universität Trier
Postfach 3825
D-5500 Trier

CATENA SUPPLEMENT 15 p. 329–342 Cremlingen 1989

SUSPENDED LOAD YIELD OF A SMALL ALPINE DRAINAGE BASIN IN UPPER BAVARIA

Michael **Becht**, München

1 Introduction

In the northern marginal Limestone Alps there are numerous occurrences of unconsolidated Pleistocene valley fills. Small tributary valleys were blocked by large valley glaciers, such as the Inn, Isar and Loisach glaciers, and partially filled by allochthonous ice masses. Sediment traps were created for the fluvially and glacially transported rock waste and large quantities of unconsolidated sediment accumulated. They have been eroded, in part, postglacially and at the present time are dissected by a large number of active erosional incisions. During flood events, the hydrographs and sediment loads of the streams depend greatly on the relative share of these erosional cuts in the drainage basin area.

Within the framework of the International Hydrological Decade, the drainage basin of the Lainbach (18.8 km^2) near Benediktbeuren (fig.1) was selected as a representative area for the investigation of the water budget at the northern margin of the Alps (HERRMANN et al. 1973). The Pleistocene unconsolidated sediments have a maximum thickness of 200 m and an areal extent of more than 7 km^2 (KARL & DANZ 1969) in the drainage basin. In the north of the basin there are forest covered flysch ridges and in the south the Wetterstein limestone forms steep summits such as the Benediktenwand (1800 m) and Glaswand (1497 m) (DOBEN 1985). The limestone is highly karstified so that runoff on the upper parts of the slopes is of only short duration, despite the high precipitation.

The drainage basin of the Lainbach has a total relief of 1125 m and can be subdivided into three subbasins (fig.2):

1. The Kotlaine (6.2 km^2) drains the northern and eastern part of the area. The major portion of the erosional cuts in the unconsolidated sediments occurs in this subbasin.

2. The drainage basin of the Schmiedlaine (9.4 km^2) includes the karstified southern area (2.9 km^2) and the area southwest of the Lainbach valley.

3. The Lainbach (3.3 km^2), in this case defined as the lower course below the junction of Kotlaine and Schmiedlaine. Flysch predominates in this area. The unconsolidated sediments that were present at the end of the last glaciation have, for

ISSN 0722-0723
ISBN 3-923381-18-2

3-923381-18-4/89/5011851/US$ 2.00 + 0.25

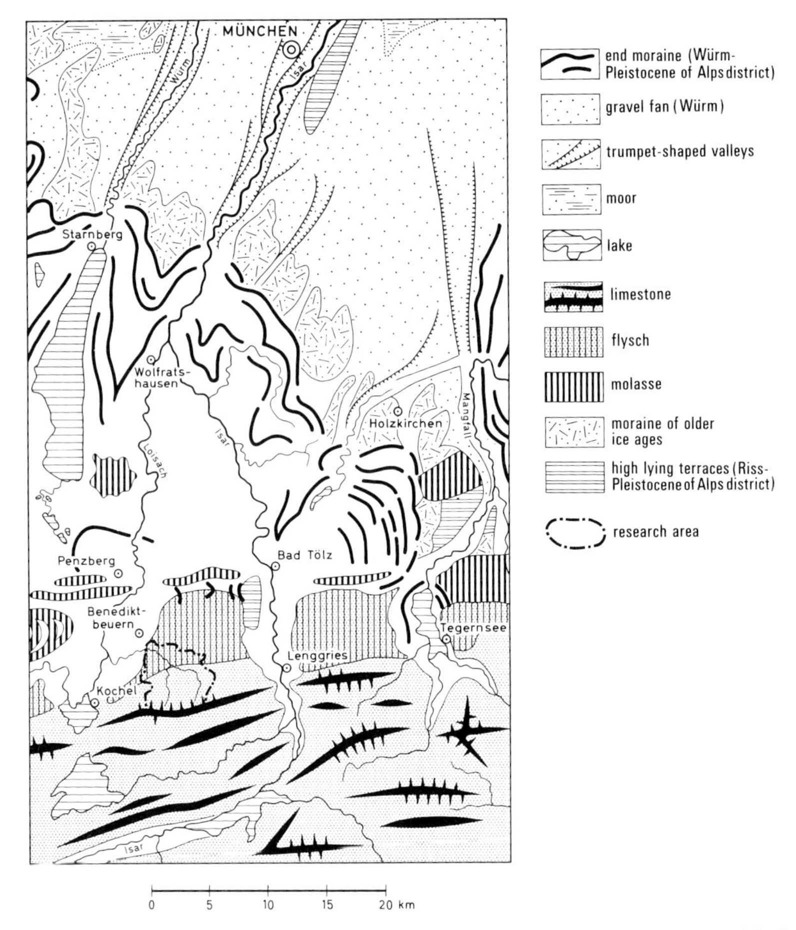

Fig. 1: *Location of the Lainbach drainage basin (after TROLL, unpublished).*

the most part, been removed by erosion.

The channels of the Lainbach, the Kotlaine and the Schmiedlaine are deeply incised into the unconsolidated sediments and flow on pre-Quaternary bedrock for long distances. Vertical and lateral erosion largely ceased after 1886 when the first erosion controls were installed.

The suspended load yield of the drainage basin has been investigated since 1984. Sampling during runoff events is carried out at the measurement points S1 on the Lainbach, S2 on the Kotlaine and S3 on the Schmiedlaine (fig.2). At the points S4 to S8 the solid load yield from small subbasins (<0.5 km^2) that have a high proportion of erosional incisions is measured. In addition, the suspended load of the karst springs in the Schmiedlaine basin is also being sampled (S9 and S10).

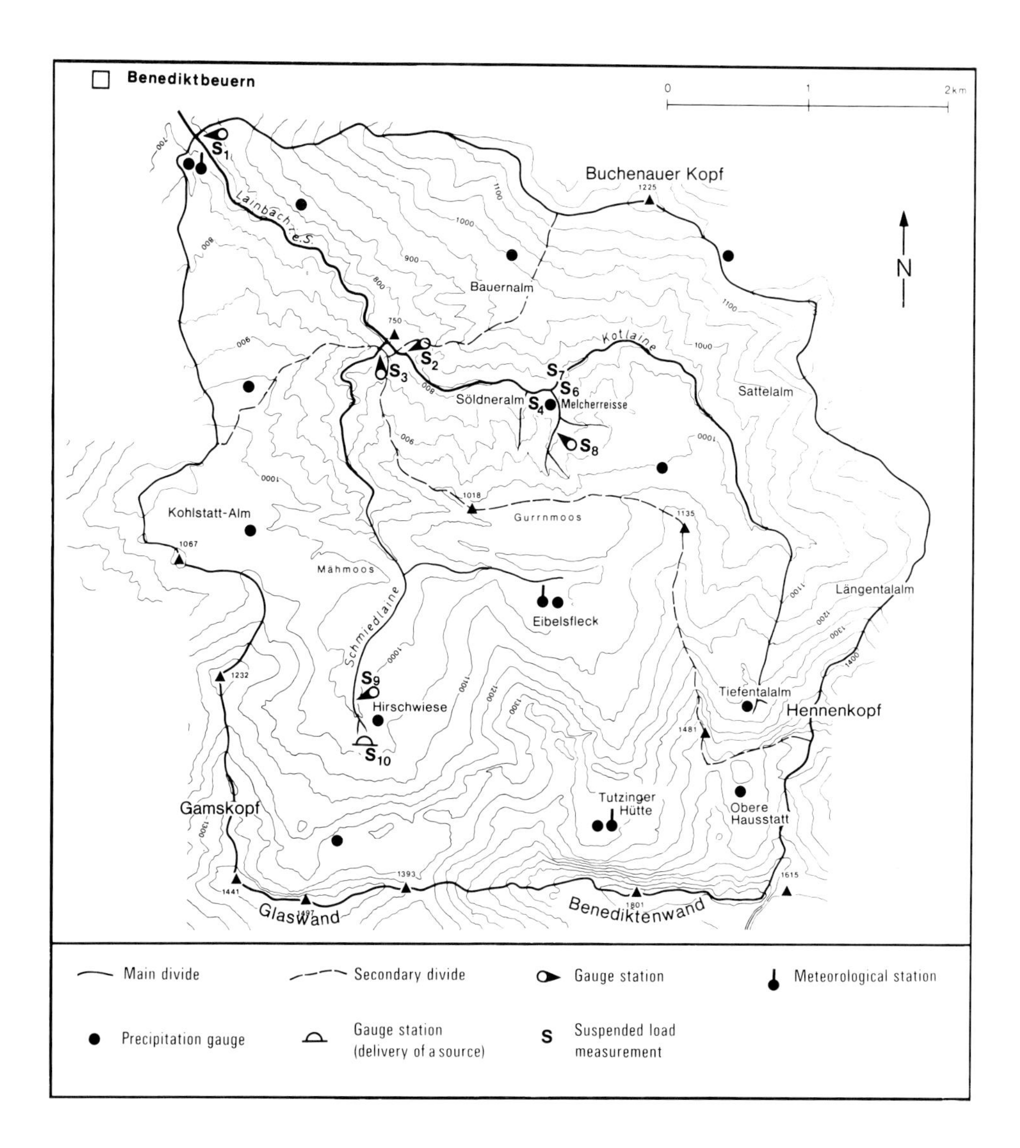

Fig. 2: *Instrumentation of the Lainbach drainage basin.*

Suspended load measurement points:
S1: Lainbach gauging station; S 2: Kotlaine gauging station; S 3: Schmiedlaine gauging station; S 4: Söldneralm; S 6: Mouth of Melcherbach; S 7: Kotlaine above the mouth of Melcherbach; S 8: Melcherbach gauging station; S 9: Schmiedlaine Bridge gauging station; S 10: Spring of the Schmiedlaine

Photo 1: *Aerial view of the Lainbach valley.*

Photo 2: *Lainbach valley and Benediktenwand (view southward). In the middle ground is the Melcherreisse.*

2 The spatial differentiation of the suspended load yield in the Lainbach valley

2.1 Suspended yield from an erosional incision (Melcherreisse)

The suspended particles transported into the streams after rainfall and meltwater runoff events originate, for the most part, in small erosion cuts with no vegetation on the banks of the streams and from larger erosional incisions (Reissen) that have developed in the areas of the Pleistocene loose sediments.

The share of the largest erosional incision, Melcherreisse, of the solid load discharge of its effluent streams, the Kotlaine and Lainbach, was determined for one intensive summer rainfall event. At the Melcherbach gauging station on July 18, 1986 after a heavy rain related to a thunderstorm, one preliminary maximum and two main maxima of suspended load concentration occurred (fig.3). These maxima of discharge and suspended load transport were the immediate reaction of this drainage basin (0.14 km^2), to rainfall events and short term rainfall intensity variations. In the upper course of the Kotlaine (S4: Söldneralm, fig.2) only two of the three suspended load maxima were present and at the gauging station Kotlaine, only one. The suspended load concentration did, however, increase from 15 to 29 kg/m^3 between Söldneralm and Kotlaine because several additional inflows with high wash loads from nearby large erosional cuts occurred in this stream segment.

A quantitative comparison of these suspended load concentrations with the measurements at the Melcherbach station was not possible because a large proportion of the sand and gravel fractions was transported in suspension in the upper course and moved as bedload at the Söldneralm and Kotlaine stations (tab.1). They could not, therefore, be included in suspended load measurements. In order to facilitate such a comparison, the share of coarse suspended load and of bedload in the total solid load transport of the Melcherbach is shown in fig.3. The share increases from 10% in the first maximum to 22% in the second and 35% in the third. During the final falling stage it is 95%.

The intensity of the suspended load removal from the Melcherreisse is influenced by the substrate. After a hot summer period the unconsolidated sediment, which contains a high proportion of silt and clay, is firmly baked together and contains desiccation cracks. The effective precipitation during the initial rain, which produced the first suspended load maximum on July 18, 1986, amounted, therefore, to only 2.2% of the total of 6 mm precipitation. During the following main rainfall of 24 mm the effective precipitation increased to 4.8%.

In spite of the small runoff quotient, suspended load particles and bedload particles were mobilized on the bare, steep surfaces, about 2 hectares in extent, in the Melcherbach drainage basin (BECHT & KOPP 1988). As the wetness of the substrate increased, larger particles were removed and the bedload transport became intensified with the increase in duration of the event. The small maximum registered at 16.25 h was caused by an increase in the coarse suspended load and the bedload during declining discharge (fig.3). Bedload particles move more slowly than the suspended load, which moves at the flow velocity of the water, and arrive at the measurement

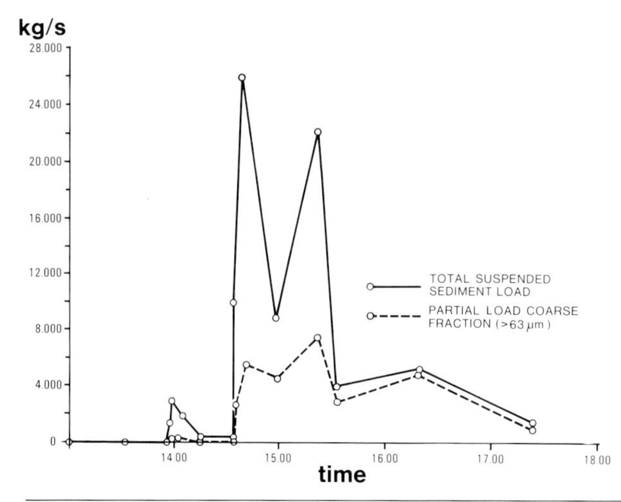

Fig. 3: *Suspended load at the Melcherbach gauging station during the flood event of July 18, 1986.*

	Total load (t)	suspended load coarse (t)	suspended load fine (t)	coarse suspended load in % of total load	fine suspended load of Melcherreisse in % of total fine suspended load
Melcherreisse	83.2	45.9	37.3	55	100
Kotlaine	70.7	7.7	63.0	11	59
Lainbach	72.1	6.8	65.3	9	57
Schmiedlaine	11.2	—	—	—	—

Tab. 1: *Suspended sediment yield at the gauging stations of the Melcherbach, Kotlaine, Schmiedlaine and Lainbach on July 18, 1986.*

station after a delay.

Since the share of the coarse suspended load decreases from 55% in the upper course in the Melcherbach to less than 10% in the lower course of the Kotlaine and the Lainbach (tab.1), only the wash load component, which is not deposited under the existing conditions of gradient and discharge, is used for a comparison of the suspended load transport at each measurement station.

During the event of July 18, 1986, the Melcherreisse drainage basin supplied 59% of the wash load of the Kotlaine and 57% of the wash load of the entire Lainbach basin. Since the wash load particles are derived solely from incisions in the Melcherbach area in which there is no vegetation cover, this means that only 0.1% of the drainage basin area produced 57% of its wash load yield. The Melcherreisse accounts for 30% of the total erosional area in the Kotlaine basin (BECHT & KOPP 1988) so that relative also to the Kotlaine, its wash load production is very large. In the Melcherreisse, the transport of material to the channel is intensified by gravitational mass movements of the loose sediments. Rotational slips, induced by groundwater seepage, and mudflows which deposit material in the channels occur only in the Melcherreisse within the Lainbach basin at the present time.

These processes are especially intensive during the spring snow melt. The discharge in the channels is still relatively low at this time compared to the rain-induced flood discharges in the summer so that only a small proportion of the slide and mudflow deposits is removed from the Lainbach valley during spring (BECHT 1987). Most of the material is redeposited temporarily after being transported a short distance in the steep upper course of the Melcherbach. In April of 1987, the mudflows advanced for the first time to the Kotlaine, the trunk stream of the Melcherbach. Although the greater part of the redeposited loose sediments is, for the most part, removed from the Lainbach valley during the flood discharges of the summer half year, measurements during the occurrences of mudflows in the spring of 1985 (BECHT 1986) showed that, within three days, the Lainbach had transported about 3000 t of suspended load from its basin.

2.2 Spatial distribution of the suspended load removal as a function of the amount and intensity of precipitation

2.2.1 The shares of the subbasins Kotlaine, Schmiedlaine and Lainbach of the total sediment removal

Because of the high density of erosional incisions, the Kotlaine drainage supplies, on average, about 72% of the entire suspended load of the Lainbach valley. During local thunder showers or during the occurrence of mudflows the proportion increases to more than 95%. The erosional area of the Melcherreisse is the dominant supplier of the suspended load to the Kotlaine and to the Lainbach.

The longer an intensive precipitation event lasts, the more the sediment removal increases from the other parts of the Lainbach basin. For example, on August 11–12, 1984 there was a suspended load removal of 3660 t from the Kotlaine and of 1880 t from the Schmiedlaine. There are only a few erosional incisions in the partial drainage basin of the Lainbach trunk stream, so that sediment production was lowest in this basin.

Slides can, however, cause temporary stream load increases. In September 1984, undercutting by lateral erosion in the lower course of the Schmiedlaine caused the banks to slide with the result that 50% of the total suspended load came from this area. A few weeks later, the load had decreased again to the prior value.

2.2.2 Hysteresis curves of suspended load concentration

The maximum of the suspended load transport frequently does not coincide with the maximum of the discharge. This phenomenon is independent of the suspended load concentration. Most often the suspended load maximum precedes the discharge maximum, so that a clockwise hysteresis (KLEIN 1984) results (fig.4). The discharge maximum is composed of the partial discharges of the entire drainage basin but the suspended load particles are mainly removed by wash denudation from erosional incisions that lie close to the main trunk channels. The particles arrive, therefore, at the measurement stations earlier than the discharge maximum.

A clockwise hysteresis results even if the maxima of the discharge and suspended load occur simultaneously, be-

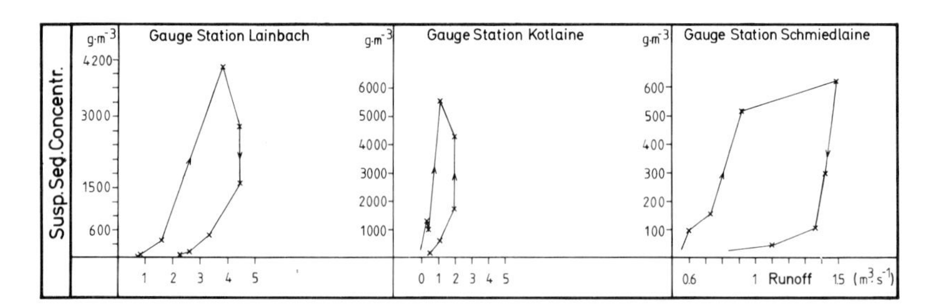

Fig. 4: *Hysteresis curves of the suspended load concentration at the Kotlaine, Schmiedlaine and Lainbach on July 1, 1985.*

cause the suspended load concentration decreases rapidly with decreasing rainfall intensity while the discharge remains longer at a high level. This is due to delayed near-surface runoff (rapid interflow).

The effect occurs during intensive rains of long duration because, at that time, the supply area of the suspended load is not limited to the erosional incisions but includes a large part of the total drainage area. The carbonate content of the suspended load also decreases in these cases from 65% to a mean of 45% since wash denudation takes place in both the calcareous unconsolidated sediments and the flysch areas. Additional information relating to the sediment from the flysch was obtained by X-ray diffractometer analysis (BECHT 1986).

Thunder showers can be very local and have high rainfall intensities. The quantities of transported suspended load and the arrival of the load maximum and the discharge maximum at a measurement station depend upon the location of the thunderstorm centre. Since thunderstorms frequently have, for orographic reasons, their centre in the area of large erosional incisions, there tends to be a coincidence of both maxima in these cases.

2.3 Grain size distribution of the suspended load as a function of runoff duration and intensity

The conventional separation between suspended load and bedload at the critical silt/sand boundary of 0.063 mm is useful in lowland streams but the suspended load in a torrent may temporarily include sand and gravel components. This also applies to upper courses of the streams in the Lainbach basin. A comparison of the suspended load transport at the measurement stations requires, therefore, a distinction between fine-grained (<0.063 mm) and coarse-grained (>0.063 mm) suspended load.

In the winter half-year, the share of coarse-grained suspended load transported in the trunk streams Lainbach, Kotlaine and Schmiedlaine remains below 5% of the total because the discharges of the snow melt period remain far below those of rain-induced summer flood events (BECHT 1987).

In the 1985 snow melt season, the share of coarse-grained suspended load in the upper course at the Melcherbach station was, after the occurrence of mudflows, 25% of the total and decreased down valley to 10% at the Söldneralm station, to 3.5% at the Kotlaine station and 0.8% at Lainbach station which lies

Date	Maximum discharge	coarse suspended load during maximum discharge
30.07.1985	0.482 m^3/s	1%
15.06.1985	2.169 m^3/s	11%
27.08.1985	3.777 m^3/s	21%
11.08.1984	22.770 m^3/s	37%

Tab. 2: *The proportion of coarse-grained suspended load of the Kotlaine during single flood events in the summer.*

at the exit of the basin (fig.2). In the summer half year the tractive force of the stream flow becomes much higher and the share of the coarse-grained suspended load rises during the high discharges of rainfall-induced floods (tab.2).

During some of the rare very intensive flood events the share of the coarse suspended load may increase considerably for a short period, an increase that also causes an increase in the overall suspended load concentration (fig.5). The concentration peak usually occurs with some delay after the passage of the peak discharge (fig.5). Such transport waves of coarse suspended load are produced by the destruction of gravel pavements in the stream bed (LEKACH & SCHICK 1983) which can take place when the tractive force exceeds a critical value. Once the bed material has been set in motion, even the discharge of the falling stage of the flood may suffice to keep the coarse-grained load in suspension. It is, therefore, possible that during long lasting flood events, waves of coarse-grained suspended sediments can occur after a discharge peak and before a further discharge increase. The threshold discharge for the occurrence of coarse-grained suspended sediment waves is about 8 m^3/s in the Lainbach and about 6 m^3/s in the Kotlaine and the Schmiedlaine.

Slides from the steep sides of the Melcherreisse into the channel are another possible source of coarse sediment waves and can cause sudden increases of the coarse-grained suspended load in the Kotlaine, if the tractive force of the stream is sufficiently high.

The high discharge of thunderstorm rainfall is of short duration compared to other rainfall events and often does not last long enough to transport the coarse load over great distances. The discharge decreases again rapidly so that after thundershowers more than 95% of the suspended load consists of silt and clay particles.

The suspended load is considerably more fine-grained during the rising stage of a flood than during the falling stage. As the rainfall intensity lessens, the supply of fines from the erosional incisions is stopped. This leads to a clockwise hysteresis of the suspended load concentration since the amount of fine-grained load decreases rapidly after the passage of the maximum and there is also a relative increase of the coarse-grained suspended load. The hysteresis of the coarse component is anti-clockwise because the discharge maximum occurs ahead of the load maximum.

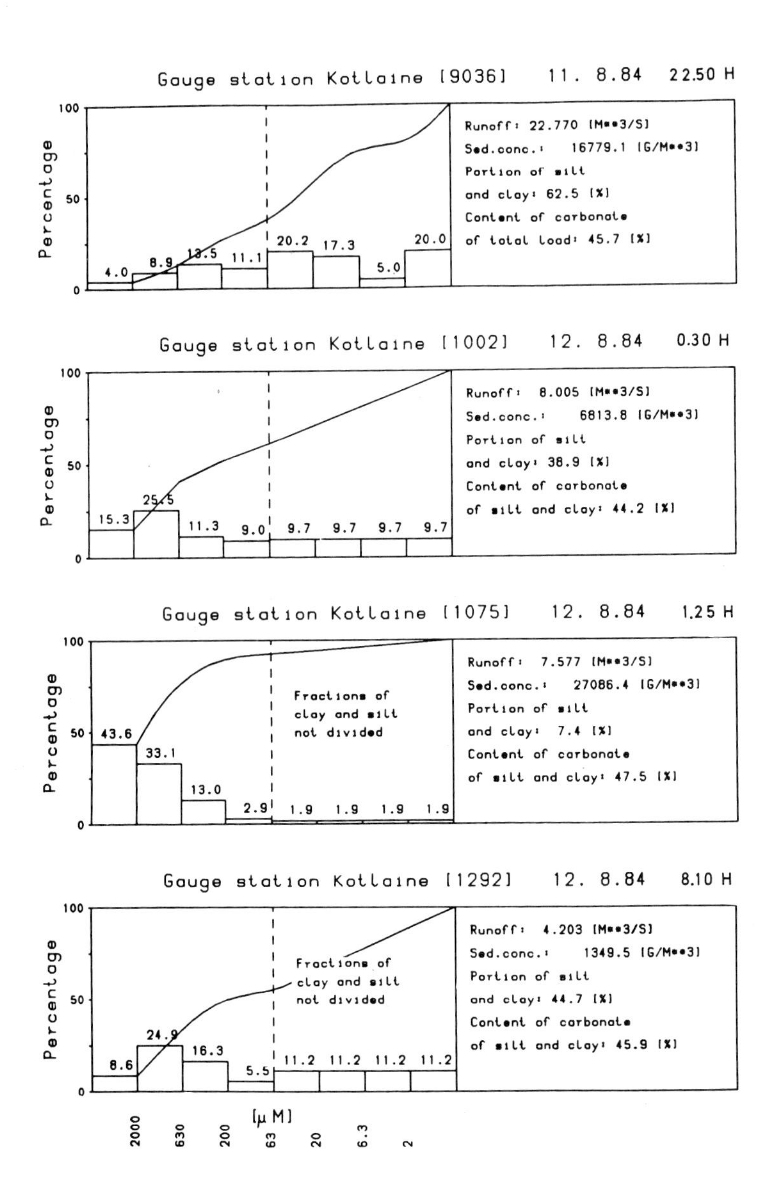

Fig. 5: *Variations in the grain size distribution of the suspended load during the flood event of August 11–12, 1984 at the Kotlaine gauging station.*

3 The annual variation of the suspended load removal in the Lainbach valley

3.1 Suspended load removal during the winter half year

In the northern marginal Limestone Alps, suspended load transport events occur in connection with rainfall events and snowmelt discharges. During periods of continuous frost, which may last for several weeks, the trunk streams do not carry suspended loads.

A spatial and temporal differentiation of the suspended load removal by snow melt runoff is due to the exposure of the large erosional incisions to solar radiation; no suspended load was supplied from surfaces under forest. During the winter, melting after new snowfalls always begins in the steep erosion incisions that are exposed to the south. The snow cover frequently slides off because of the steepness of the slopes and only a thin snow layer remains through which the solar radiation penetrates into the soil. The resulting increase in ground temperature causes the snow to melt within a few days. The suspended load transported by the trunk streams during this winter period is supplied almost exclusively from these south facing slopes. The amount of the load is apparently related to the radiation input which indicates that exposure to radiation is the dominating factor for snow melt processes on slopes exposed to the south. Additional measurements are needed to test this relationship statistically (BECHT 1987).

By contrast, on slopes that are exposed to the north, the snow melt takes place as a result of an increase in air temperature and higher sun angle in early spring, usually in the last third of March, and the supply of suspended load comes from all erosional incisions in the Lainbach, regardless of their exposure. In the winter and early spring of 1985, the total removal of suspended load from the Lainbach basin by snow melt runoff was nearly 300 t, of which about 150 t were supplied before the general spring thaw and 120 t mainly from north facing erosional incisions, after the beginning of the general thaw. In addition, rainfall events at the lower and middle elevations of the area during the winter of 1984/85 induced flood events which removed an additional 600 t of suspended load. The total production during this period from snow melt and rainfall runoff was, therefore, about 1000 t.

Not included are the effects of mudflows which were observed for the first time in 1985. Mudflows in the area of the Melcherreisse, triggered by saturation of the substrate and rising groundwater level during snow melt, contributed 3000 t load to the Lainbach system within the three days from April 3 to April 6, 1985, about three times the supply from rainfall and snow melt in the entire basin during the winter half year.

3.2 Suspended load transport in the summer half year

3.2.1 The influence of individual flood events on the monthly suspended load yield

The quantity of suspended load transported in the streams increases exponentially with the discharge (BECHT 1986). During the summer most of the suspended load is transported during the few days of particularly high floods. During the period of observation from 1984 to 1986, more suspended load was transported in a few hours on August

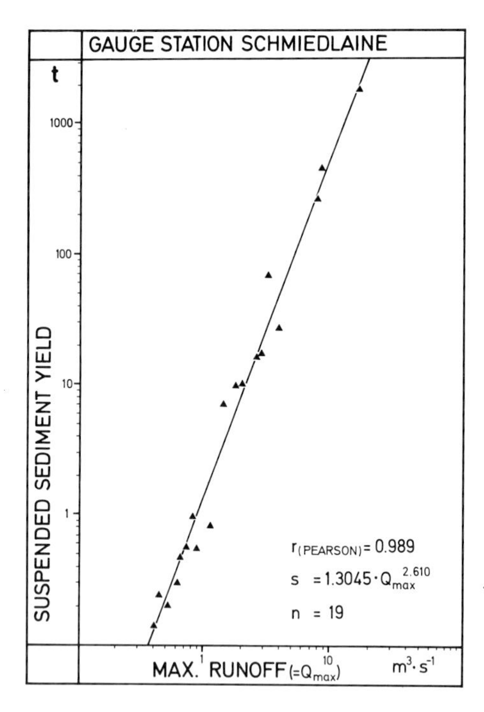

Fig. 6: *The statistical relationship between suspended load and peak discharge during flood events at the Schmiedlaine gauging station (based on measurements of the suspended load in 1984 and 1985).*

11 and 12, 1984 than in all other summer floods of 1984 combined or in all summer floods of 1985 or of 1986.

3.2.2 Calculation of the suspended load removal in the summer half year

An attempt to estimate total suspended load removal on the basis of the statistical relationship between load concentration and runoff in the Lainbach valley (BECHT 1986) encountered the difficulties pointed out by WALLING (1978), particularly because the hysteresis of the suspended load concentration differs for different events. Separation of rising and falling stages also did not produce satisfactory results. Because of the great number of flood events sampled, however, it was possible to relate the suspended load during a flood event to the peak discharge of that event with sufficient certainty (fig.6).

Since the calculation of the suspended sediment yield is also based on the total runoff amount of the flood event, a partial correlation test was made to determine whether the high correlation coefficient for yield = f (peak discharge) was caused by a relationship between total runoff and peak discharge. A high correlation resulted, even after the effect of total runoff was eliminated. For example, at the Lainbach station the correlation coefficient decreased from 0.978 to 0.866 after the elimination of the total runoff effect, for n = 21 pairs of data. Significant differences between expected and measured values may occur after local slides or mudflows, which cause a disproportionately high amount of suspended load.

After several successive flood events the suspended load removal becomes lower than the expected value. This is due to the decrease of available fine-grained material on the land surface and also to the increasing share of the base flow, which does not bring suspended load with it, in the total discharge.

The suspended load removal can be estimated for the summer half years 1972 to 1986 for the Lainbach basin for which runoff records are available (FELIX et al. 1988). The mean value per summer half year is 10,000 t. The values of individual summers can deviate considerably from this mean, as the data for 1984 to 1986 show; the standard deviation is 5780 t. It is evident from the fourteen years of data that the months of June, July and August contain the majority of intensive rainfall events and have, therefore, the highest values of suspended load removal. In spring and autumn, extreme flood events are rare and there is less suspended load at this time of year.

4 Areal denudation in the Lainbach basin

If the mudflows that have occurred since 1985 are considered to be singularities and are left out of calculations of long term trends, the sediment removal takes place predominantly in the summer half year. Without the mudflows the mean share of the winter half year is about 10% of the total annual loss.

The mean annual suspended load production per unit area is 580 t/km^2. The extreme values measured are 150 t/km^2 in 1976 and 1070 t/km^2 in 1974. Inclusion of the mudflow events would increase the mean annual production to 740 t/km^2.

On the basis of the suspended load removal, the mean denudation of the area is 0.38 mm/a. If the mean production of the basin is related only to the 12 ha of active erosional incisions from which most of the suspended load is probably derived (BECHT & KOPP 1987), the mean denudation rate on these 12 ha is approximately 60 mm/a. Since about half the Quaternary fill, the main source of the solid load, consists of gravel-sized particles which are transported as bedload, the total removal of solid load in the Lainbach valley is about twice as high as the value calculated only on the basis of the suspended load removal. No Holocene accumulation of these coarse components has been found in the area, indicating that they must be removed and transported during flood events.

The results of the suspended load measurements in the Lainbach valley agree with data of SOMMER (1980) from the Dürrache valley in the northern Limestone Alps where there was a mean annual rate of 777 t/km$^2 \cdot a$ from 1976 to 1978. PETERS-KÜMMERLY (1973) had a denudation rate that ranged from 0.01 to 0.51 mm/a in the Swiss Alps, an indication of the great variation in Alpine denudation rates. Streams in the Limestone Alps are burdened with glacial debris and carry, therefore, heavy sediment loads, in contrast to the central Alps where the suspended load removal is lower by about a power of ten (SOMMER 1980, VORNDRAN 1979).

The availability of fine-grained particles appears to be the limiting factor in humid middle latitudes. In upland areas outside the Alps much lower loads are transported than in the streams of the Limestone Alps. NIPPES (1983), for example, estimated a mean value of 39 t/km$^2 \cdot$a for the Dreisam river in the Black Forest.

References

BECHT, M. (1986): Die Schwebstofführung der Gewässer im Lainbachtal bei Benediktbeuren/Obb. Münchener Geogr. Abh. Bd. B2, Reihe B.

BECHT, M. (1987): Auswirkungen der Schneeschmelze auf die Schwebstofführung von Wildbächen. Beitr. z. Hydrologie, Jg. 11, H.2, (in print).

BECHT, M. & KOPP, M. (1988): Aktuelle Geomorphodynamik in einem randalpinen Wildbacheinzugsgebiet und deren Beeinflussung durch die Wirtschaftsweise des Menschen. 46. Deutscher Geographentag München, Tagungsbericht und wissenschaftliche Abh., Stuttgart, 526–534.

DOBEN, K. (1985): Geologische Karte von Bayern 1:25,000, Erläuterungen zum Blatt Nr. 8334 Kochel am See. München.

FELIX, R., PRIESMEIER, K., WAGNER, O., VOGT. H. & WILHELM, F. (1988): Abfluß in Wildbächen. Untersuchungen im Einzugsgebiet des Lainbachtals bei Benediktbeuren/Oberbayern. Münchener Geogr. Abh., Reihe B, Bd. B6.

HERRMANN, A., PRIESMEIER, K. & WILHELM, F. (1973): Wasserhaushaltsuntersuchungen im Niederschlagsgebiet des Lainbaches bei Benediktbeuren/Obb. DGM, 17. Jg., H. 3, 65–73.

KARL, J. & DANZ, W. (1969): Der Einfluß des Menschen auf die Erosion im Bergland. Schriftenreihe der Bayer. Landesstelle f. Gewässerkunde, H. 1, München.

KLEIN, M. (1984): Anti-clockwise hysteresis in suspended sediment concentration during individual storms: Holbeck catchment, Yorkshire, England. CATENA **11**, 251–257.

LEKACH, J. & SCHICK, A.P. (1983): Evdidence for transport of bedload in waves: Analysis of fluvial sediment samples in a small upland stream channel. CATENA **10**, 267–279.

MANGELSDORF, J. & SCHEURMANN, K. (1980): Flußmorphologie. München/Wien, 262 pp.

MÜLLER-DEILE, G. (1940): Geologie der Alpenrandzone beiderseits vom Kochelsee in Oberbayern. Mitt. Reichsanst. Bodenforsch. **34**, München.

NIPPES, K.-H. (1983): Erfassung von Schwebstofftransporten in Mittelgebirgsflüssen. Geoökodynamik Bd. **4**, 105–124.

PETERS-KÜMMERLY, B. (1973): Untersuchungen über Zusammensetzung und Transport von Schwebstoffen in einigen Schweizer Flüssen. Geogr. Helv., Jg. 28/3, 137–151.

SOMMER, N. (1980): Untersuchungen über die Geschiebe- und Schwebstofführung und den Transport von gelösten Stoffen in Gebirgsbächen. Intern. Symp. Interpraevent, Bd. 2, 69–94, Bad Ischl.

VORNDRAN, G. (1979): Geomorphologische Massenbilanzen. Augsburger Geogr. Hefte, 1.

WALLING, D.E. (1978): Reliability considerations in the evaluation of an analysis of river loads. Z. f. Geomorph., Supp. Bd. **29**, 29–43.

Address of author:
Dr. Michael Becht
Institut für Geographie der Universität München
Luisenstraße 37
8000 München 2
Federal Republic of Germany

CATENA SUPPLEMENT 15 p. 343–347 Cremlingen 1989

GEOMORPHOLOGICAL MAPPING IN THE FEDERAL REPUBLIC OF GERMANY THE GMK 25 AND THE GMK 100

Dietrich **Barsch**, Heidelberg
Gerhard **Stäblein**, Bremen

1 The GMK program

At the XVIIIth International Geographical Congress in Rio de Janeiro (1956), the IGU Commission on Geomorphological Mapping initiated the research work that led to the publication of possible detailed legends for geomorphological maps at various scales (DEMEK 1972, DEMEK & EMBLETON 1978). Based on these suggestions and on geomorphological mapping methods developed in other countries (German Democratic Republic, Switzerland, the Netherlands, France, Poland and Canada), geomorphologists in the Federal Republic of Germany sought to devise a generally applicable system to represent the landforms in the Federal Republic on geomorphological maps and to develop data relevant to their interpretation (LESER 1985). From 1976 to 1986 this work was contained in a Priority Program (Schwerpunktprogramm) of the German Research Foundation (Deutsche Forschungsgemeinschaft) to develop geomorphological maps (GMK). A large number of geomorphologists have participated in this program.

The purpose of the program was to prepare and publish a series of geomorphological map sheets at the scales 1:25,000 (GMK 25) and 1:100,000 (GMK 100) of areas in the Federal Republic. Fig.1 shows the location of the completed map sheets. They include areas of all major landform types present in central Europe: the coastal marshes, the old and young moraine regions, the uplands and scarplands, the Alpine Foreland and the Alps.

Each map coincides with a topographic map quadrangle and with the corresponding geological and soil maps so that the contents of the GMK can be directly related to these other maps. The program has been described in detail in several publications, some in English (see publications by BARSCH, LESER, LIEDTKE, MÄUSBACHER and STÄBLEIN in the list of references). Only a brief review of the concept and scope of the program is possible here.

2 Concept and map legends

Landforms are shown, as far as possible, in terms of form elements such as slopes, crests and scarps, and in terms of form attributes such as slope angle,

ISSN 0722-0723
ISBN 3-923381-18-2

3-923381-18-4/89/5011851/US$ 2.00 + 0.25

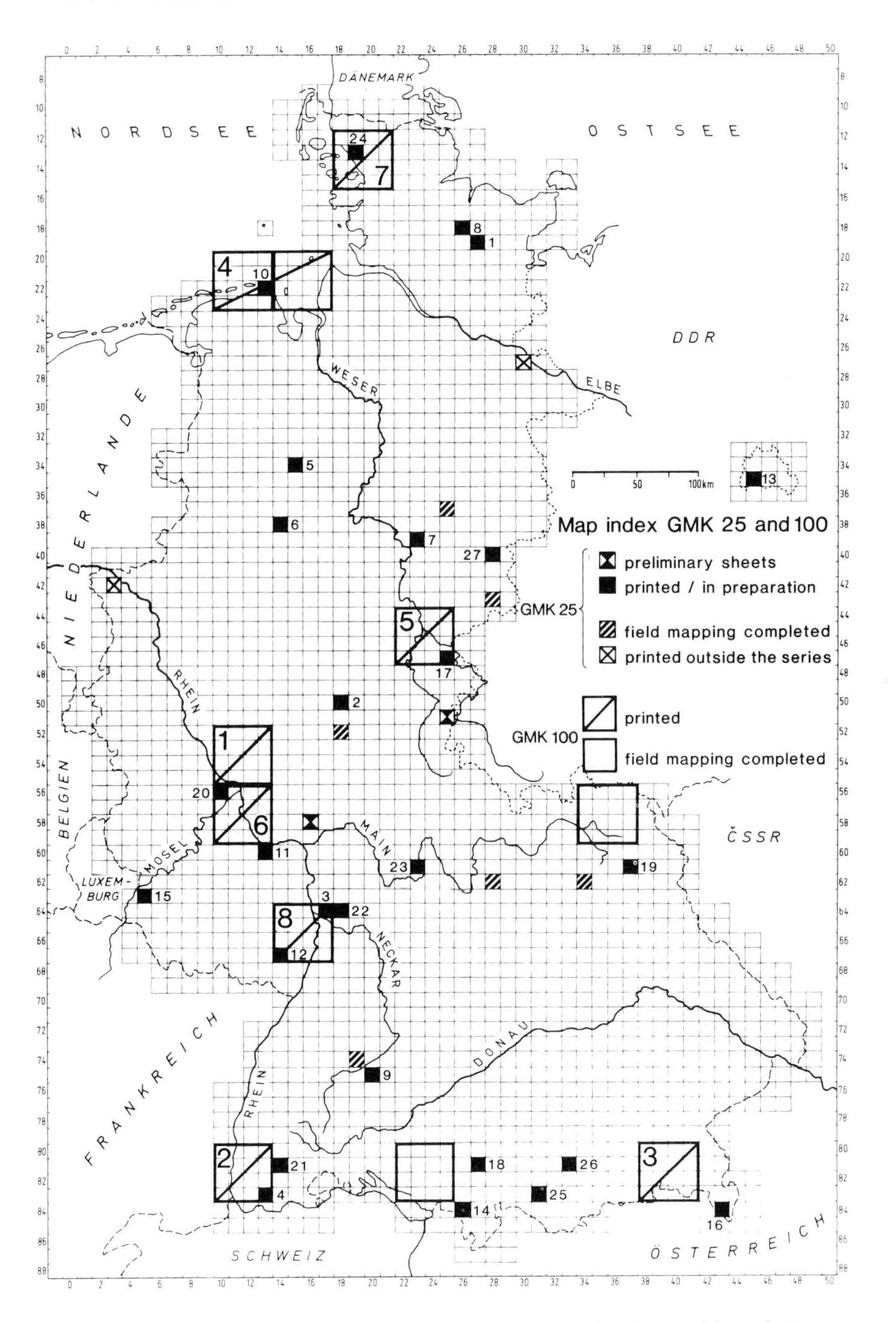

Fig. 1: *Map index for the GMK 25 and 100 of the Federal Republic of Germany.*

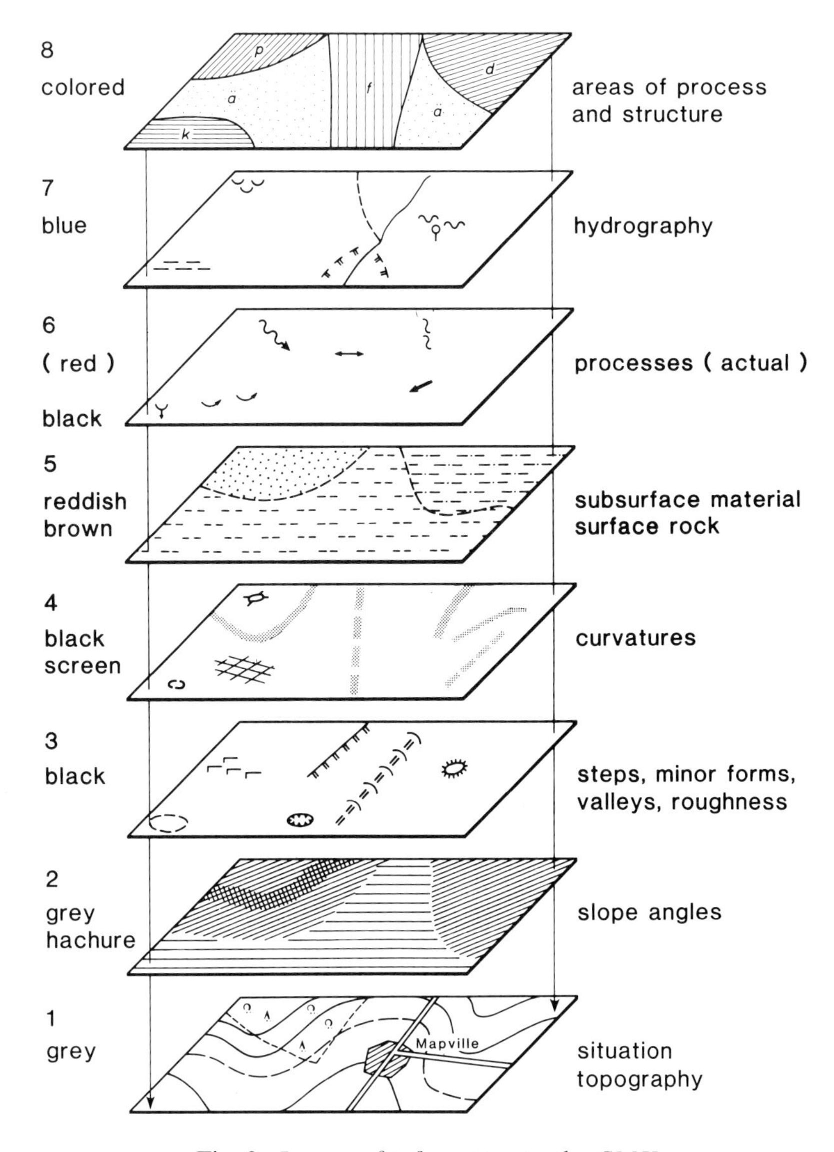

Fig. 2: *Layers of information in the GMK.*

surface materials and morphogenetic information (STÄBLEIN 1978). As a result, the legends for both the GMK 25 (LESER & STÄBLEIN 1975) and the GMK 100 (FRÄNZLE et al. 1979) contain patterns for individual form elements and attributes but not for all their possible combinations. Features or surface characteristics which cover an area larger than 2×4 mm on the map, that is, a ground area of 50x100 m for maps of 1:25,000 or of 200×400 m for maps of 1:100,000, are shown true to scale. Forms smaller than this lower limit of cartographic resolution are shown by symbols. Not all minor landforms can be included; preference is given to forms that are characteristic or genetically relevant for the area.

The geomorphological map has several layers of information (fig.2). The base, printed in grey, is the topographic map with grid coordinates, contours and other information that identifies location and topography. Different classes of slope angles are shown by grey hachuring of graduated intensity. Morphographic elements (minor landforms, surface roughness, axes of curvature) are shown by black symbols, the near-surface materials, including exposures of bedrock, by open reddish-brown patterns. Selected past processes and process indications (scars, materials) are indicated by black symbols, present-day processes in red. The hydrographic information of the topographic map is augmented by special symbols in blue. Areas in which the landforms have resulted from a dominant past or present process system are shown in solid colours, for example, fluvial landforms in light green, aeolian landform areas in yellow and (Pleistocene) periglacial landform areas in lilac. In spite of the many layers and categories of information, care has been taken to ensure that patterns that overlap remain recognizable and that symbols do not interfere with one another. Considerable cartographic skill and a high standard of printing techniques has been required to achieve this.

3 Applicability of the GMK

Because the legend is standardized for all GMK map sheets, the contents of different sheets can be easily compared with one another. This applies to the individual layers of information as well as to the representation of complex landform systems on the maps by combinations of symbols and colours. In addition, each sheet of the GMK 25 and the GMK 100 is accompanied by a detailed explanatory text (for example, BARSCH & MÄUSBACHER 1979, LESER 1983). Special maps and data assemblages that are of value for site evaluation and in regional planning can be produced by extracting information from GMK sheets (MÄUSBACHER 1985). The GMK concept has been found usable for mapping all relief types present in central Europe and elsewhere (BARSCH, FISCHER & STÄBLEIN 1987, BARSCH & LESER 1987).

Recent evaluations of the GMK 25 and GMK 100 mapping systems and of their usefulness for research and practical applications in geomorphology and other fields have been made by GELLERT (1986, 1988).

The GMK 25 and GMK 100 sheets that have been published may be ordered from GEOCENTER, Postfach 800830, 7000 Stuttgart.

References

BARSCH, D., FISCHER, K. & STÄBLEIN, G. (1987): Geomorphological Mapping of High Mountain Relief, Federal Republic of Germany (with geomorphological map Königssee, scale 1:25,000). Mountain Research and Development **7 (4)**, Boulder, 361–374.

BARSCH, D. & LESER, H. (eds.) (1987): Regionale Bespiele zur geomorphologischen Kartierung in verschiedenen Maßstäben 1:5 000 bis 1:200 000). Beiträge zum GMK-Schwerpunktprogramm VI. Berliner Geogr. Abh. **42**, 1–76.

BARSCH, D. & LIEDTKE, H. (1980): Principles, scientific value and practical applicability of the geomorphological map of the Federal Republic of Germany at the scale of 1:25 000 (GMK 25) and 1:100 000 (GMK 100). Z. Geomorph. N.F. Suppl. **36**, 296–313.

BARSCH, D. & LIEDTKE, H. (eds.) (1985): Geomorphological Mapping in the Federal Republic of Germany. Contributions to the GMK-Priority program IV. Berliner Geogr. Abh. **39**, 1–89.

BARSCH, D. & MÄUSBACHER, R. (1979): Erläuterungen zur Geomorphologischen Karte 1:25 000 der Bundesrepublik Deutschland GMK 25 Blatt 3, 6417 Mannheim-Nordost. Berlin, 56 pp.

BARSCH, D. & STÄBLEIN, G. (eds.) (1982): Erträge und Fortschritte der geomorphologischen Detailkartierung. Beiträge zum GMK-Schwerpunktprogramm III. Berliner Geogr. Abh. **35**, 1–134.

DEMEK, J. (ed.) (1972): Manual of detailed geomorphological mapping. Prague, 344 pp.

DEMEK, J. & EMBLETON, C. (1978): Guide to medium-scale geomorphological mapping. Stuttgart, 348 pp.

FRÄNZLE, O. et al. (1979): Legendenentwurf für die geomorphologische Karte 1:100 000 (GMK 100). Heidelberger Geogr. Arb. **65**, 1–18.

GELLERT, J. F. (1986): Die geomorphologische Detailkartierung (GMK 25) in der Bundesrepublik Deutschland, eine analytische Betrachtung. Petermanns Geogr. Mitt. **130**, 63–68.

GELLERT, J.F. (1988): Die geomorphologische Karte 1:100 000 (GMK 100) der BRD — Analyse und Probleme. Petermanns Geogr. Mitt. **132**, 293–297.

LESER, H. (1985): Incorporation of the GMK 25 BRD in the international development of geomorphological maps. Berliner Geogr. Abh. **39**, 9–16.

LESER, H. (1983): Geographisch-landeskundliche Erläuterungen der topographischen Karte 1:100 000 des Raumordnungsverbandes Rhein-Neckar. Forsch. z. dt. Ldke, Trier, 221.

LESER, H. & STÄBLEIN, G. (eds) (1975): Geomorphologische Kartierung, Richtlinien zur Herstellung geomorphologischer Karten 1:25 000 ("grüne Legende"). unveränderte Auflage, Berliner Geogr. Abh, Sonderheft, 1–39.

LIEDTKE, H. (1984): Geomorphological mapping in the Federal Republic of Germany at scales of 1:25 000 and 1:100 000; a priority programme supported by the German Research Foundation (DFG). Bochumer Geogr. Arb. **44**, 67–73.

MÄUSBACHER, R. (1985): Die Verwendbarkeit der geomorphologischen Karte 1:25 000 (GMK 25). Berliner Geogr. Abh. **40**, 1–97.

STÄBLEIN, G. (ed.) (1978): Geomorphologische Detailaufnahme, Beiträge zum GMK-Schwerpunktprogramm I. Berliner Geogr. Abh. **30**, 1–95.

STÄBLEIN, G. (1980): Die Konzeption der Geomorphologischen Karten GMK 25 und GMK 100 im DFG-Schwerpunktprogramm. Berliner Geogr. Abh. **31**, 13–30.

Address of authors:
Prof. Dr. Dietrich Barsch
Geographisches Institut der Univesität
Im Neuenheimer Feld
D-6900 Heidelberg
Prof. Dr. Gerhard Stäblein
Universität Bremen
Physiogeographie & Polargeographie
FB 5 Geowissenschaften
Klagenfurter Straße
D-2800 Bremen 33

VORANKÜNDIGUNG

Erscheinungstermin: August 1989

Claus Dalchow

VORLESUNGSAUSWERTUNGEN HEINRICH ROHDENBURG: GEOÖKOLOGIE - GEOMORPHOLOGIE

CATENA paperback

1989/broschiert/ca. 180 Seiten/DM 29.50/ US$ 17.50

ISBN 3-923381-21-2

BESTELLSCHEIN

Ich bestelle hiermit .... Exemplare: Claus Dalchow, **VORLESUNGSAUSWERTUNGEN HEINRICH ROHDENBURG: GEOÖKOLOGIE - GEOMORPHOLOGIE**, CATENA paperback. 1989. Broschiert DM 29.50/US$ 17.50. ISBN 3-923381-21-2

Name ....

Anschrift ....

Datum/ Unterschrift ....

Bitte belasten Sie meine Kreditkarte: ☐ Eurocard/Mastercard/Access ☐ Visa ☐ Diners ☐ American Express

Kartennummer .... gültig bis .... Unterschrift: ....

CATENA VERLAG, Brockenblick 8, D-3302 Cremlingen-Destedt, West Germany, tel.05306-1530, fax 05306-1560

SOIL TECHNOLOGY

A Cooperating Journal of **CATENA**

This quarterly journal is concerned with applied research and field applications on

- **soil physics,**
- **soil mechanics,**
- **soil erosion and conservation,**
- **soil pollution,**
- **soil restoration,**
- **drainage and irrigation,**
- **land evaluation.**

The majority of the articles will be published in English but original contributions in French, German or Spanish, with extended summaries in English will occasionally be considered according to the basic principles of the publisher CATENA whose name not only represents the link between different disciplines of soil science but also symbolizes the connection between scientists and technologists of different nations, different thoughts and different languages.

Editorial Office SOIL TECHNOLOGY: Coordinator:
Dr. D. Gabriels
Department of Soil Physics, Faculty of Agricultural Sciences, State University Gent
Coupure Links 653, B-9000 Gent, Belgium
Tel. 32-91-236961